Automotive Engine Repair and Rebuilding

Third Edition

By  Chek-Chart

Richard K. DuPuy, *Editor*
Michael Monaghan, *Managing Editor*

Automotive Engine Repair and Rebuilding, Third Edition

Classroom Manual and Shop Manual

Copyright © 1999 by Chek-Chart Publications

International Standard Book Number: 1-57932-189-5
Library of Congress Catalog Card Number: 98-73080

Chek-Chart Publications

320 Soquel Way
Sunnyvale, CA 94086
408-739-2435

"Chek" us out on the web at www.chekchart.com.

Edit
Richa...

Managing Editor
Michael Monaghan

Manufacturing Coordinator
Christine Moos

Production Team Supervisor
Steve Pool

Production Coordinator
Kristy Nash

Production Team
Svetlana Dominguez
Maribeth Echard
Carl Pierce
Mary Ellen Stephenson

ACKNOWLEDGEMENTS

In producing this series of textbooks for automotive technicians, Chek-Chart has drawn extensively on the technical and editorial knowledge of the nation's vehicle manufacturers and suppliers. Automotive design is a technical, fast-changing field, and we gratefully acknowledge the help of the following companies and organizations in allowing us to present the most up-to-date information and illustrations possible. These companies and organizations are not responsible for any errors or omissions in the instructions or illustrations, or for changes in procedures or specifications made by the vehicle manufacturers or suppliers, contained in this book or any other Chek-Chart product:

ABS Products, South Gate, CA
Audi of America, Inc.
BHJ Products, Inc.
Bayco Division, VSB, Inc., Sacramento, CA
Champion Spark Plug Co.
Chaves Performance Engines, San Jose, CA
Chrysler Motors Corporation
Competition Cams
Clayton Industries
DCM Tech, Inc.
DeAnza College, Cupertino, CA
Evergreen Valley College, San Jose, CA
Federal-Mogul Corporation
Fel-Pro, Inc.
Ford Motor Company
Furtado's Machine Shop, San Jose, CA
General Motors Corporation
  AC-Delco Division
  Buick Motor Division
  Cadillac Motor Division
  Chevrolet Motor Division
  Oldsmobile Division
  Pontiac Motor Division

Griffin Auto Parts, Inc., Santa Clara, CA
Hines Industries, Inc.
Hopper Shop Equipment Sales, Stanton, CA
The Hotsy Corporation, Engelwood, CO
Iskenderian Racing Cams
Honda Motor Co., Inc.
Kent-Moore Tool Group
MAC Tools, Inc.
Mazda Motor Corporation
Melling Automotive Products, Jackson, MI
Mobil Oil Corporation
Nissan Motor Corporation
Norton/TRW Ceramics
Penniman & Richards, San Jose, CA
Porsche Cars North America, Inc.
Rottler Manufacturing
Sim-Test Division of Hawk Instruments, Inc.
Snap-On Tools Corporation
Storm Vulcan
Sunnen Products Company
Total Seal

Toyota Motor Sales, U.S.A., Inc.
TRW, Inc.
Volkswagen of America

The authors have made every effort to ensure that the material in this book is as accurate and up-to-date as possible. However, neither Chek-Chart nor any related companies can be held responsible for mistakes or omissions, or for changes in procedures or specifications made by the manufacturers or suppliers.

The comments, suggestions, and assistance of the following reviewers were invaluable:

Bob Weber, Parcellville, VA
Ronald M. Davis, Stockton, CA

Original art and photographs were produced by John Badenhop, Dave Douglass, Gerald A. McEwan, Richard K. DuPuy, and William J. Turney. The project is under the direction of Roger L. Fennema.

Contents

On the Covers:
Front — The Lexus 3-Liter, 24-valve engine.
Rear — The Volkswagen 1.8-Liter, turbocharged,
4-cylinder engine; and the Pontiac 2.3-Liter,
DOHC Quad 4-cylinder engine.

Introduction to Automotive Engine Repair and Rebuilding

Automotive Engine Repair and Rebuilding is part of the Chek-Chart Automotive Series. The package for each course has two volumes, a *Classroom Manual* and a *Shop Manual*.

Other titles in this series include:

- Automatic Transmissions and Transaxles
- Automotive Brake Systems
- Automotive Heating, Ventilation, and Air Conditioning
- Engine Performance, Diagnosis, and Tune-Up
- Fuel Systems and Emission Controls
- Automotive Electrical and Electronic Systems
- Automotive Steering, Suspension, and Wheel Alignment.

Each book is written to help the instructor teach students to become competent and knowledgeable professional automotive technicians. The two-manual texts are the core of a complete learning system that leads a student from basic theories to actual hands-on experience.

The entire series is job-oriented, especially designed for students who intend to work in the automotive service profession. A student will be able to use the knowledge gained from these books and from the instructor to get and keep a job. Learning the material and techniques in these volumes is a giant leap toward a satisfying, rewarding career.

The books are divided into *Classroom Manuals* and *Shop Manuals* for an improved presentation of the descriptive information and study lessons, along with representative testing, repair, and overhaul procedures. The manuals are to be used together: The descriptive material in the *Classroom Manual* corresponds to the application material in the *Shop Manual*.

Each book is divided into several parts, and each book is complete by itself. Instructors will find the chapters to be complete, readable, and well thought-out. Students will benefit from the many learning aids included, as well as from the thoroughness of the presentation.

The series was researched and written by the editorial staff of Chek-Chart Publications. For 70 years, Chek-Chart has provided vehicle and equipment manufacturers' service specifications to the automotive service field. Chek-Chart's complete up-to-date automotive data bank was used extensively to prepare this textbook series.

Because of the comprehensive material, the hundreds of high-quality illustrations, and the inclusion of the latest automotive technology, instructors and students alike will find that these books will keep their value over the years. In fact, they will form the core of the master technician's professional library.

How to Use This Book

Why Are There Two Manuals?

Unless you are familiar with the other books in this series, *Automotive Engine Repair and Rebuilding* will not be like any other textbook you've ever used before. It is actually two books, the *Classroom Manual* and the *Shop Manual*. They have different purposes, and should be used together.

The *Classroom Manual* teaches you what you need to know about how the automotive engine operates and why the relationship between engine parts is so important. The *Classroom Manual* will be valuable in class and at home, for study and for reference. You can use the text and illustrations for years to refresh your memory — not only about the basics of automotive engine repair, but also about related topics in automotive history, physics, and technology.

In the *Shop Manual*, you learn test procedures, troubleshooting, and how to overhaul the systems and parts you read about in the *Classroom Manual*. The *Shop Manual* provides the practical hands-on information you need to work on automotive engines. Use the two manuals together to fully understand how engines work and how to fix them when they do not work.

What Is in These Manuals?

These key features of the *Classroom Manual* make it easier for you to learn, and to remember what you learn:

- Each chapter is divided into self-contained sections for easier understanding and review. The organization shows you clearly which parts make up which systems, and how various parts or systems that perform the same task differ or are the same.
- Most parts and processes are fully illustrated with drawings or photographs. Important topics appear in several different ways, to make sure you can see other aspects of them.
- Important words in the *Classroom Manual* text are printed in **boldface type** and are defined on the same page and in a glossary at the end of the manual. Use these words to build the vocabulary you need to understand the text.
- Review questions are included for each chapter. Use them to test your knowledge.
- Every chapter has a brief summary at the end to help you to review for exams.
- Every few pages you will find sidebars — short blocks of related information — in addition to the main text.

The *Shop Manual* has detailed instructions on overhaul, repair, and rebuilding procedures for modern automotive engines. These are easy to understand, and often have step-by-step explanations to guide you through the procedures. This is what else you'll find in the *Shop Manual*:

- Helpful information that tells you how to use and maintain shop tools and test equipment
- A thorough coverage of the metric system units needed to work on modern engines
- Safety precautions
- Test procedures and troubleshooting hints will help you work better and faster
- Tips the professionals use that are presented clearly and accurately
- There is a sample test at the back of the *Classroom Manual*, similar to those given for Automotive Service Excellence (ASE) certification. Use it to help you study and prepare yourself when you are ready to be certified as an expert in one of several areas of automotive technology.

Where Should I Begin?

If you already know something about automotive engines and know how to repair them, you will find that this book is a helpful review. If you are just starting in vehicle repair, then the book will give you a solid foundation on which to develop professional-level skills.

Your instructor will design a course to take advantage of what you already know, and what facilities and equipment are available to work with. You may be asked to read certain chapters of these manuals out of order. That's fine. The important thing is to really understand each subject before you move on to the next.

Study the vocabulary words in boldface type. Use the review questions to help you understand the material. When you read the *Classroom Manual*, be sure to refer to your *Shop Manual* to relate the descriptive text to the service procedures. And when you are working on actual vehicle systems and components, look back to the *Classroom Manual* to keep the basic information fresh in your mind. Working on such a complicated piece of equipment as a modern vehicle isn't always easy. Use the information in the *Classroom Manual*, the procedures in the *Shop Manual*, and the knowledge of your instructor to help you.

The *Shop Manual* is a good book for work, not just a good workbook. Keep it on hand while you're working on equipment. It folds flat on the workbench and under the car, and can stand quite a bit of rough handling.

When you perform test procedures and overhaul equipment, you will need a complete and accurate source of manufacturers' specifications, and the techniques for pulling computer trouble codes. Most shops have either the vehicle manufacturers' annual shop service manuals, which lists these specifications, or an independent guide, such as the Chek-Chart *Car Care Guide*. This unique book, with multi-year coverage, is updated each year to give you service instructions, capacities, and troubleshooting tips that you need to work on specific vehicles.

PART ONE

Automotive Engine Fundamentals

1

Engine Operation and Construction

This is a book about the repair and rebuilding of modern automobile engines. Automobile engines have been developed and improved continuously for over 100 years, since Gottlieb Daimler and Wilhelm Maybach installed a four-stroke engine in a modified horse-drawn carriage in 1886. To know enough about an engine to service it, you must understand its design, construction, and operation. To understand engine operation completely, you must understand not only the internal engine parts, but the intake, exhaust, and ignition systems as well.

Part One of this Classroom Manual, the first five chapters, summarizes engine design, operation, and construction. Most of this material is a preview of subjects covered in greater detail later in the book. You may already have studied some of the subjects covered in Part One. If so, these chapters will be a quick review to refresh your knowledge.

If this material is new to you, study it thoroughly. In any case, the information in Part One is a necessary foundation for the detailed descriptions of engine components in Part Two.

ENGINE OPERATION

All engines use some kind of fuel to produce mechanical power. The oldest "engine" known to man is the simple lever. Food "fuels" the muscle pushing the lever to move objects that the muscle alone could never budge. In a similar way, the automotive engine uses fuel to perform work, figure 1-1.

It is common to think of the automotive engine as a gasoline engine. Most of the automotive engines in the world are fueled by gasoline. But the correct name for the automotive engine is **internal combustion engine**. It can be designed to run on any fuel that **vaporizes** easily or on any flammable gas.

The automotive engine is called an internal combustion engine because the fuel it uses is burned inside the engine. **External combustion engines** burn the fuel outside the engine. A common example of an external combustion engine is the steam engine. Fuel is burned to produce heat to make steam, but this burning takes place anywhere from a few feet to several miles away from the engine. Figure 1-2 shows the basic differences between internal and external combustion engines.

The internal combustion engine burns its fuel inside a combustion chamber. One side of this chamber is open to a piston. When the fuel burns, the hot gases expand very rapidly and push the piston away from the combustion chamber. This basic action of heated gases expanding and pushing is the source of power for all internal combustion engines. This includes piston, rotary, and turbine engines.

Figure 1-1. Today's automotive engines produce high power with low fuel consumption and low exhaust emissions — not an easy design task.

Compression and Combustion

Gasoline by itself will not burn; it must be mixed with oxygen in the air. If fuel burns in the open air, it produces no power because it is not confined. If the same amount of fuel is enclosed and burned, it will expand with some force. To get the most force from the burning of a liquid fuel, it must be vaporized, mixed with air, and compressed to a small volume before it is burned. This compression and combustion is the most efficient way of releasing the energy stored in the air-fuel mixture.

In a piston engine, a piston moving in a cylinder provides compression. An example of this type of compression can be found in a two-section mailing tube with metal ends, figure 1-3. Push the inside tube in very quickly, and it will compress the air inside. Release the inside tube quickly and it will fly out. In a similar way, the piston compresses the air-fuel mixture in the cylinder.

Internal combustion engines are designed to compress and burn the vaporized air-fuel mixture in a sealed chamber, figure 1-4. Here, the combustion energy can work on the movable piston to produce mechanical energy. When the heat from the burning fuel causes the fuel vapor, air, and exhaust gases inside the cylinder to expand, it produces much more power than was required to compress it. The burning,

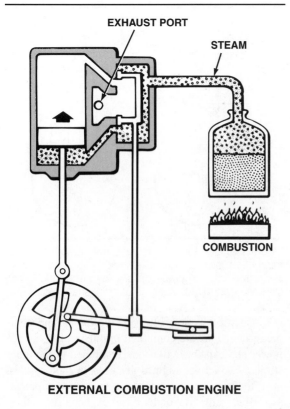

EXTERNAL COMBUSTION ENGINE

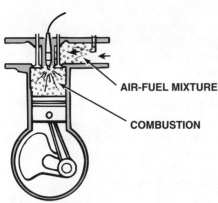

INTERNAL COMBUSTION ENGINE

Figure 1-2. The fuel for an internal combustion engine is burned inside the engine. The fuel for an external combustion engine is burned outside the engine.

expanding gases push the piston to the other end of the cylinder.

Vacuum

The air-fuel mixture enters the combustion chamber past an intake valve. The suction or vacuum that pulls the mixture into the cylinder is created by the descending piston. You can create this same suction in the example of the two-piece mailing tube shown in figure 1-3. Draw the assembled tubes apart quickly

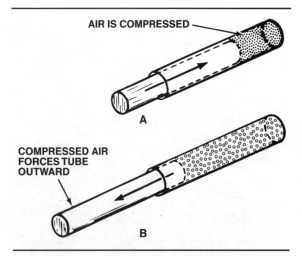

Figure 1-3. Push the inner tube in rapidly, and air is compressed (A). Release the inner tube quickly, and the compressed air forces it out (B).

and let go. The suction tends to pull the tubes back together. If an open intake valve were in the end of the outer tube, pulling the inside tube would draw air past the valve. This suction is known as engine vacuum. Suction exists because of a difference in air pressure between the two areas. We will study vacuum and air pressure in more detail in later chapters.

The Four-Stroke Cycle

The piston creates a vacuum by moving down through the cylinder **bore**. The movement of the piston from one end of the cylinder to the other is called a **stroke**, figure 1-5. After the piston reaches the end of the cylinder, it will move back to the other end. As long as the engine is running, the piston continues to move, or stroke, back and forth in the cylinder.

Internal Combustion Engine: An engine, such as a gasoline or diesel engine, in which fuel is burned inside the engine.

Vaporize: To change from a solid or liquid into a gaseous state.

External Combustion Engine: An engine, such as a steam engine, in which fuel is burned outside the engine.

Bore: The diameter of an engine cylinder; to enlarge or finish the surface of a drilled hole.

Stroke: One complete top-to-bottom or bottom-to-top movement of an engine piston.

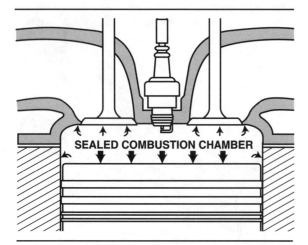

Figure 1-4. For combustion to produce power in an engine, the combustion chamber must be sealed.

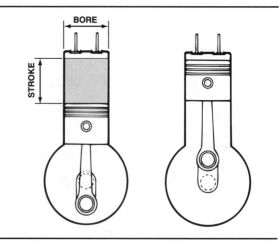

Figure 1-5. One top-to-bottom or bottom-to-top movement of the piston is called a stroke. One piston stroke performs 180 degrees of crankshaft rotation; two strokes perform 360 degrees of crankshaft rotation.

An internal combustion engine must go through four separate actions to complete one operating sequence, or cycle. Depending on the type of reciprocating engine, a complete operating cycle may require either two or four strokes. In the **four-stroke engine**, four strokes of the piston in the cylinder are needed to complete one full operating cycle. Each stroke is named after the action it performs—intake, compression, power, and exhaust—in that order, figure 1-6.

1. Intake stroke: As the piston moves down, the mixture of vaporized fuel and air is drawn into the cylinder past the open intake valve.
2. Compression stroke: The intake valve closes, the piston returns up, and the mixture is compressed within the combustion chamber.
3. Power stroke: The mixture is ignited by a spark, and the expanding gases of combustion force the piston down in the cylinder. The exhaust valve opens near the bottom of the stroke.
4. Exhaust stroke: The piston returns up as the exhaust valve opens, and the burned gases are pushed out to prepare for the next intake stroke. The intake valve usually opens just before the top of the exhaust stroke. This four-stroke cycle is continuously repeated in every cylinder as long as the engine is running.

Engines that use the four-stroke sequence are known as four-stroke engines. This four-stroke cycle engine is also called the Otto-cycle engine after its inventor, Dr. Nikolaus Otto, who built the first successful four-stroke engine in 1876. Most automobile engines are four-stroke, spark-ignition engines. Other types of engines include two-stroke and compression-ignition (diesel) engines.

A **two-stroke engine** also goes through intake, compression, power, and exhaust actions to complete one operating cycle. However, the intake and compression actions are combined in one stroke, and the power and exhaust actions are combined in the other stroke. Two-stroke engines are used in motorcycles, lawn mowers, heavy trucks, construction equipment, as well as ships. Beyond a basic explanation of their operation, they are not covered in this text.

Diesel engines do not use a spark to ignite the air-fuel mixture. The heat from their high compression ignites the fuel. Diesel engine operation is explained later in this chapter. Aside from the differences in ignition and fuel systems, diesel and gasoline engines are physically quite similar. Although this text does not feature diesel engine repair and rebuilding, much of the information on gasoline engines is typical of light-duty diesel engine service.

Reciprocating Engine

Except for the Wankel rotary engine, all production automotive engines are **reciprocating**, or piston type.

Four-Stroke Engine: The Otto-cycle engine. An engine in which a piston must complete four strokes to make up one operating cycle. The strokes are intake, compression, power, and exhaust.

Two-Stroke Engine: An engine in which a piston makes two strokes to complete one operating cycle.

Reciprocating Engine: Also called a piston engine. An engine in which the pistons move up and down, or back and forth, as a result of combustion of an air-fuel mixture at one end of the piston cylinder.

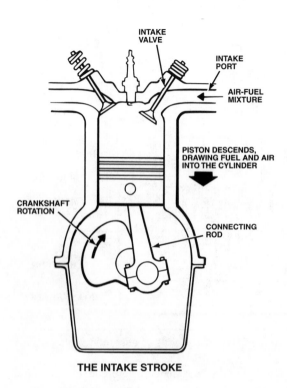

THE INTAKE STROKE

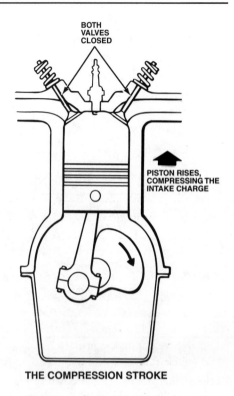

THE COMPRESSION STROKE

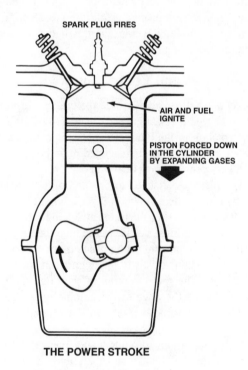

THE POWER STROKE

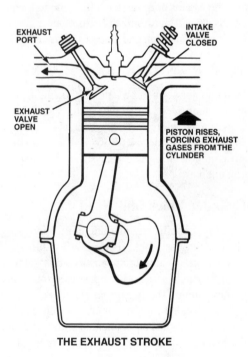

THE EXHAUST STROKE

Figure 1-6. The downward movement of the piston draws the air-fuel mixture into the cylinder through the intake valve on the intake stroke. On the compression stroke, the mixture is compressed by the upward movement of the piston with both valves closed. Ignition occurs at the beginning of the power stroke, and combustion drives the piston downward to produce power. On the exhaust stroke, the upward-moving piston forces the burned gases out the open exhaust valve.

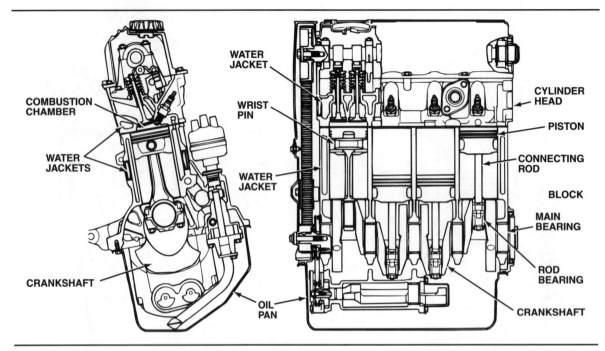

Figure 1-7. This 4-cylinder Chrysler engine has a typical engine block and cylinder head assembly.

Reciprocating means up and down or back and forth action of a piston in a cylinder. Power is produced by the in-line motion of a piston in a cylinder. However, this *linear* motion must be changed to rotating motion to turn the wheels of a car or truck. In the following paragraphs we will show how the parts of an engine produce reciprocating motion and change it to rotating motion.

MAJOR ENGINE COMPONENTS

Of the major parts of an automobile engine, so far we have mentioned only the pistons and cylinders. It takes many more parts, however, to build a complete engine that will do useful work. The following parts are common in typical four-stroke internal combustion engines.

Cylinder Block and Head

Most automobile engines are built upon a cylinder block, or engine block, figure 1-7. The block is usually an iron or aluminum casting that contains the engine cylinders as well as passages for coolant and oil circulation. The top of the block is covered by the cylinder head, which has more coolant passages and forms most of the combustion chamber. The bottom of the block is covered with an oil pan, or an oil sump.

Crankcase

All piston engines have a crankcase, figure 1-8. It is a housing that supports or encloses the crankshaft. Early automotive designs used a separate crankcase bolted

to the cylinders. Most modern automotive engines use a crankcase that is cast in one piece with the cylinder block. The entire casting of block and crankcase is known as the cylinder block, or simply the engine block. The term crankcase is still used to describe the open space around the crankshaft, which also usually includes the oil pan.

Crankshaft and Connecting Rod

The crankshaft revolves inside the crankcase portion of the engine block, figures 1-7 and 1-8. Main bearing caps bolt to the block and hold the crankshaft in place. Large shell bearings, called main bearings, are used between the crankshaft and caps.

The piston is attached to one end of a connecting rod by a pin called a piston pin, or a wrist pin, figure 1-9. The other end of the rod is attached to the crankshaft. Rod bearings, similar to the main bearings, are used between the connecting rod and the crankshaft.

As the piston strokes in the cylinder, it rotates the crankshaft. The rotary motion of the crankshaft can be used to turn the wheels of an automobile, rotate the blades of a lawnmower, or turn the propeller of an airplane.

It is important to understand the relationship between the revolving crankshaft and the stroking piston. The piston always makes two strokes for each revolution of the crankshaft. The complete four-stroke cycle requires two crankshaft revolutions.

Because only one of the four strokes is a power stroke, the crankshaft must coast for one and one-half

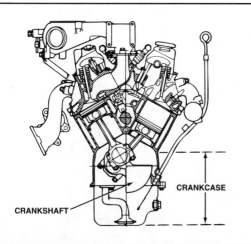

Figure 1-8. The crankcase of this Chrysler V6 engine, like most modern engines, is the lower portion of the cylinder block casting and the oil pan.

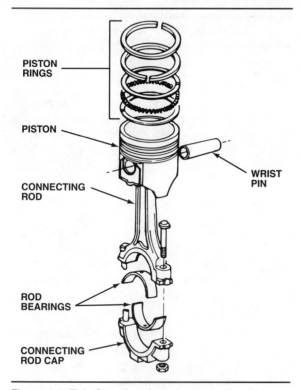

Figure 1-9. This Chrysler piston is attached to the connecting rod by a wrist pin, which lets it pivot as the engine runs.

revolutions in a single-cylinder engine. It does this because of the **inertia** of its rotating parts, particularly the flywheel.

Flywheel

Because the power in a piston or rotary engine is applied in impulses, the engine tends to jerk or pulse.

This tendency is reduced by the flywheel. The flywheel is a large, heavy disc of metal attached to the end of the crankshaft, figure 1-10. The flywheel works on the principle of inertia. The inertia of the flywheel resists any change in speed. When there is a power impulse, the heavy flywheel resists a rapid increase in engine speed, and during the coasting period (the exhaust, intake, and compression strokes),

■ **Nikolaus August Otto (1832 to 1891) — Creator of the Four-Stroke Internal Combustion Engine**

Nikolaus Otto was born in 1832, in a small hamlet in Germany near the Rhine River. Poor economic conditions forced him to drop out of secondary school to become a grocery clerk, and he ended up as a salesman of tea, sugar, and kitchenwares.

Otto was intrigued by the Lenoir internal combustion engine, introduced in 1860 as the first internal combustion engine commercially available. Handicapped by his lack of education, Otto spent three years and all his own money (and much of the money of his friends) trying to improve the Lenoir. In 1864, Otto formed a company with Eugen Langen, a technologically-minded speculator who provided much-needed capital. Their company, today called Klöckner-Humboldt-Deutz AG, is the first and oldest internal combustion engine manufacturing company in the world. The company and its refined engines were immensely successful.

The first four-stroke engine was not built until 1876. All previous internal combustion engines — including Otto's — were non-compression, in that fuel and air were drawn into the cylinder during part of a piston's downward stroke and then ignited. The expanding gases then pushed the piston down the remainder of its stroke. Many inventors used this design to make the pistons double-acting, with a power stroke each way. Otto's new engine used a downward stroke of the piston to draw in an intake charge, and a second upward stroke to compress it. It also required two more strokes to extract power and push out exhaust gases—the four-stroke engine cycle. Otto's competitors were dubious, believing that to waste three piston strokes for a single power stroke must surely outweigh any advantages of compressing the mixture.

They were quite wrong. Comparing Otto's new engine with his earlier best-seller showed that the four-stroke weighed one-third as much, could run almost twice as fast, and needed only 7 percent of the cylinder displacement to produce the same horsepower, with almost identical fuel consumptions. Within 10 years, a four-stroke engine powered the first motorcycle, and soon after that the engine appeared in what would be called the horseless carriage.

Inertia: The tendency of an object at rest to remain at rest and of an object in motion to remain in motion.

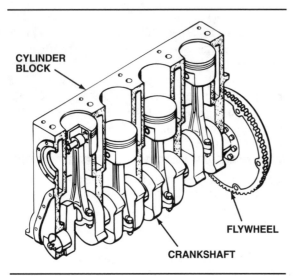

Figure 1-10. Flywheel inertia helps smooth out the impulses of the firing strokes.

the flywheel keeps the engine turning because it also resists a decrease in speed.

When an automatic transmission is used, a torque converter is bolted to the flywheel. Because the torque converter is heavy, the flywheel can be much lighter. The flywheel used with a torque converter is often called a "flexplate." The total weight of torque converter and flexplate equals the weight of the flywheel and clutch on a manual transmission engine.

Engine rotation

The front of the engine is commonly considered to be the end opposite the flywheel. When you look at the front of an engine as just defined, the crankshaft and flywheel of most engines rotate clockwise.

This was the routine definition of an engine's physical features and direction of rotation in the simpler days when nearly all cars had front-mounted engines and rear-wheel-drive. Now that many cars with front-wheel-drive have transversely mounted engines, there is some confusion over which end of the engine is the front. In this book, no matter where the engine is situated, front or rear, in-line or transverse, the front is always the end opposite the flywheel.

Camshaft

The camshaft controls the opening and closing of the valves, and is driven by the crankshaft. Lobes on the cam-shaft push each valve open as the shaft rotates, figure 1-11. A spring closes each valve when the lobe is not holding it open.

Once during each revolution of the camshaft, the lobe will push the valve open. The timing of the valve opening is critical to the operation of the engine. Intake valves must be opened just before the beginning

of the intake stroke. Exhaust valves must be opened just before the beginning of the exhaust stroke. Because the intake and exhaust valves open only once during every two revolutions of the crankshaft, the camshaft must run at half the crankshaft speed.

Turning the camshaft at half the crankshaft speed is accomplished by using a gear or sprocket on the camshaft that is twice the diameter of the crankshaft gear or sprocket, figure 1-11. If you count the teeth on each gear or sprocket, you will find exactly twice as many on the camshaft as on the crankshaft.

Valves

All modern automotive piston engines use **poppet valves**, figure 1-12. These are valves that work by linear motion. Most water faucet valves operate with a circular motion. The poppet valve is opened simply by pushing on it. The poppet valve must have a seat on which to rest and from which it closes off a passageway. There also must be a spring to hold the valve against the seat. In operation, a push on the end of its stem opens the valve, and when the force is removed, the spring closes the valve.

Valve arrangement

Intake and exhaust valves on modern engines are located in the cylinder head. Because the valves are "in the head," this basic arrangement is called an I-head design. In the past, poppet valves have been arranged in three different ways, figure 1-13.

- The L-head design positions both valves side-by-side in the engine block. Because the cylinder head is rather flat and contains only the combustion chamber, water jacket, and spark plugs, L-head engines also are called "flatheads." Still very common on lawn mowers, this valve arrangement has not been used in a domestic automotive engine since the mid-1960s.

- The F-head design positions the intake valve in the cylinder head and the exhaust valve in the engine block. A compromise between the L-head and I-head designs, the F-head was last used in the 1971 Jeep.

- The I-head design, in both overhead-valve or overhead-camshaft form, positions both the intake and the exhaust valves in the cylinder head. All modern automotive engines use this design.

In the overhead-valve engine, the camshaft is in the engine block and the valves are opened by valve lifters, pushrods, and rocker arms, figures 1-11 and 1-12. In the overhead-camshaft engine, the camshaft is mounted in the head, either above or to one side of

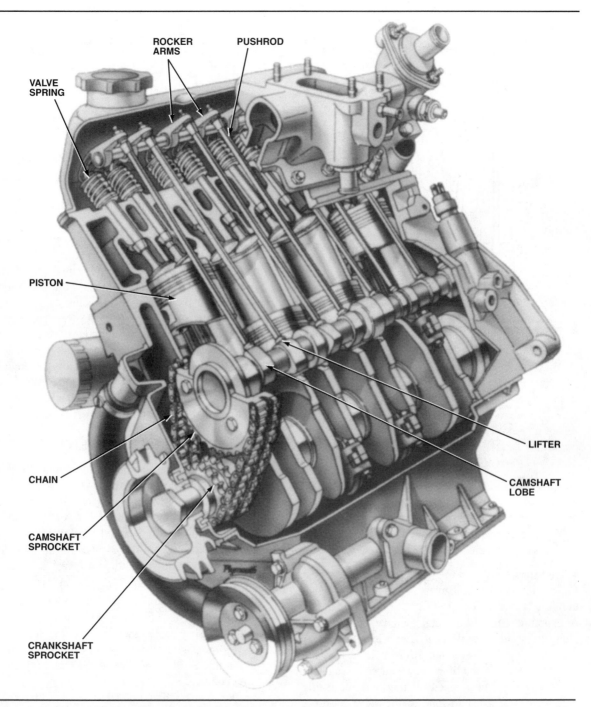

Figure 1-11. Each valve is opened by a lobe on the camshaft and closed by a spring. The camshaft sprocket has twice as many teeth as the crankshaft sprocket, allowing the camshaft to rotate at one-half crankshaft speed.

the valves, figure 1-14. This improves valve action at higher engine speeds. The valves may open directly by means of valve lifters or camshaft followers, or through rocker arms. The double overhead camshaft engine has two camshafts, one on each side of the valves. One camshaft operates the intake valves; the other operates the exhaust valves.

Poppet Valve: A valve that plugs and unplugs its opening by linear movement.

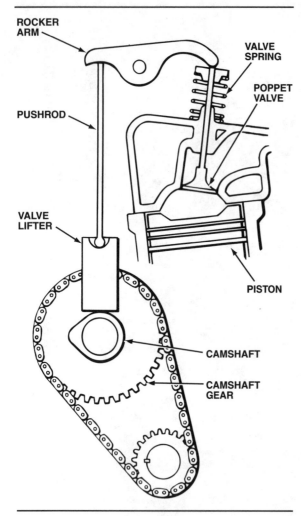

Figure 1-12. Modern automotive engines use poppet valves.

Valve lifters

Valve lifters can be mechanical or hydraulic, as shown in figure 1-15. A mechanical valve lifter is solid metal. A hydraulic lifter is a metal cylinder containing a plunger that rides on oil. A chamber below the plunger fills with engine oil through a feed hole and, as the camshaft lobe lifts the lifter, the chamber is sealed by a check valve. The trapped oil transmits the lifting motion of the camshaft lobe to the valve pushrod. Hydraulic lifters are generally quieter than mechanical lifters and do not normally need to be adjusted, because the amount of oil in the chamber varies to keep the valve adjustment correct.

Number of valves

As you saw in figure 1-6, most automobile engines have one intake and one exhaust valve per cylinder. This means that a 4-cylinder engine has 8 valves, a 6-cylinder engine has 12 valves, and a V8 has 16 valves.

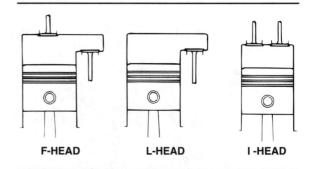

Figure 1-13. Historically, designers have arranged valves for four-stroke engines in these three ways.

Many engines have been built, however, with more than two valves per cylinder. Engines with three valves per cylinder are currently used in several Japanese production cars. Engines with four valves per cylinder have been used in racing engines since the 1912 Peugeot Grand Prix cars. Street motorcycle engines with five valves per cylinder have been built in recent years. In spite of performance advantages, the higher costs and greater complexity of engines with more than two valves per cylinder kept such designs from being common in production engines until the mid-1980s, figure 1-16.

CYLINDER ARRANGEMENT

While single-cylinder engines are common in motorcycles, outboard motors, and small agricultural machines, automotive engines have more than one cylinder. Most car engines have 4, 6, or 8 cylinders, although engines with 3 and 12 cylinders are also being produced. Within the engine block, the cylinders are arranged in one of three ways:

- In-line engines have a single bank of cylinders arranged in a straight line. The cylinders do not have to be vertical, as shown in figure 1-17. They can be inclined to either side. Most in-line engines have 4 or 6 cylinders, but many in-line engine have been built with 3, 5, and 8 cylinders.
- V engines, figure 1-18, have two banks of cylinders, usually inclined either 60 degrees or 90 degrees from each other. Most V-type engines have 6 or 8 cylinders, but V2 (or V-twin), V4, V12, and V16 engines have been built.
- Horizontally opposed, "flat", or "boxer" engines have two banks of cylinders 180 degrees apart, figure 1-19. These engine designs are often air-cooled, and are found in the original VW Beetle and some Ferrari, Porsche, and Subaru models. Ferrari's and Subaru's designs are liquid cooled. Volkswagen vans use a version

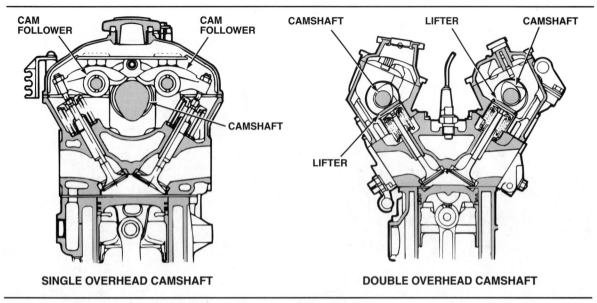

CAM FOLLOWER   CAM FOLLOWER   CAMSHAFT   LIFTER   CAMSHAFT

CAMSHAFT

LIFTER

SINGLE OVERHEAD CAMSHAFT   **DOUBLE OVERHEAD CAMSHAFT**

Figure 1-14. Single-overhead camshafts usually require an additional component, such as a rocker arm, to operate all the valves. Double-overhead-camshaft engines actuate the valves directly.

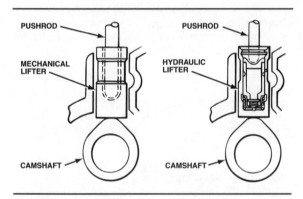

PUSHROD   PUSHROD

MECHANICAL LIFTER   HYDRAULIC LIFTER

CAMSHAFT   CAMSHAFT

Figure 1-15. Mechanical lifters are solid metal. Hydraulic lifters use engine oil to take up clearance and transmit motion.

of the traditional air-cooled VW horizontally opposed engine with liquid-cooled cylinder heads. Most opposed engines have 2, 4, or 6 cylinders, but flat engines have been built with 8, 12, and 16 cylinders.

Engine Balance

When an engine has more than one cylinder, the crankshaft is usually made so that the firing impulses are evenly spaced. Engine speed is measured in revolutions per minute, or in degrees of crankshaft rotation. It takes two crankshaft revolutions, or 720 degrees of crankshaft rotation, to complete the 4-stroke sequence. If a 4-stroke engine has two cylinders, the firing impulses and crankshaft throws can be spaced so that there is a power impulse every 360

degrees. On an in-line 4-cylinder engine, the crankshaft is designed to provide firing impulses every 180 degrees. An in-line 6-cylinder crankshaft is built to fire every 120 degrees. In an 8-cylinder engine, either in-line or V, the firing impulses occur every 90 degrees of crankshaft rotation.

As you can see, the more cylinders an engine has, the closer the firing impulses will be. On 6- and 8-cylinder engines, for example, the firing impulses are close enough that power strokes overlap slightly. In other words, a new power stroke begins before the power stroke that preceded it ends. This provides a smooth transition from one firing pulse to the next. A 4-cylinder engine has no overlap of its power strokes, which makes it a relatively rough-running engine compared to engines with more cylinders. Therefore, an 8-cylinder engine runs smoother than a 4-cylinder engine.

Figure 1-20 shows common crankshaft arrangements and firing impulse frequencies. Other arrangements are possible, such as V4, opposed 4-cylinder, and several combinations of 2-cylinder layouts. The cylinder arrangement, cylinder numbering order, and crankshaft design all determine the firing order of an engine, which we will study at the end of this chapter.

ENGINE DISPLACEMENT AND COMPRESSION RATIO

In any discussion of engines, the term "engine size" comes up often. This does not refer to the outside dimension of an engine, but to its displacement. As the piston strokes in the cylinder, it moves through,

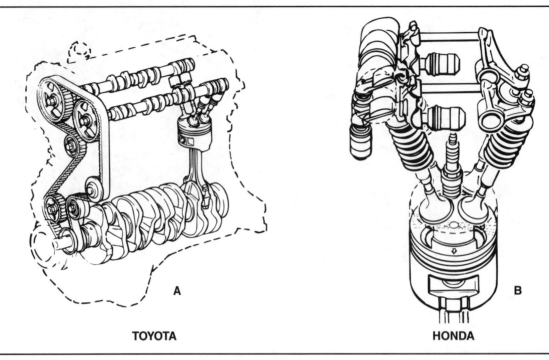

TOYOTA   **HONDA**

Figure 1-16. All 4-valve-per-cylinder production engines use overhead camshafts. Most have separate intake and exhaust camshafts as in the Toyota example (A). Some have a single camshaft, such as the Honda Acura V6 design (B) that operates the exhaust valves through short pushrods.

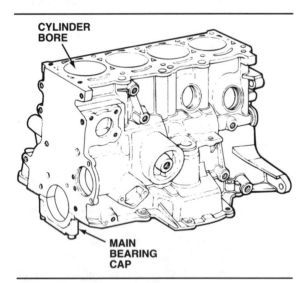

Figure 1-17. Most in-line engines position the cylinders vertically, like this Chrysler 4-cylinder.

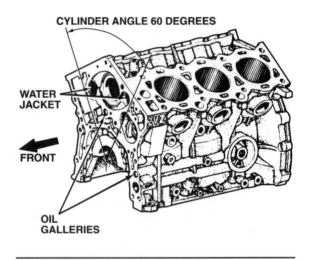

Figure 1-18. This Lexus V cylinder block has its cylinders inclined at 60 degrees.

or displaces, a specific volume. Another important engine measurement term is compression ratio. Displacement and compression ratio are related to each other, as you will learn in the following paragraphs.

Engine Displacement

Engine **displacement** is a measurement of engine volume. The number of cylinders is a factor in determining

Displacement: A measurement of engine volume. It is calculated by multiplying the piston displacement of one cylinder by the number of cylinders. The total engine displacement is the volume displaced by all the pistons.

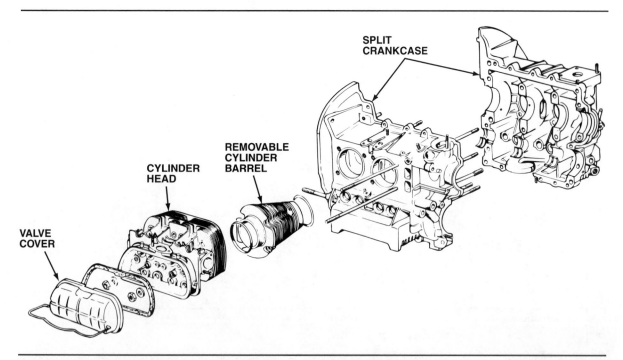

Figure 1-19. This Volkswagen horizontally opposed engine is built with a split crankcase. Individual cylinder castings bolt to the split crankcase, and cylinder heads attach to each pair of cylinders.

■ Isaac de Rivaz and His Self-Propelled Carriage

The first self-propelled vehicles using internal combustion engines were those built by Isaac de Rivaz, a Swiss engineer and government official. De Rivaz built the first version in 1805, but his improved model of 1813 was more impressive. This "great mechanical chariot" was 17 feet long by 7 feet wide (5.2 m by 21 m) and weighed 2,100 pounds (950 kg). Its top speed was about three miles per hour (5 km/hr) on a level road, and it could climb a 12 percent grade.

The engine that drove this marvel used a single cylinder, open at the top, with a bore and stroke of 36.5 cm by 150 cm (14.4 inches by 59 inches). A long rod

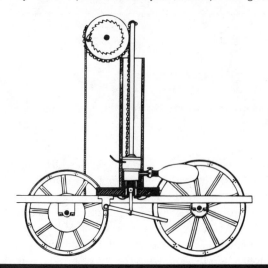

attached to the piston was connected by a chain to a drum outside the cylinder. When the piston descended, the chain would rotate the drum, turning the drive axle through a rope and pulley. When the piston rose, a ratchet let the rope and pulley freewheel.

Coal gas (a burnable mixture of hydrogen and methane) was stored in a collapsible leather bladder, pumped into a mixing chamber in a carburetor by a bellows for each stroke of the piston. To start an engine cycle, the driver pulled on a lever, which first dropped the floor of the cylinder a few inches to draw in an intake charge, and next closed an electric circuit to cause a spark to jump a gap between two wires in the combustion chamber. The burning, expanding gases shot the piston up to the top of its stroke, as the ratchet mechanism freewheeled. After combustion stopped, the gases in the cylinder cooled and contracted, and atmospheric pressure pushed the piston down. This engaged the ratchet and rolled the carriage forward about 16 to 20 feet (5 to 6 m). Exhaust was expelled when the driver raised the floor of the cylinder again in preparation for the next stroke of the engine.

De Rivaz's carriage was not very practical, as trips were limited to the two-mile (3 km) fuel capacity of the leather bag, and the driver had to pull and push the lever for each piston stroke. However, his use of a fuel-mixing carburetor, spark ignition, and a portable fuel tank were all engineering feats that would eventually become common on the internal combustion engines in modern day "carriages."

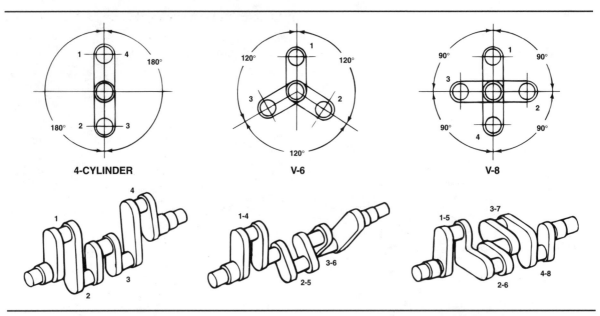

Figure 1-20. The more cylinders an engine has, the closer together the firing impulses are. Here are common crankshaft designs for 4-, 6-, and 8-cylinder engines.

displacement, but the arrangement of cylinders is not. Engine displacement is calculated by multiplying the piston displacement of one cylinder by the number of cylinders. The total engine displacement is the volume displaced by all the pistons.

We learned earlier that the bore and stroke are important engine dimensions. Both are necessary measurements for calculating engine displacement. The displacement of one cylinder is the volume through which the piston's top surface moves as it travels from the bottom of its stroke (**bottom dead center** or **BDC**) to the top of its stroke (**top dead center** or **TDC**), figure 1-21. Piston displacement is computed as follows:

$$\text{Piston displacement} = \left(\frac{\text{Bore}}{2}\right)^2 \times 3.1416 \times \text{Stroke}$$

1. Divide the bore (cylinder diameter) by two. This will give you the radius of the bore.
2. Square the radius (multiply it by itself).
3. Multiply the square of the radius by 3.1416 to find the area of the cylinder cross section.
4. Multiply the area of the cylinder cross section by the length of the stroke.
5. You now know the piston displacement for one cylinder. Multiply this by the number of cylinders to determine the total engine displacement.

For example, to find the displacement of a 6-cylinder engine with a 3.80-inch bore and a 3.40-inch stroke:

1. $\frac{3.80}{2} = 1.90$

2. $1.90 \times 1.90 = 3.61$
3. $3.61 \times 3.1416 = 11.3412$

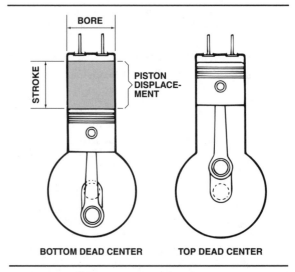

Figure 1-21. The bore and stroke of a piston are used to calculate an engine's displacement.

4. $11.3412 \times 3.40 = 38.56$
5. $38.56 \times 6 = 231.36$

The displacement is 231 cubic inches. Fractions of an inch are usually not included.

Metric displacement specifications

When stated in U.S. customary values, displacement is given in cubic inches. The engine's cubic inch displacement is abbreviated cid. When stated in metric values, displacement is given in cubic centimeters (cc) or in liters (one liter equals 1,000 cc). To convert engine displacement specifications from one value to another, use the following formulas:

- To change cubic centimeters to cubic inches, multiply by 0.061 (cc × 0.061 = cu in.).
- To change cubic inches to cubic centimeters, multiply by 16.39 (cid × 16.39 = cc).
- To change liters to cubic inches, multiply by 61.02 (liters × 61.02 = cid).

Our 231-cid engine from the previous example is also a 3,786-cc engine (231 × 16.39 = 3,786). When expressed in liters, this figure is rounded up to 3.8 liters.

Metric displacement in cc can be calculated directly with the displacement formula, using centimeter measurements instead of inches. Here is how it works for the same engine with a bore equaling 96.52 mm (9.652 cm) and a stroke equaling 86.36 mm (8.636 cm):

1. $\frac{9.652}{2}$ = 4.826
2. 4.826 × 4.826 = 23.29
3. 23.29 × 3.1416 = 73.17
4. 73.16 × 8.636 = 631.90
5. 631.81 × 6 = 3,791 cc

This figure is a few cubic centimeters different from the 3,786-cc displacement we got by converting 231 cubic inches directly to cubic centimeters. This is due to rounding. Again, the engine displacement can be rounded up to 3.8 liters.

Compression Ratio

The **compression ratio** compares the total cylinder volume when the piston is at BDC to the volume of the combustion chamber when the piston is at TDC, figure 1-22. Total cylinder volume may seem to be the same as piston displacement, but it is not. Total cylinder volume is the piston displacement plus the combustion chamber volume. The combustion chamber volume with the piston at TDC is sometimes called the **clearance volume**.

Compression ratio is the total volume of a cylinder divided by the clearance volume. If the clearance volume is ⅛ of the total cylinder volume, the compression ratio is 8 to 1. The formula is as follows:

Bottom Dead Center (BDC): The exact bottom of a piston stroke.

Top Dead Center (TDC): The exact top of a piston stroke. Also a specification used when tuning an engine.

Compression Ratio: A ratio or the total cylinder volume when the piston is at BDC to the volume of the combustion chamber when the piston is at TDC.

Clearance Volume: The combustion chamber volume with the piston at TDC.

■ William Cecil's Internal Combustion Engine

Internal combustion engines are by no means new. The first internal combustion engine that would operate continuously by itself was built in 1820 by William Cecil.

Cecil's engine was constructed with three cylinders arranged in a "T," and connected at their intersection by a rotating valve. The vertical cylinder contained a piston which descended to draw in a charge of air mixed with hydrogen — the two horizontal cylinders remained empty during the intake stroke. At the bottom of the intake stroke, the central valve rotated to briefly open a passage connecting the cylinder to a small flame, igniting the hydrogen. The valve then rotated again, permitting the burning, expanding gases to fill up the two cylinders forming the crosspiece of the "T," pushing out the air they contained through flapper exhaust valves. As the burned gases cooled and contacted, they pulled the flapper valves shut. Atmospheric pressure then pushed the piston back up into the cylinder for the power stroke, and the cycle repeated.

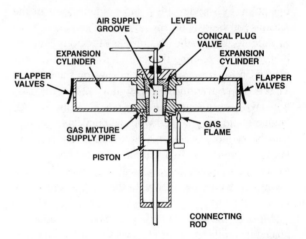

Cecil's engine had a cylinder capacity of about 30 cid (500 cc), and used a flywheel that weighed fifty pounds (23 kg). It would run evenly on a fuel-air mixture of 1:4, although best power was obtained with a richer mixture of 1.25:1. Because of the flame-type ignition, top speed was limited to about 60 RPM — above this speed the flame could not light the hydrogen reliably.

Cecil demonstrated a working model of his engine to the Cambridge Philosophical Society in 1820, but chose to not pursue its development. He was ordained two years after inventing the engine, and spent the remainder of his life as a clergyman with the Church of England.

$$\frac{\text{Total volume}}{\text{Clearance volume}} = \text{Compression ratio}$$

To determine the compression ratio of an engine in which each piston displaces 510 cc and which has a clearance volume of 65 cc:

510 + 64 = 574 cc (total cylinder volume)

$$\frac{574}{64} = 8.968$$

The compression ratio is 8.986 to 1. This would be rounded and expressed as a compression ratio of 9 to 1. This can also be written 9:1.

A higher compression ratio is desirable because it increases the efficiency of the engine, making the engine develop more power from a given quantity of fuel. A higher compression ratio increases cylinder pressure, which pacts the fuel molecules more tightly together. The flame of combustion then travels more rapidly and across a shorter distance.

ENGINE STRUCTURE

We have now covered the basic parts of the engine. At this point, let's review them and examine some of the reasons for the way an engine is put together.

After looking at the illustrations of various engines, it is natural to assume that cylinder heads are always placed on top and crankshafts always on the bottom. This is the conventional way, figure 1-23, but it does not have to be done that way, nor is it always. An engine could be designed with the crankshaft on top, and the cylinder heads and carburetor on the bottom. However, you would have to crawl under the vehicle to service the carburetor, and the air going into the carburetor would be filled with the dust kicked up by the tires.

Most of the components on a modern engine are placed in similar positions by every manufacturer. For example, it would be theoretically possible to build an engine with the combustion chamber between the piston and the crankshaft, with the connecting rod passing right through the center of the chamber. Engine design is a compromise. When designing an engine, manufacturing costs, accessibility, and serviceability have to be considered. As a technician, you can appreciate those last two qualities.

Consider the flywheel. Its purpose is to provide inertia. It could, conceivably, be placed at the front of the crankshaft, or even in the middle. But then it would not provide a convenient place for attaching a clutch or torque converter.

Cooling systems usually have the radiator at the front of the car to take advantage of the airflow. But it does not have to be that way either. The radiator could be anywhere there is enough airflow to remove heat from the coolant.

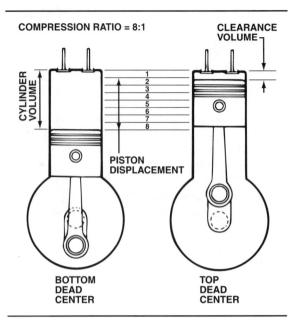

Figure 1-22. Compression ratio is the ratio of the total cylinder volume to the clearance volume.

Exhaust systems, like many parts of the vehicle, have also evolved. The best place to get rid of the exhaust from a vehicle is at the rear. But large trucks exhaust straight up in the air and some vehicles exhaust to the side.

At this point, you should have a basic understanding of the internal parts of an engine. It should be clear to you how pistons, cylinders, crankshafts, camshafts, valves, induction and exhaust all work together to form an engine. We cannot overemphasize the importance of knowing the basic four-stroke cycle thoroughly. It is the foundation upon which the whole engine operates. Knowledge of it is necessary to understand what follows.

ENGINE SYSTEMS

So far we have covered the basic structure of a modern four-stroke automobile engine. However, there are a number of support systems that are essential to proper engine operation. These systems feed the engine its air and fuel, cool the engine, keep it lubricated, and exhaust the burned, spent air-fuel mixture. Some of these systems help the engine to run, while others play a supporting role, by reducing heat and friction so that the engine does not destroy itself, or by quieting the exhaust and reducing air pollution.

Intake System

Because an automotive piston engine burns air and fuel, there must be passages to route the air-fuel mixture to the combustion chamber. The intake system consists of a filter for the incoming air and a manifold that allows the air and fuel to mix before

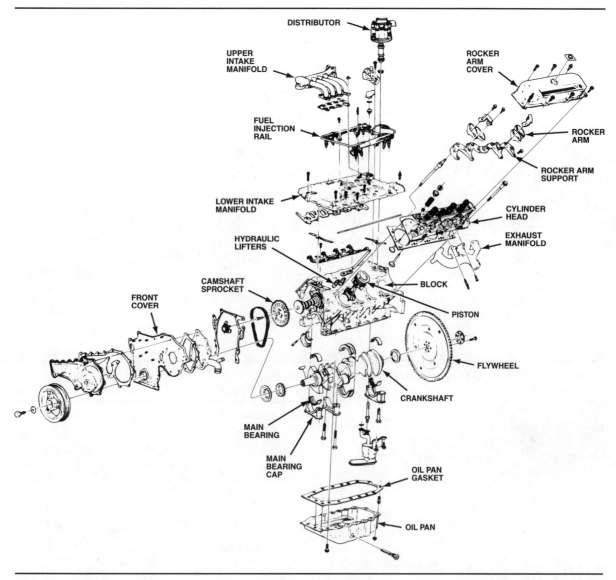

Figure 1-23. Not all the engines that you service will be Cadillac V8s, but they will all share most of the parts shown here.

delivering the mixture to the combustion chamber, figure 1-24. The manifold also supports the carburetor or fuel injection system. We detail the intake system later in the classroom manual.

Fuel System

The fuel system stores the fuel and delivers it to the engine. It consists of a tank, a fuel pump, lines, filters, and a fuel delivery system — the carburetor or fuel injection, figure 1-25. Fuel lines route fuel from the tank to the fuel delivery system where it is used by the engine. One or more filters remove dirt and contaminants from the fuel during its flow through the lines. The fuel pump provides the pressure that moves the fuel through the lines.

Cooling System

The great heat of combustion would rapidly destroy an engine if the cooling system did not remove much of it. The cooling system consists of a radiator, hoses, water pump, and a liquid coolant, figure 1-26. The water pump forces the coolant through the engine, where it absorbs heat along the way, and then flows through the radiator. As the coolant passes through the radiator, heat dissipates into the air. We detail the cooling system later in the classroom manual.

Lubrication System

The moving parts of a reciprocating engine cause a great deal of friction-generated heat as they roll and mesh together. Without proper lubrication they would

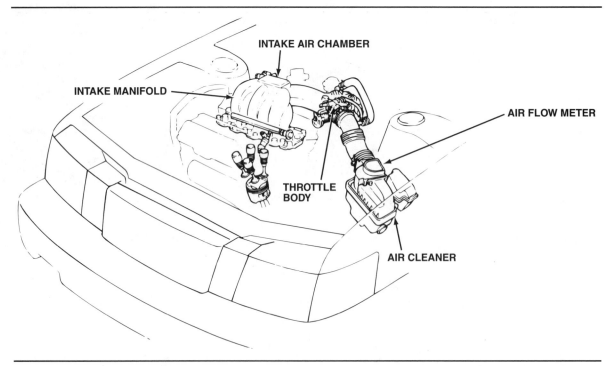

Figure 1-24. This Lexus intake system features a remote-mounted air cleaner and fuel-injection.

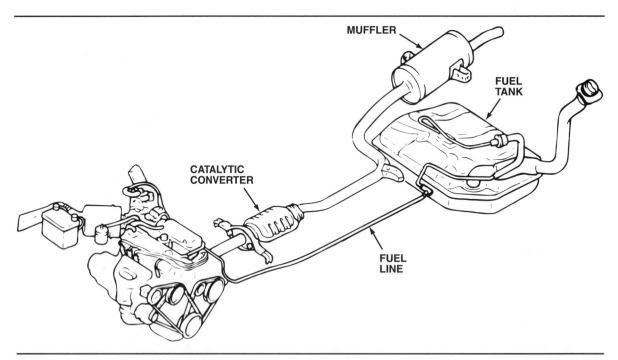

Figure 1-25. This Mitsubishi Precis fuel system is similar to many that mount the fuel pump inside the fuel tank.

quickly weld themselves solid. The automotive lubrication system consists of a motor oil, an oil pan for its storage, a filter, and an oil pump to force the oil through passageways in the engine so the oil reaches critical moving parts, figure 1-27. We cover the lubrication system later in the classroom manual.

Exhaust System

The main parts of the exhaust system, the exhaust manifolds and pipes, direct the burned exhaust gases out of the engine and to the rear of the vehicle. Early in the development of the automobile, manufacturers added mufflers to quiet the exhaust. In 1975, manufacturers

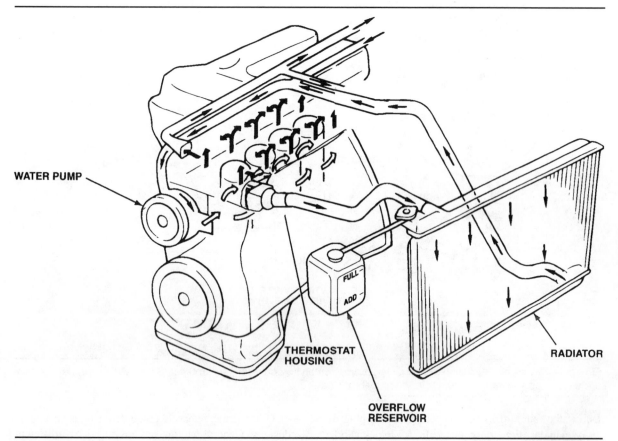

Figure 1-26. The coolant is forced through the engine, hoses, and radiator by a pump. Air flowing over the radiator reduces coolant and, therefore, engine temperature.

added catalytic converters to reduce tailpipe emissions, figure 1-28. We cover the exhaust system in more detail later in the classroom manual.

THE IGNITION SYSTEM

After the air-fuel mixture is drawn into the cylinder and compressed by the piston, it must be ignited. The ignition system creates a high electrical potential or voltage. This voltage jumps a gap between two electrodes in the combustion chamber. The arc (spark) between the electrodes ignites the compressed mixture. The two electrodes are part of the spark plug, which is a major part of the ignition system.

Ignition Interval

The timing of the spark is critical to proper engine operation. It must occur near the start of the power stroke. If the spark occurs too early or too late, full power will not be obtained from the burning air-fuel mixture.

As we have seen, every two strokes of a piston rotate the crankshaft 360 degrees, and there are 720 degrees of rotation in a complete four-stroke cycle.

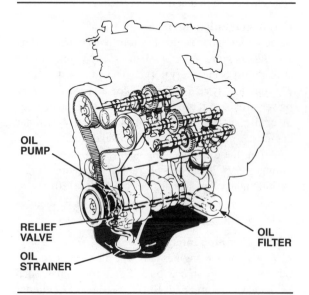

Figure 1-27. Modern lubrication systems pull oil from the pan, send it through the oil filter and feed it under pressure to important rotating parts like the camshaft and crankshaft bearings.

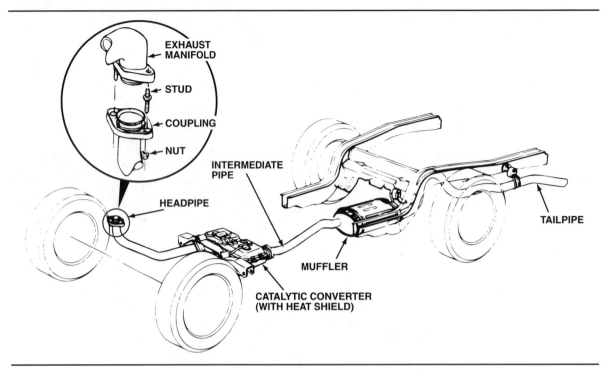

EXHAUST
MANIFOLD

STUD

COUPLING

NUT

INTERMEDIATE
PIPE

HEADPIPE

TAILPIPE

MUFFLER

CATALYTIC CONVERTER
(WITH HEAT SHIELD)

Figure 1-28. This exhaust system has a single pipe, catalytic converter, and muffler. Some V engines have a separate pipe, converter, and muffler for each cylinder bank.

During the four strokes of the cycle, the spark plug for each cylinder fires only once. In a single-cylinder engine, there would be only one spark every 720 degrees. These 720 degrees are called the **ignition interval**, or **firing interval**, which is the number of degrees of crankshaft rotation that occur between ignition sparks.

Common ignition intervals

Since a 4-cylinder engine has four power strokes during 720 degrees of crankshaft rotation, one power stroke must occur every 180 degrees (720 ÷ 4 = 180). The ignition system must produce a spark for every power stroke, so it produces a spark every 180 degrees of crankshaft rotation. This means that a 4-cylinder engine has an ignition interval of 180 degrees.

An in-line 6-cylinder engine has six power strokes during every 720 degrees of crankshaft rotation, for an ignition interval of 120 degrees (720 ÷ 6 = 120). An 8-cylinder engine has an ignition interval of 90 degrees (720 ÷ 8 = 90).

Unusual ignition intervals

Most automotive engines have 4, 6, or 8 cylinders, but other engines are in use today. Some companies, such as Jaguar, Ferrari, and BMW, produce 12-cylinder engines, with a 60-degree firing interval. Audi and Mercedes vehicles with 5-cylinder engines have a 144-degree firing interval; Suzuki produces a 3-cylinder engine with a 240-degree firing interval used in the Chevrolet Metro.

Other firing intervals result from unusual engine designs. General Motors has produced two different V6 engines from V8 engine blocks. The Buick version, developed in the 1960s, has alternating 90- and 150-degree firing intervals, figure 1-29. The uneven firing intervals resulted from building a V6 with a 90-degree crankshaft and block. This engine was modified in mid-1977 by redesigning the crankshaft to provide uniform 120-degree firing intervals, as in an in-line six. In 1978, Chevrolet introduced a V6 engine that fires at alternating 108- and 132-degree intervals.

Spark frequency

In a spark-ignition engine, each power stroke begins with a spark igniting the air-fuel mixture. Each power stroke needs an individual spark. An 8-cylinder engine, for example, requires four sparks per engine revolution (remember that there are two 360-degree engine revolutions in each 720-degree operating cycle). When the engine is running at 1,000 rpm, the ignition system must deliver 4,000 sparks per minute. At high speed, about 4,000 rpm, the ignition system must deliver 16,000 sparks per minute. The ignition system must perform precisely to meet these demands.

Firing Order

The order in which the air-fuel mixture is ignited within the cylinders is called the **firing order**, and it varies with different engine designs. Firing orders are

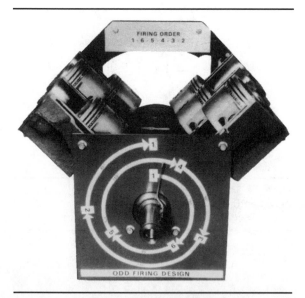

Figure 1-29. Buick's early V6 engine required uneven firing intervals to accommodate a 90-degree angle between the cylinder banks.

designed to reduce the vibration and imbalance created by the power strokes of the pistons. Engine designers number the cylinders for identification. However, the cylinders seldom fire in the order in which they are numbered.

The ignition system must deliver a spark to the correct cylinder at the correct time. To get the correct firing order, the spark plug cables must be attached to the distributor cap, or to the coil on distributorless ignition systems, in the proper sequence, figure 1-30.

In-line engines
Straight or in-line engines are numbered from front to rear, figure 1-31. The most common firing order for both domestic and imported 4-cylinder engines is 1-3-4-2. That is, the number one cylinder power stroke is followed by the number three cylinder power stroke, then the number four power stroke and, finally, the number two cylinder power stroke. Then, the next number one power stroke occurs again. Some in-line 4-cylinder engines have been built with firing orders of 1-2-4-3. One of these two firing orders is necessary due to the geometry of the engine.

The 5-cylinder, spark-ignited engine built by Audi and the 5-cylinder diesel engine built by Mercedes are also numbered front to rear. They share the same firing order of 1-2-4-5-3.

The cylinders of an in-line 6-cylinder engine are numbered from front to rear. The firing order for all in-line 6-cylinder engines, both domestic and imported, is 1-5-3-6-2-4.

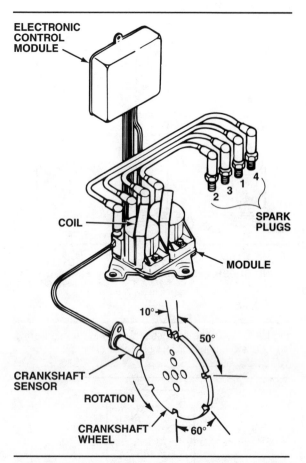

Figure 1-30. The spark plug wires must be connected in the proper sequence, or the engine will run poorly or not at all.

V engines
The V engine structure allows designers greater freedom in selecting a firing order and still producing a smooth-running powerplant. Consequently, there is great variety of cylinder numbering and firing orders for V engines, too many to cover completely. Figures 1-32 through 1-36 show a representative sampling of common cylinder numbering styles and firing orders for these engines.

Ignition Interval (Firing Interval): The number of degrees of crankshaft rotation between ignition sparks. Sometimes called firing interval.

Firing Order: The sequence by cylinder number in which combustion occurs in the cylinders of an engine.

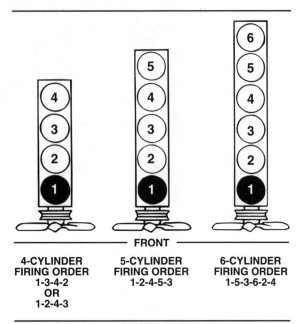

FRONT

4-CYLINDER
FIRING ORDER
1-3-4-2
OR
1-2-4-3

5-CYLINDER
FIRING ORDER
1-2-4-5-3

6-CYLINDER
FIRING ORDER
1-5-3-6-2-4

Figure 1-31. These are customary cylinder numbering and possible firing orders of in-line engines.

FIRING ORDERS

DOMESTIC
1-6-5-4-3-2

JAPANESE
1-2-3-4-5-6

Figure 1-32. The domestic 90-degree V6 engines and several Japanese 60-degree engines are numbered this way. However, they use different firing orders.

FIRING ORDER
1-2-3-4-5-6

Figure 1-33. General Motors 60-degree V6 engines and many Japanese V6 engines are numbered and fired this way.

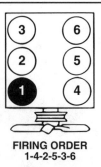

FIRING ORDER
1-4-2-5-3-6

Figure 1-34. Ford and Acura V6 engines are numbered and fired this way.

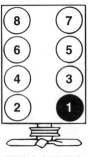

FIRING ORDERS

DOMESTIC AND LEXUS
1-8-4-3-6-5-7-2

INFINITI
1-8-7-3-6-5-4-2

Figure 1-35. Most Chrysler and General Motors V8s as well as the 1990 Lexus V8s are numbered and fired this way. The 1990 Infiniti V8 is numbered this way, but fires differently.

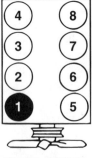

FIRING ORDERS

FORD
1-3-7-2-6-5-4-8
OR
1-5-4-2-6-3-7-8

AUDI AND MERCEDES-BENZ
1-5-4-8-6-3-7-2

Figure 1-36. Ford V8 engines are numbered this way, but use two different firing orders, depending on the engine. Audi and Mercedes-Benz V8 engines are numbered like the Fords, but use a third firing order.

ENGINE-IGNITION SYNCHRONIZATION

During the engine operating cycle, the intake and exhaust valves open and close at specific times. The ignition system delivers a spark when the piston is near the top of the compression stroke and both valves are closed. These actions must all be coordinated, or engine damage can occur.

Distributor Drive

The distributor must supply one spark to each cylinder during each cylinder's operating cycle. The distributor cam has as many lobes as the engine has cylinders, or in an electronic ignition system, the trigger wheel has as many teeth as the engine has cylinders. One revolution of the distributor shaft will deliver one spark to each cylinder. Since each cylinder needs only one spark for each two crankshaft revolutions, the distributor shaft must turn at only one-half engine crankshaft speed. Therefore, the distributor is driven by the camshaft, which also turns at one-half crankshaft speed.

On vehicles equipped with a distributorless ignition system (DIS), a DIS fires the spark plugs using a multiple coil pack containing two or three separate ignition coils, according to the number of cylinders. A computer called the ignition control module (ICM) discharges each coil separately in sequence, with each coil serving two cylinders 360 degrees apart in the firing order.

Crankshaft Position

As you learned earlier, the exact bottom of the piston stroke is called bottom dead center (BDC). The exact top of the piston stroke is called top dead center (TDC). The ignition spark occurs near TDC, as the compression stroke is ending. As the piston approaches the top of its stroke, it is said to be before top dead center (BTDC). A spark that occurs BTDC is called an advanced spark, figure 1-37. As the piston passes TDC and starts down, it is said to be after top dead center (ATDC). A spark that occurs ATDC is called a retarded spark.

Burn Time

The instant the air-fuel mixture ignites until its combustion is complete is called the burn time. For pump gasoline, burn time requires only a few milliseconds. The exact duration of burn time varies depending on a number of factors, the most important of which include fuel type, compression ratio, bore size, spark plug location, and combustion chamber shape.

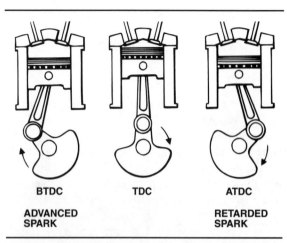

Figure 1-37. Piston position is identified in terms of crankshaft rotation.

As its name implies, burn time is a function of time and not of piston travel or crankshaft degrees. The ignition spark must occur early enough so that the combustion pressure reaches its maximum just when the piston is beginning its downward power stroke. Combustion should be completed by about 10 degrees ATDC. If the spark occurs too soon, BTDC, the rising piston will be opposed by combustion pressure. If the spark occurs too late, the force on the piston will be reduced. In either case, power will be lost. In extreme cases, the engine could be damaged. Ignition must start at the proper instant for maximum power and efficiency. Engineers perform many tests with running engines to determine the proper time for ignition to begin.

Engine Speed

As engine speed increases, piston speed increases. If the air-fuel ratio remains relatively constant, the fuel burning time will remain relatively constant. However, at greater engine speed, the piston will travel farther during this burning time. The spark must occur earlier to ensure that maximum combustion pressure occurs at the proper piston position. Making the spark occur earlier is called spark advance or ignition advance.

For example, consider an engine, figure 1-38, that requires 0.003 seconds for the fuel charge to burn and that achieves maximum power if the burning is completed at 10 degrees ATDC.

- At an idle speed of 625 rpm, position A, the crankshaft rotates about 11 degrees in 0.003 seconds. Therefore, timing must be set at 1 degree BTDC to allow ample burning time.
- At 1,000 rpm, position B, the crankshaft rotates 18 degrees in 0.003 seconds. Ignition should begin at 8 degrees BTDC.

• At 2,000 rpm, position C, the crankshaft rotates 36 degrees in 0.003 seconds. Spark timing must be advanced to 26 degrees BTDC.

INITIAL TIMING

As we have seen, ignition timing must be set correctly for the engine to run at all. This is called the engine's initial, or base, timing. Initial timing is the correct setting at a specified engine speed. In figure 1-38, initial timing was l degree BTDC. Initial timing is usually within a few degrees of top dead center. For many years, most engines were timed at the specified slow-idle speed for the engine. However, some engines built since 1974 require timing at speeds either above or below the slow-idle speed.

Timing Marks

We have seen that base timing is related to crankshaft position. To properly time the engine, we must be able to determine crankshaft position. The crankshaft is completely enclosed in the engine block, but most cars have a pulley and vibration damper bolted to the front of the crankshaft figure 1-39. This pulley rotates with the crankshaft and can be considered an extension of the shaft.

Marks on the pulley show crankshaft position. For example, when a mark on the pulley is aligned with a mark on the engine block, the number 1 piston is at TDC.

Timing marks vary widely, even within a manufacturer's product line. There are two common types of timing marks, figure 1-40:

• A mark on the crankshaft pulley, and marks representing degrees of crankshaft position, on the engine block, position A.
• Marks on the pulley, representing degrees of crankshaft position, and a pointer on the engine block, position B.

Some vehicles may also have a notch on the engine flywheel and a scale on the transmission cover or bellhousing. Some vehicles, in addition to a conventional timing mark, have a special test socket for electromagnetic engine timing. Technicians time these vehicles with a conventional timing light, or a special test probe that fits into the socket that was also used on the assembly line in the manufacture of the vehicle.

OTHER ENGINE TYPES

Other engine types besides the four-stroke engine have been installed in automobiles over the years, but only three have been used with any real success — the two-stroke, the diesel, and the rotary engines.

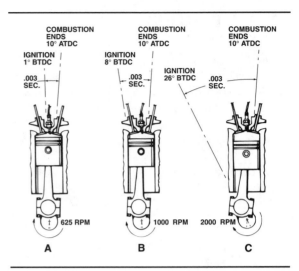

Figure 1-38. As engine speed increases, ignition timing must be advanced.

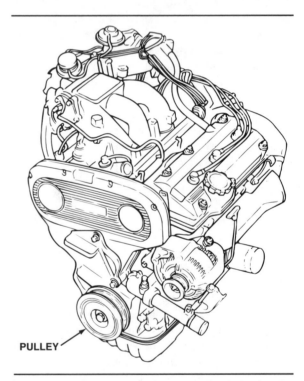

Figure 1-39. Most engines have a pulley bolted to the front end of the crankshaft.

The Two-Stroke Engine

While four-stroke engines develop one power stroke for every two crankshaft revolutions, two-stroke engines produce a power stroke for each revolution. Poppet valves are not used in most automobile and motorcycle two-stroke engines. The valve work is done instead by the piston, which uncovers intake and exhaust ports in the cylinder as it nears the bottom of its stroke.

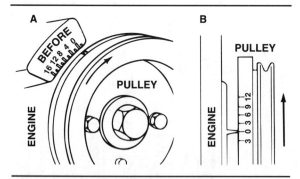

Figure 1-40. These are common types of timing marks.

Two-stroke gasoline engines used in motorcycles and some small cars draw the air-fuel mixture into the crankcase where it is partially compressed for delivery to the cylinders. As the piston moves up, pressure increases in the cylinder above it while pressure decreases in the crankcase below it. By using the crankcase to pull in the air-fuel mixture, the two-stroke engine combines the intake and compression strokes. Also, by using exhaust ports in the cylinder wall, the power and exhaust strokes are combined. The operating cycle of a two-stroke engine shown in figure 1-41 works like this:

1. Intake and Compression: As the piston moves up, a low-pressure zone is created in the crankcase. A **reed valve** or **rotary valve** opens, and the air-fuel mixture is drawn into the crankcase. At the same time, the piston compresses a previous air-fuel charge in the cylinder. Ignition occurs in the combustion chamber when the piston is near top dead center.

2. Power and Exhaust: Ignited by the spark, the expanding air-fuel mixture forces the piston downward in the cylinder. As the piston travels down, it uncovers exhaust ports in one side of the cylinder. The pressure in the cylinder forces most of the exhaust through the ports and out of the engine. As the piston continues down, it compresses the air-fuel charge in the crankcase. The intake reed valve or rotary valve closes to hold the charge in the crankcase. After uncovering the exhaust ports, the piston uncovers similar intake ports nearer the bottom of the stroke.

The compressed mixture in the crankcase moves through a transfer port to the intake ports and flows into the cylinder. The force of the incoming mixture also helps to drive remaining exhaust gases from the cylinder. In some engines a ridge on the top of the piston deflects the intake mixture upward in the cylinder so it will not flow directly out the exhaust ports.

The piston then begins another upward compression stroke, closing off the intake and the exhaust

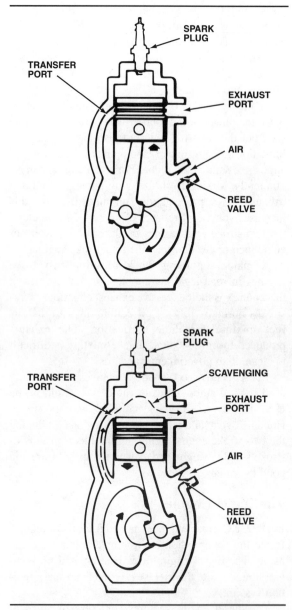

Figure 1-41. Two-stroke cycle engine operation produces a power stroke on every revolution of the crankshaft.

Reed Valve: A one-way check valve. A reed, or flap, opens to admit a fluid or gas under pressure from one direction, while closing to deny movement from the opposite direction.

Rotary Valve: A valve that rotates to cover and uncover the intake port of a two-stroke engine at the proper time. Rotary valves are usually flat discs driven by the crankshaft. Rotary valves are more complex than reed valves, but are effective in broadening a two-stroke engine's power band.

ports, and the process begins again. Because the crankcase of a two-stroke gasoline engine is used for air-fuel intake, it cannot be used as a lubricating oil reservoir. Therefore, engine oil for a two-stroke gasoline engine must be mixed with the fuel or injected directly into the crankcase.

In theory, the two-stroke engine should develop twice the power of the four-stroke engine of the same size, but the two-stroke design also has its practical limitations. With intake and exhaust occurring at almost the same time, it does not breathe or take in air efficiently. Mixing fresh fuel with unburned fuel and exhaust gases preheats the mixture. Because this increases volume, it reduces efficiency. Nevertheless, the two-stroke engine does produce more power per cubic inch or per liter than the four-stroke engine.

A major stumbling block to using two-stroke engines in road-legal automobiles or motorcycles in this country is their excessive exhaust emissions. Two-strokes normally rely on oil coming in with the air-fuel mixture for their lubrication. The exhaust produced from this "dirty" intake mixture is difficult to clean up to government standards.

Consequently, no two-stroke engines are currently used in road automobiles or motorcycles sold in the U.S. However, Chrysler, General Motors, Ford, Honda, Toyota, and others are working on eliminating the two-stroke engine's emission problems. The two-stroke's light weight and power for its size are too good to ignore.

The Diesel Engine

In 1892, a German engineer named Rudolf Diesel perfected the compression-ignition engine that bears his name. The diesel engine uses heat created by compression to ignite the fuel, so it requires no spark ignition system.

The diesel engine requires compression ratios of 16:1 and higher. Incoming air is compressed until its temperature reaches about 1,000°F (540°C). As the piston reaches the top of its compression stroke, fuel is injected into the cylinder, where it is ignited by the hot air, figure 1-42. As the fuel burns, it expands and produces power.

Diesel engines differ from gasoline-burning engines in other ways. Instead of a carburetor to mix the fuel with air, a diesel uses a precision **injection pump** and individual fuel injectors. The pump delivers fuel to the injectors at a high pressure and at timed intervals. Each injector measures the fuel exactly, spraying it into the combustion chamber at the precise moment required for efficient combustion, figure 1-43. The injection pump and injector system thus perform the tasks of the carburetor and distributor in a gasoline engine.

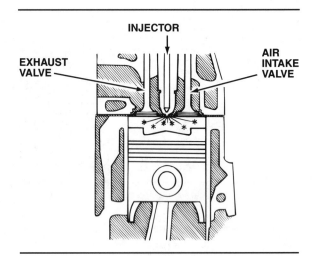

Figure 1-42. Diesel combustion occurs when fuel is injected into the hot, highly compressed air in the cylinder.

The air-fuel mixture of a gasoline engine remains nearly constant — changing only within a narrow range — regardless of engine load or speed. But in a diesel engine, *air* remains constant and the amount of *fuel* injected is varied to control power and speed. The air-fuel mixture of a diesel can vary from as lean as 85:1 at idle, to as rich as 20:1 at full load. This higher air-fuel ratio and the increased compression pressures make the diesel more fuel-efficient than a gasoline engine.

Like gasoline engines, diesel engines are built in both two-stroke and four-stroke versions. The most common two-stroke diesels are the truck and industrial engines made by the Detroit Diesel. In these engines, air intake is through ports in the cylinder wall. Exhaust is through poppet valves in the head. A blower box blows air through the intake port to supply air for combustion and to blow the exhaust gases out of the exhaust valves. Crankcase fuel induction cannot be used in a two-stroke diesel.

For many years, diesel engines were used primarily in trucks and heavy equipment. Mercedes-Benz, however, has built diesel cars since 1936, and the energy crises of 1973 and 1979 focused attention on the diesel as a substitute for gasoline engines in automobiles.

In the late 1970s, General Motors and Volkswagen developed diesel engines for cars. They were followed quickly by most major vehicle manufacturers in offering optional diesel engines for their vehicles. By the early 1980s many vehicle manufacturers were predicting diesel power for more than 30 percent of the domestic auto population. These predictions did not come true.

In the mid 1980s, increasing gasoline supplies and lower prices in combination with the disadvantages of

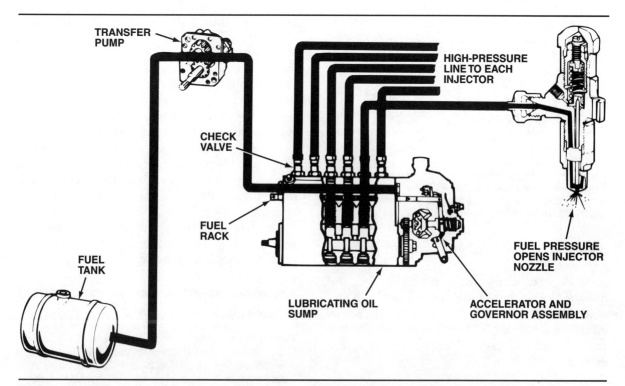

Figure 1-43. This is a typical automotive diesel fuel injection system.

noise and higher purchasing costs reduced the incentives for customers to buy diesel automobiles. Stringent diesel emission regulations added even more to manufacturing costs and made the engines harder to certify for sale. A few automobile diesel engines, such as the Oldsmobile V8, were derived from gasoline powerplants and suffered reliability problems. In the late 1980s, diesel engine use in the U.S. and Canada was mostly limited to truck and industrial applications, although late-model diesel automobiles, such as those made by Volkswagen, continue to sell in the United States and in Europe, where gasoline costs remain high.

The Rotary (Wankel) Engine

The reciprocating motion of a piston engine is both complicated and inefficient. For these reasons, engine designers have spent decades attempting to devise engines in which the working parts would all rotate on an axis. The major problem with this rotary concept has been the sealing of the combustion chamber. Of the various solutions proposed, only the rotary design of Felix Wankel — as later adapted by NSU, Curtiss-Wright, and Toyo Kogyo (Mazda) — has proven practical.

Although the same sequence of events occur in both a rotary and a reciprocating engine, the rotary is quite different in design and operation. A curved triangular rotor moves on an **eccentric**, or off-center,

geared portion of a shaft within an elliptical chamber, figure 1-44. As it turns, seals on the rotor's corners follow the housing shape. The rotor thus forms three separate chambers whose size and shape change constantly during rotation. The intake, compression, power, and exhaust functions occur within these chambers, figure 1-45. Wankel engines can be built with more than one rotor. Mazda production engines, for example, were 2-rotor engines.

One revolution of the rotor produces three power strokes, or pulses, one for each face of the rotor. In fact, each rotor face can be considered the same as one piston. Each pulse lasts for about three-quarters of a rotor revolution. The combination of rotary motion and longer, overlapping power pulses results in a smooth-running engine.

Nearly equivalent in power output to that of a 6-cylinder piston engine, a 2-rotor engine is only one-third to one-half the size and weight. With no pistons, connecting rods, valves, lifters, and other reciprocating

Injection Pump: A pump used on diesel engines to deliver fuel under high pressure at precisely timed intervals to the fuel injectors.

Eccentric: Off center. A shaft lobe which has a center different from that of the shaft.

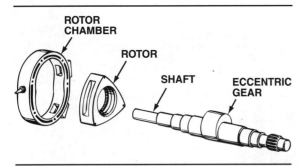

Figure 1-44. The main parts of a Wankel rotary engine are the rotor chamber, the three-sided rotor, and the shaft with an eccentric gear.

parts, the rotary engine has 40 percent fewer parts than a piston engine.

While the rotary overcomes many of the disadvantages of the piston engine, it has its own disadvantages. It is basically a very "dirty" engine. In other words, it gives off a high level of emissions and requires additional external devices to clean up the exhaust.

SUMMARY

Most automobile engines are internal combustion, reciprocating four-stroke engines. An air-fuel mixture is drawn into sealed combustion chambers by a vacuum created by the downward stroke of a piston. The mixture is ignited by a spark.

Valves at the top of the cylinder open and close to admit the air-fuel mixture and release the exhaust. These valves are driven by a camshaft and synchronized with engine rotation. The sequence in which the cylinders fire is the firing order. Several valve designs have been used, including I-head, F-head, and L-head designs. Most engine cylinders have two valves, but some engines have three, four, or five valves per cylinder.

Displacement and compression ratio are two frequently used engine specifications. Displacement indicates engine size, and compression ratio compares total cylinder volume to compression chamber volume.

The ignition interval is the number of degrees between ignition sparks. A 4-cylinder engine commonly has an ignition interval of 180 degrees, a V8—90 degrees, and an in-line 6-cylinder—120 degrees (although many intervals have been used over the years). Still, each cylinder fires once every 720 degrees of crankshaft rotation. The ignition spark is provided by the distributor or electronic ignition, and is synchronized with the crankshaft rotation.

Correct ignition timing is essential for the engine to operate. Timing marks on the front of the engine block or the pulley indicate crankshaft position, and can be used to alter the timing from the engine's base or initial timing.

The most successful automobile engine types used besides four-stroke gasoline engines are the two-stroke, the diesel, and the rotary (Wankel).

■ Rudolf Diesel

The theory of a diesel engine was first set down on paper in 1893 when Rudolf Christian Karl Diesel wrote a technical paper, "Theory and Construction of a Rational Heat Engine." Diesel was born in Paris in 1858. After graduating from a German technical school, he went to work for the refrigeration pioneer, Carl von Linde. Diesel was a success in the refrigeration business, but at the same time he was developing his theory on the compression ignition engine. The theory is really very simple. Air, when it is compressed, gets very hot. If you compress it enough, say on the order of 20:1, and squirt fuel into the compressed air, it will ignite.

At first, the diesel engine was used to power machinery in shops and plants. By 1910, it was used in ships and locomotives and a 4-cylinder engine was even used in a delivery van. The engine proved too heavy at that time to be practical for automobile use. It was not until 1927, when Robert Bosch invented a small fuel injection mechanism, that the use of the diesel engine in trucks and cars became practical.

Unlike many of the early inventors, Diesel did make a great deal of money from his invention, but in 1913 he disappeared from a ferry crossing the English Channel and was presumed to have committed suicide.

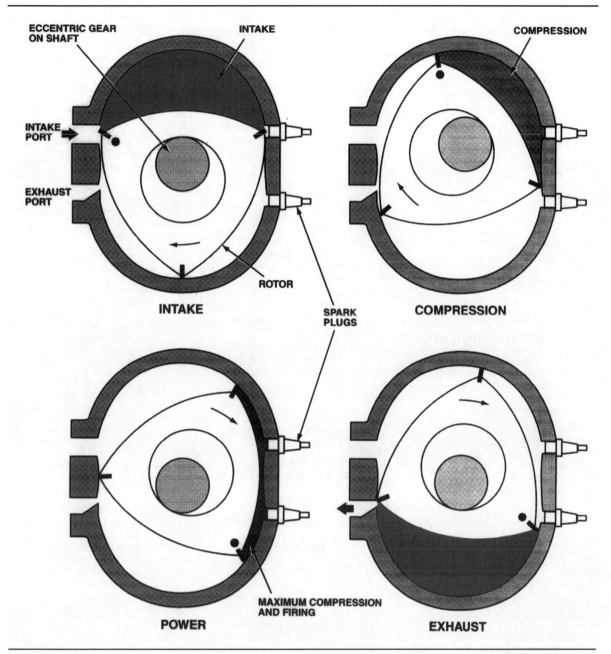

Figure 1-45. These are the four stages of rotary operation. They correspond to the intake, compression, power, and exhaust strokes of a 4-stroke reciprocating engine. The sequence is shown for only one rotor face, but each face of the rotor goes through all four stages during each rotor revolution.

Review Questions

Choose the single most correct answer.
Compare your answers with the correct answers on Page 328.

1. An automotive internal combustion engine:
   a. Uses energy released when a compressed air-fuel mixture is ignited
   b. Has pistons which are driven downward by explosions in the combustion chambers
   c. Is better than an external combustion engine
   d. Burns gasoline

2. A spark plug fires near the end of the:
   a. Intake stroke
   b. Compression stroke
   c. Power stroke
   d. Exhaust stroke

3. The flywheel:
   a. Keeps the engine turning during the "coasting" or non-power stroke periods of the engine
   b. Provides a mounting plate for the clutch or torque converter
   c. Smoothes out increases and decreases in engine power output, providing continuous thrust
   d. All of the above

4. The camshaft rotates at:
   a. The same speed as the crankshaft
   b. Double the speed of the crankshaft
   c. Half the speed of the crankshaft
   d. It depends on the engine design

5. Which of the following is never in direct contact with the engine valves?
   a. Pushrods
   b. Valve springs
   c. Rocker arms
   d. Valve seats

6. The bore is the diameter of the:
   a. Connecting rod
   b. Cylinder
   c. Crankshaft
   d. Combustion chamber

7. The four-stroke cycle operates in which order?
   a. Intake, exhaust, power, compression
   b. Intake, power, exhaust, compression
   c. Compression, power, intake, exhaust
   d. Intake, compression, power, exhaust

8. Diesel engines:
   a. Have no valves
   b. Produce ignition by heat of compression
   c. Have low compression
   d. Use special carburetors

9. How many strokes of a piston are required to turn the crankshaft through 360 degrees?
   a. One
   b. Two
   c. Three
   d. Four

10. A "retarded spark" is one that occurs:
    a. At top dead center
    b. Before top dead center
    c. After top dead center
    d. At bottom dead center

11. An internal combustion engine is an efficient use of available energy because:
    a. The combustion of the fuel produces more energy than is required to compress and fire it
    b. Gasoline is so easy to burn
    c. Internal combustion engines turn with so little friction
    d. Compression ratios are high enough to vaporize gasoline

12. Which of the following shows the typical ignition interval of an in-line 6-cylinder engine?
    a. 600 degrees ÷ 6 = 100 degrees
    b. 360 degrees ÷ 3 = 120 degrees
    c. 90 degrees × 6 = 540 degrees
    d. 720 degrees ÷ 6 = 120 degrees

13. How many sparks (spark plug firings) per crankshaft revolution are required by a typical 8-cylinder engine?
    a. 8
    b. 6
    c. 4
    d. 2

14. The firing order of an engine is
    a. The same on all V8s
    b. Stated in reference books and on the cylinder block
    c. Deduced by common sense
    d. None of the above

15. Ignition timing:
    a. Requires a basic setting specified in degrees BTDC for a particular engine
    b. Varies while engine is running, synchronizing combustion with proper piston position
    c. Is indicated by the alignment of markings on the crankshaft pulley or flywheel with other markings or a pointer fixed to the cylinder block
    d. All of the above

16. In a four-stroke engine, the piston is driven down in the cylinder by expanding gases during the _____ stroke.
    a. Compression
    b. Exhaust
    c. Power
    d. Intake

17. A two-stroke engine:
    a. Produces a power stroke for each crankshaft revolution
    b. Has intake and exhaust ports
    c. May use the crankcase for fuel induction
    d. All of the above

18. The ignition interval of an engine is the number of degrees of crankshaft rotation that:
    a. Occur between ignition sparks
    b. Are required to complete one full stroke
    c. Take place in a 4-stroke engine
    d. All of the above

19. The rotary engine:
    a. Operates without pistons, connecting rods, or poppet valves
    b. Is not a reciprocating engine
    c. Produces three power strokes per revolution of its rotor
    d. All of the above

20. Which of the following is not used in calculating engine displacement?
    a. Stroke
    b. Bore
    c. Number of cylinders
    d. Valve arrangement

21. To change cubic centimeters to cubic inches, multiply by:
    a. 0.061
    b. 16.39
    c. 61.02
    d. 1000

22. Compression ratio is
    a. Piston displacement plus clearance volume
    b. Total volume times number of cylinders
    c. Total volume divided by clearance volume
    d. Stroke divided by bore

2

Engine Physics and Chemistry

In this chapter, we first detail the principles of physics and chemistry that are applied in an internal combustion engine. Next, we look at an engine's air and fuel requirements. Finally, we examine the basic problems of air pollution and how the automobile engine is involved in the problem.

ENGINE PHYSICS

We sometimes talk about engines in terms of their horsepower and torque ratings. We also know that engines burn fuel to release energy and do work. But what do these terms mean? A simple review of key terms will help you understand why engines of different sizes and designs produce different amounts of power and can do different amounts of work.

Matter, Mass, and Weight

All the substances from which things in the world are made are composed of **matter**. In general, matter is anything that can be touched. Matter is found on the earth's surface in one of three states: solid, liquid, or gas.

We measure the amount of matter in an object by its **mass**. Mass is what gives an object weight. Mass also gives an object inertia, which we defined in Chapter 1.

On the Earth, the words mass and weight can be used interchangeably, but they are not really the same thing. Mass is the amount of matter in an object. The **weight** of the object is the force it exerts against a surface because of the Earth's gravity — against the ground, for instance. An astronaut who weighs 150 pounds (68 kg) on Earth will weigh only 25 pounds (11 kg) standing on the surface of the moon, figure 2-1. The astronaut's *mass* is the same in both cases, but the moon's gravitational field is only 1/6 that of the Earth, so the astronaut's *weight* is only 1/6 as much.

This distinction becomes important to automotive technicians when dealing with metric measurements of torque and horsepower. In the U.S. Customary system, mass and force are often measured with the same units: pounds. In the metric system, mass is measured in grams (kilograms), and force is more often measured in different units called newtons. We will explain these units more fully in later sections when we deal with torque and horsepower.

Energy

The ability to do work is called **energy**. Heat, light, sound, and electricity are all forms of energy. These things do not have weight or occupy space, but they all have the ability to do work. The energy of both food and fuel is commonly measured in terms of how

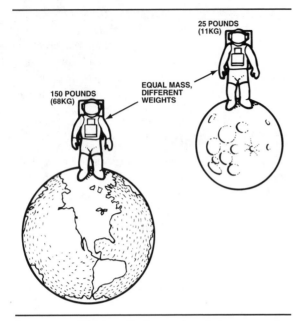

Figure 2-1. Weight will change in relation to gravitational pull, but mass remains constant.

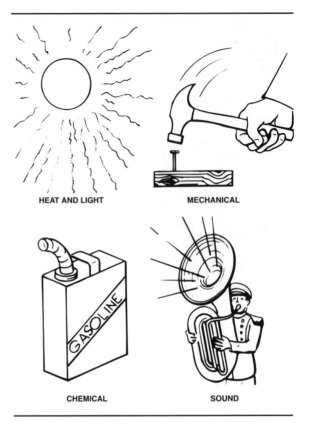

Figure 2-2. Energy, the ability to perform work, exists in many forms.

much heat it can produce, in calories. A **calorie** is the amount of heat required to raise the temperature of one gram of water by one degree Celsius. This is not very much heat, so most people discuss heat in terms of kilocalories (1000 calories). The **Calorie** values listed in nutrition books are actually kilocalories, and the word is usually capitalized to distinguish it from the smaller calorie.

You may occasionally see references in older books and manuals to another way of measuring heat, the **British Thermal Unit (Btu)**. The Btu is the amount of heat needed to raise the temperature of one pound (454 g) of water by one degree Fahrenheit (0.556 degrees Celsius). One Btu is the same amount of heat as 252 calories.

Energy can be stored in many forms, figure 2-2. Energy stored and available to do work but not actually being used is called **potential energy**. A car parked on a hill top and a stick of dynamite both have potential energy. A fundamental form of mechanical energy is **kinetic energy** — the energy of mass in motion. A moving car and a spinning crankshaft possess kinetic energy. The amount of that energy is determined by the object's mass and speed. The greater the mass of an object and the faster it moves, the more kinetic energy it possesses.

ENGINE TORQUE

A push or pull acting on an object is called a **force**. Forces may start, stop, or change the direction of motion. To apply force we need energy. In a car engine, it is combustion of the air-fuel mixture in the

Matter: All substances are made from this material.

Mass: The measure of the amount of matter in an object.

Weight: The measure of the earth's gravitational pull on an object.

Energy: The ability to do work by applying force.

calorie: The amount of heat required to raise the temperature of one gram of water by one degree Celsius. When capitalized, it means 1,000 calories.

British Thermal Unit (Btu): The amount of heat required to raise the temperature of one pound of water one degree Fahrenheit.

Potential Energy: Energy stored but not being used. A wound-up spring and gallon can of gasoline both have potential energy.

Kinetic Energy: The energy of mass in motion. All moving objects possess kinetic energy.

Force: A push or pull acting on an object; it may cause motion or produce a change in position. Force is measured in pounds in the U.S. system and in newtons in the metric system.

■ Learn the Metric System!

The United States is slowly but surely converting to the metric system. Many measurements are already in metrics. For instance, cameras and film generally use millimeters (mm) for measuring the size of the lenses or the film. In electronics, we have always used the metric system of seconds, volts, watts, amperes and hertz (cycles per second). The pharmaceutical industry changed to the metric system more than 25 years ago, and the nation's vehicle manufacturers not only use liter measurement for displacement, but build some of their own engines and vehicles to metric dimensions. This nation's tire manufacturers use kilopascals as well as pounds per square inch to indicate tire inflation. Gasoline is sold by the liter in some parts of the country. And the fasteners that hold most domestic vehicles together are a mixture of metric and SAE sizes.

Technicians need to know and understand both the metric and U.S. Customary systems, and they must be able to use simple mathematics to convert from one system to the other. Technicians must also own wrenches and sockets that fit both metric and U.S. Customary sizes.

Standard Metric Prefixes and Abbreviations

Prefix	Magnitude		Abbreviation
Giga	1,000,000,000	or billion	G
Mega	1,000,000	or million	M
Kilo	1,000	or thousand	k
Hecto	100	or hundred	h
Deka	10	or ten	da
Deci	1/10	or one-tenth	d
Centi	1/100	or one-hundredth	c
Milli	1/1,000	or one-thousandth	m
Micro	1/1,000,000	or one-millionth	μ
Nano	1/1,000,000,000	or one-billionth	n

Standard Conversions from the U.S. Customary to the Metric System

Desc.	unit	x conversion factor	to get
length	inch	25.4	millimeters (mm)
	foot	0.3048	meters (m)
	yard	0.9144	meters (m)
	mile	1.609	kilometers (km)
volume	gallon	3.7854	liters (L)
	fluid ounce	29.57	milliliters (mL)
	cubic inch	16.387	cubic centimeters (cc)
	cubic inch	0.01639	liters (L)
weight	pound	0.4536	kilograms (kg)
	ounce	28.3495	grams (g)
temperature	F	(F [-]32) 0.556	Celsius (C)
pressure	psi	6.895	kilopascal (kPa)
vacuum	in-Hg	25.4	millimeters of mercury (mm-Hg)
force	pound	4.448	newton (N)
	ounce	0.278	newton (N)
torque	in-lb	0.11298	newton-meter (Nm)
	ft-lb	1.3558	newton-meter (Nm)
power	hp	0.746	kilowatts (kW)
velocity	mph	1.6093	kilometers/hour (km/h)
fuel consumption	mpg	0.4251	kilometers/liter (km/L)

Standard Conversions from the Metric to the U.S. Customary System

Desc.	unit	x conversion factor	to get
length	mm	0.0394	inches (in)
	m	3.281	feet (ft)
	m	1.094	yards (yd)
	km	0.6215	miles
volume	L	0.2642	gallons
	mL	0.0338	fluid ounces
	L	61.024	cubic inches (cid)
	cc	0.6102	cubic inches (cid)
weight	kg	2.2046	pounds (lbs)
	g	0.0353	ounces (oz)
temperature	C	(C [+]32) 1.798	Fahrenheit (F)
pressure	kPa	0.1450	pounds per square inch (psi)
force	N	0.2248	pounds (lbs)
	N	0.0141	ounces (oz)
torque	Nm	8.851	inch-pounds (in-lbs)
	Nm	0.7376	foot-pounds (ft-lbs)
power	kW	1.34	horsepower (hp)
vacuum	mm-Hg	0.0394	inches of mercury (in-Hg)
velocity	km/h	0.6214	miles per hour (mph)
fuel consumption	km/l	2.3524	miles per gallon (mpg)

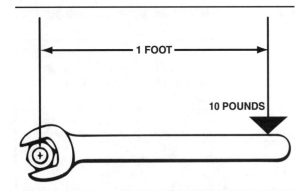

Figure 2-3. Torque is a twisting force equal to the distance to the axis times the force applied.

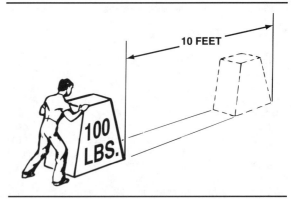

Figure 2-4. Work is calculated by multiplying force times distance. If you push 100 pounds 10 feet, you have done 1,000 foot-pounds of work.

cylinder that releases the energy in the fuel that exerts force on the piston. As you learned in Chapter 1, the downward, linear motion of the piston is changed to rotating, or turning motion in the crankshaft. The measure of the twisting or turning force the crankshaft exerts is called **torque**.

Torque is measured as the amount of force multiplied by the length of the lever through which it acts. When you put a 1-foot (30 cm)(*0.3 meter*) wrench on a bolt and apply 10 pounds (44 N) of force to the end of the wrench to turn the bolt, you exert 10 foot-pounds (14 Nm) of torque, figure 2-3.

In the metric system, force is measured in units called newtons. One newton equals the same amount of force as 0.225 foot-pounds, so one foot-pound is the same as 4.44 newtons. Torque in the metric system is measured as the amount of force (in newtons) multiplied by the distance through which the force acts (in meters). The unit is the **Newton-meter** (Nm), and is commonly specified in automotive service manuals for torque wrench settings. In some manuals, the metric units are the *only* ones presented. Most modern torque wrenches have both a U.S. Customary and a metric scale. Newton-meters and foot-pounds can be converted back and forth, using the following formulas:

- Foot-pounds × 1.3558 = Newton-meters
- Newton-meters × 0.7376 = foot-pounds

Engine torque is measured at the end of the rotating crankshaft or the flywheel. Therefore, a device to measure engine torque while the engine runs is necessary. Imagine a pulley attached to an engine's crankshaft, with a clutch to connect and disconnect it from the engine. Around the pulley is a rope attached to a weight on the floor. If we start the engine and slowly engage the clutch, the pulley will turn and lift the weight. If we keep adding weights until the engine finally cannot lift them, the total weight will stop the engine. In this way, we can measure the amount of torque an engine generates.

We can do this with any engine. Some engines are relatively strong and can lift very heavy loads. The amount of weight lifted by different engines gives us a standard of comparison. This simple example is a handy way to compare torque output of simple, stationary engines. However, it is not practical for use with an automobile engine that must deliver torque continuously to drive a vehicle.

Now let's attach an arm to the flywheel clutch, or friction brake, on our engine and connect the other end of the arm to a scale. If we engage the clutch, the arm will push on the scale and give a reading in pounds. But to measure torque in foot-pounds, the length of the arm must be considered. If the length of the arm is two feet and the reading on the scale is 100 pounds (45 kg), we have a torque measurement of 200 foot-pounds (271 Nm).

Torque is a common measurement for comparing engines. It measures the amount of force an engine can exert, and is related to how much work the engine can perform. Torque is exerted whether or not any motion is produced, but when engine torque produces motion, we say **work** is accomplished. Work is calculated by multiplying the applied force by the distance the object moves. If you push 100 pounds 10 feet, you have done 1,000 foot-pounds of work, figure 2-4. These foot-pounds are a measure of work, not of force, and are not at all the same as the foot-pounds used to measure torque. Sometimes automobile manuals or specification sheets will give torque values in pounds-feet (lbs-ft), instead of foot-pounds (ft-lbs). This is done to avoid confusion between torque and work.

ENGINE HORSEPOWER

Torque may help us rank an engine's ability to do work, but it does not indicate how fast an engine can do work. **Power** is the measure of the rate or speed at which the work is done. As a comparison, torque measures an engine's ability to drive a vehicle up a hill, while power measures how rapidly the car can drive up the hill. A simple equation shows the relationship between power, work, and time:

$$\text{Power} = \frac{\text{Work}}{\text{Time}}$$

In the U.S., engines are commonly rated in **horsepower**. We use horsepower as a unit of measurement because horses were a common form of power used to do work immediately before the industrial revolution. Therefore, horses were something that could easily be compared to machines used to do similar work. In 1783, James Watt calculated that a horse could lift a 200-pound (91-kg) hopper of coal 165 feet (50 m) up in a mine shaft in one minute, figure 2-5. When we multiply 200 pounds (91 kg) by 165 feet (50 m), we get 33,000 foot-pounds (44,741 Nm). The standard definition of one horsepower then is the power needed to do 33,000 foot-pounds (44,741 Nm) of work in one minute. Two horsepower can do the same amount of work in half the time; four horsepower in one-quarter the time. In the metric system, engine power is measured in **kilowatts**. Horsepower and kilowatts can be converted back and forth, using the following formulas:

- Horsepower × 0.746 = Kilowatts (kW)
- Kilowatts × 1.34 = Horsepower.

To find an engine's horsepower, if we know the torque at a given rpm, we can use this formula:

$$\text{Horsepower} = \frac{\text{torque} \times 2 \text{ pi} \times \text{rpm}}{33,000}$$

Because pi equals 3.1416, this can be shortened to:

$$\text{Horsepower} = \frac{\text{torque} \times \text{rpm}}{5252}$$

When we translate engine torque into engine horsepower, we pretend the engine is moving a weight around a circle. The weight is equal to the number of pounds indicated on the scale at the end of the torque arm. If the weight were allowed to rotate, the arm would make a circle with a diameter twice the length of the arm.

To derive the formula for horsepower, we go from torque to work to power. For example, let's say that the torque scale still reads 100 pounds (45 kg) at the end of a 2-foot (0.6-meter) arm, so we still have a torque measurement of 200 foot-pounds (271 Nm). Now the engine is rotating, however, so the distance we use to compute work in foot-pounds is no longer just the length of an arm. It is the circumference of the circle described by the end of the rotating arm. Here is how we calculate the circumference of that circle:

2 × radius × pi = the circumference

or

2 × 2 feet × 3.1416 = 12.57 feet.

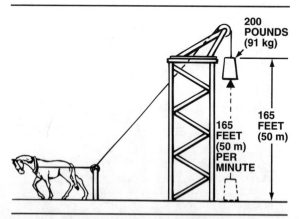

Figure 2-5. One horsepower equals 33,000 foot-pounds (200 x 165) of work per minute.

Now multiply that distance (the circumference) by 100 pounds to get the total work done during one engine revolution:

12.57 feet × 100 pounds = 1,257 foot-pounds.

We know the amount of torque exerted and the amount of work done by the engine, but we still don't know how fast the work is done and how much power is delivered. For that, we must know the speed of the engine in revolutions per minute (rpm). Let's assume the engine speed is 2,500 rpm. If we multiply the foot-pounds of work by the speed, we get the number of foot-pounds per minute:

1,257 ft-lbs × 2,500 rpm = 3,142,500 ft-lbs per minute

We know that one horsepower equals work done at the rate of 33,000 foot-pounds per minute. Therefore,

---

Torque: The tendency of a force to produce rotation around an axis. It is equal to the distance to the axis times the amount of force applied expressed in foot-pounds or inch-pounds (U.S. Customary system) or in Newton-meters (metric system).

Newton-meter (Nm): The metric unit of torque. One Newton-meter equals 1.356 foot pounds. One foot-pound equals 0.736 Newton-meter.

Work: The application of energy through force to produce motion.

Power: The rate or speed of doing work.

Horsepower: A measure of the rate at which work is done, equal to 33,000 foot-pounds of work per minute, or 0.746 kilowatts.

Kilowatt (kW): A measure of the rate at which work is done, equal to 1.3405 horsepower.

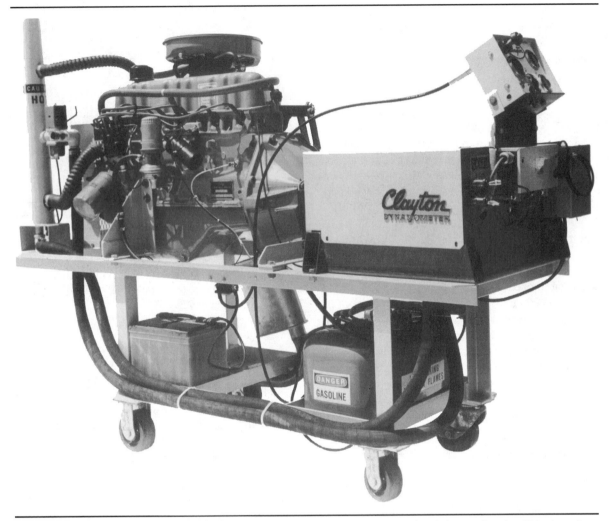

Figure 2-6. This small, portable Clayton dynamometer uses a water pump to load the engine. It will test engines with up to 250 horsepower (187 kW).

we can compute the horsepower delivered by the engine at 2,500 rpm as follows:

3,142,500 ft-lbs per minute ÷ 33,000 ft-lbs per minute = 95.23 hp.

Torque and Horsepower Relationships

The device used to measure engine torque on a scale is a **dynamometer**. Early dynamometers, called prony brakes, used a clutch or friction brake that was tightened to reduce the speed of the engine and produce a reading on the scale. Most modern dynamometers, figure 2-6 and figure 2-7, still act as a brake, but most use an electric generator or a water pump to absorb power.

In operation, the dynamometer never stops the engine; it must be kept running. Torque can be measured on an engine dynamometer at any rpm. Readings are usually taken at wide-open throttle so that the torque reading will be the maximum available at a given speed. Most street engines develop their

peak torque at a relatively low engine speed. Some street engine develop peak torque at engine speeds as low as 2,000 rpm, while other street engines do not develop peak torque until an rpm as high as 5,000 rpm. Larger street engines usually develop peak torque at a lower rpm than smaller street engines.

Peak horsepower is always developed at a higher rpm than peak torque. Some street engines develop their peak horsepower at engine speeds as low as 4,000 rpm, while other street engines develop peak horsepower at speeds of 7,000 rpm or more. Some automobile race engines develop peak horsepower at speeds as high as 14,000 rpm.

If we were to take readings of engine torque from idle to several thousand rpm, we would find torque increasing up to the rpm where the engine develops peak torque. Above the maximum figure, torque slowly falls off. This is because the faster an engine goes, the more trouble it has getting enough air.

Figure 2-7. This large Clayton dynamometer is also a water brake. It has the capacity to test engines with up to 1,000 horsepower (746 kW).

Without sufficient air to combine with fuel, combustion cannot produce force on the piston. However, as speed continues to increase, horsepower continues to rise. Remember, horsepower is a product of both torque and rpm. Horsepower will continue to increase as engine speed increases, as long as torque does not fall off too rapidly. At the upper end of an engine's speed range, torque drops off more rapidly. Horsepower then begins a similar drop.

Engine designers specify camshaft timing, intake and exhaust port sizes, and other engine tuning parameters to optimize breathing over a particular rpm range. The rpm range is chosen based on the size of the engine and the type of work the engine is required to do. Later chapters will detail how engine tuning affects horsepower and torque output.

Figure 2-8 shows typical torque and horsepower curves for a large V8 street engine, while figure 2-9 shows typical figures for a small 4-cylinder street engine. The large V8 develops its peak torque at a low 1,600 rpm, while the small 4-cylinder engine does not produce maximum torque until 4,000 rpm, more than twice the engine speed required of the larger engine. The smaller engine sacrifices breathing efficiency at low rpm so that it can breathe well at high rpm and make sufficient horsepower. Since its torque output is

relatively limited by its small size, it *needs* the rpm to build usable horsepower to do the work of driving the vehicle.

The torque curve of the large V8 engine starts high and stays high until its breathing becomes less efficient and torque falls off. Its large cylinders are able to bring in a sizable intake charge and produce great amounts of torque and horsepower even at low rpm. The designers of the V8 chose tuning specifications to best exploit this characteristic.

Brake horsepower

The horsepower measured on a dynamometer at the engine flywheel is called **brake horsepower**. The Society of Automotive Engineers (SAE) defines brake horsepower as the power available at the flywheel for doing useful work. However, there are different kinds of brake horsepower. For years, vehicle manufacturers

---

Dynamometer: A device used to measure the power of an engine or motor.

Brake Horsepower: The power available at the flywheel of an engine for doing useful work, as measured on a dynamometer.

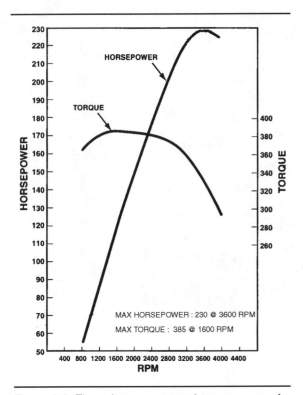

Figure 2-8. These horsepower and torque curves for the 7.4-liter General Motors Mark V light truck engine are representative of large V8 engines.

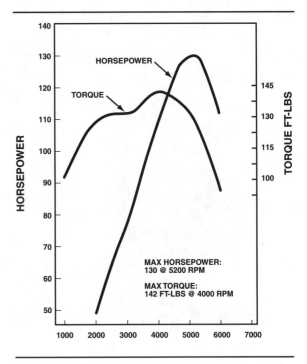

Figure 2-9. These horsepower and torque curves for the 2.2-liter Honda Accord EX are representative of small 4-cylinder engines.

advertised their engines in terms of **gross brake horsepower**. This is the power output of a basic engine, equipped only with the built-in accessories needed for its operation. These include the fuel, oil, and water pumps and any built-in emission controls. The gross brake horsepower is the maximum power available from an engine. In the days when high horsepower numbers made good advertising material, these were the numbers that the manufacturers emphasized.

A more realistic indication of the power available from an engine to drive a car is **net brake horsepower**. Net brake horsepower is the output of a fully equipped engine with accessories such as the air filter and exhaust systems, the cooling system, the alternator, and all bolt-on emission control equipment in place and operating. Since the early 1970s, vehicle manufacturers have been publishing the net brake horsepower of their engines, which is considerably lower than the gross power output.

Indicated horsepower and actual horsepower

From this brief comparison between gross and net brake horsepower, you can see that all of the power developed by an engine is not available at the flywheel for useful work. Part of an engine's power is used to drive accessories or is used in other systems.

For example, mechanical power is used to develop electrical energy in the alternator or to run the air-conditioning compressor.

We explained earlier that heat is a form of energy. Any heat lost is energy lost that could otherwise be used to produce power. A fairly large amount of power is dissipated as heat energy through the cooling system and out of the exhaust pipe.

Another part of an engine's power is used to overcome the friction between its moving parts. Friction takes away from the usable brake horsepower available at the flywheel. Through careful engine design and proper lubrication between moving parts, horsepower losses through internal friction are kept to a minimum.

Were it not for these mechanical, heat, and frictional losses, any engine could produce much more horsepower. This theoretical maximum power, called the indicated power, is the power produced in an engine's cylinders and it can be determined in a laboratory. An instrument called an indicator measures the gas pressure developed in a cylinder during combustion.

The total amount of power absorbed by mechanical, heat, and frictional losses varies from engine to engine. Understand, however, that the actual, usable brake horsepower is the indicated horsepower minus these losses.

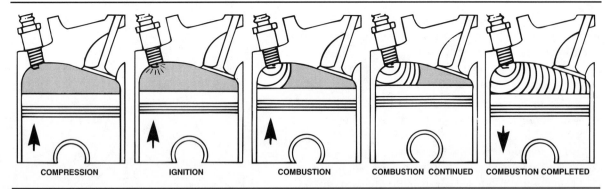

| COMPRESSION | IGNITION | COMBUSTION | COMBUSTION CONTINUED | COMBUSTION COMPLETED |

Figure 2-10. Normal combustion is a smooth, controlled burning of the air-fuel mixture at the correct time.

CHEMISTRY AND COMBUSTION

A gallon of conventional pump gasoline contains about thirty-one million calories of energy. This much energy represents a lot of explosive power. For that reason, it is easy to suppose that gasoline explodes in the cylinders to make an engine run. At one point in the development of the internal combustion engine, it was called an explosive engine. Now it is known that when an engine is functioning properly, gasoline does not explode, but undergoes controlled burning. When the spark plug ignites the air-fuel mixture, a flame front spreads out across the combustion chamber to consume the mixture, figure 2-10.

Gasoline and Combustion

Gasoline is a complex blend of chemical compounds, chiefly composed of hydrogen and carbon. The burning of gasoline is a chemical process. When gasoline burns, it combines with oxygen in the air. The result of combining the carbon part of gasoline with oxygen is carbon dioxide (CO_2). When the hydrogen combines with oxygen, water (H_2O) is formed.

The combustion process is an **oxidation reaction**. In this case, that means oxygen is combined with gasoline to form new chemical compounds. Combustion is not a simple process:

$$HC + O_2 \longrightarrow CO_2 + H_2O + Heat$$

Hydrocarbons Oxygen Carbon dioxide Water Energy

Not all of the carbon combines with oxygen. Some of the carbon is released from the gasoline to form soot, or hard carbon, that is deposited in the combustion chamber. Some of the gasoline clings to the chamber walls and does not burn. Sometimes the gasoline is burned with too little oxygen, and some of it forms carbon monoxide (CO) instead of carbon dioxide (CO_2).

If gasoline were a simple chemical compound, it would probably burn cleanly and go out the tailpipe as nothing more than CO_2 and H_2O. But gasoline is very complex. The carbon and hydrogen in gasoline are in the form of paraffins, olefins, napthenes, and aromatics, all with different chemical structures. To say that gasoline is nothing more than carbon and hydrogen is an oversimplification.

Thermal energy

The gasoline engine is really a heat engine. The more heat produced inside the combustion chamber, the more power produced by the engine. For this reason, engineers talk about gasoline as having a certain amount of thermal energy—as stated earlier, approximately thirty-one million calories per gallon. This thermal energy is converted into power through combustion.

Under some conditions, part of the thermal energy in gasoline may not be released to produce power. Power depends not only on the energy in the gasoline, but on the timing of combustion and on the amount of pressure in the combustion chamber. To produce the highest level of usable pressure in the combustion chamber, the spark is timed so that combustion begins before the piston passes through top dead center. This gives the most push on the piston.

Gross Brake Horsepower: The flywheel horsepower of a basic engine, without accessories, measured on a dynamometer.

Net Brake Horsepower: The flywheel horsepower of a fully equipped engine, with all accessories in operation, measured on a dynamometer.

Oxidation Reaction: A chemical reaction in which electrons are added to a compound, such as when oxygen molecules combine with other molecules to form a new compound.

Detonation

For combustion to be controlled and occur properly, a number of tuning parameters and engine parts must perform within a suitable range. The following elements affect the combustion of *all* engines:

- *The air-fuel mixture.* It must not be too lean.
- *The cooling system.* It must rid the engine of heat efficiently.
- *The compression ratio.* It must not be too high. Higher compression ratios require higher-quality fuel. Carbon deposits in the combustion chamber or on top of the piston may artificially raise the compression ratio.
- *The ignition timing.* Over-advanced timing can raise combustion chamber pressures drastically, and raise combustion chamber temperatures, figure 2-11.
- *The octane rating of the fuel.* Once an engine is designed, built, and tuned, it must be fueled with gasoline that has a sufficient octane rating. The gasoline must be blended to hold up under high combustion chamber pressures and temperatures.

If any of the first four elements are not within suitable limits, combustion chamber temperatures may be too high. If the temperatures are so high that the limitations of the fifth element, gasoline octane, are exceeded, there may be a secondary explosion soon after ignition called **detonation**, figure 2-12.

Detonation occurs when the expanding flame front further compresses and heats a portion of the air-fuel mixture in a corner of the combustion chamber that is already at a critically high temperature. That small portion of mixture then explodes before the flame front can reach it. This explosion creates a second pressure wave in the combustion chamber that causes the cylinder pressures to rise at an uncontrolled rate, raising cylinder temperatures above acceptable limits.

The second pressure wave also collides with the cylinder wall, combustion chamber, or piston top. The collision makes these parts vibrate, or ring like a bell, causing the characteristic "knocking" or "pinging" sound.

The effects of detonation vary with its intensity. Mild detonation causes a loss of power and efficiency because part of the thermal energy in the gasoline produces no useable power and is wasted. Some testing indicates that mild detonation also increases the rate of engine wear. The higher cylinder pressures caused by even mild detonation may increase the loads on the pistons, rings, connecting rods, crankshaft, and bearings.

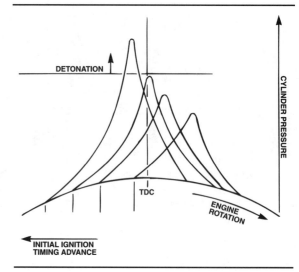

Figure 2-11 As the initial ignition timing setting is advanced, cylinder pressure rises. On this imaginary engine, advancing the spark up to 12 degrees BTDC increases usable cylinder pressure and improves power. Above 12 degrees BTDC, too much pressure increase occurs before the piston reaches TDC, fighting the piston's ascent. At 16 degrees BTDC, cylinder pressures have risen above an acceptable level and detonation is a major problem.

Anything more than mild detonation produces intense heat that can quickly lead to more severe detonation. Uncontrolled detonation can create sufficient pressure to easily break pistons or rings, figure 2-13. The cylinder pressures of severe detonation can blow a head gasket or crack the main bearing caps, cylinder head, or engine block. Such drastic damage rarely happens in a street engine unless it is operated with sustained extreme detonation. Most drivers wisely back off the throttle when they hear the first sounds of detonation.

Preignition

Another form of abnormal combustion is called **preignition**. It is the result of the air-fuel mixture being ignited too early, before the spark plug fires. This is the principal difference between detonation and preignition. Detonation always occurs after the spark plug fires.

There is another important difference between the two. Unlike detonation, which makes a characteristic noise, preignition itself is silent, despite what is often said or written in other books. The confusion developed because preignition and detonation often occur together. Preignition can raise pressure and heat in the cylinder to the point of causing detonation. If the point of preignition occurs at a place in the combustion chamber far from the spark plug, two flame fronts consume the intake charge, figure 2-14. The unburned

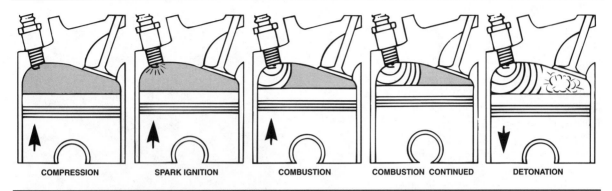

| COMPRESSION | SPARK IGNITION | COMBUSTION | COMBUSTION CONTINUED | DETONATION |

Figure 2-12. Detonation is a secondary ignition of the air-fuel mixture, caused by high cylinder temperatures. It is commonly called "pinging" or "knocking."

portion of the mixture, trapped between the two advancing flame fronts, is compressed and superheated, causing it to detonate. Also, prolonged detonation may increase the temperature of valves or combustion chamber deposits enough to cause preignition.

Typical causes of preignition include:

- Hot spots inside the combustion chamber. Common examples are spark plug threads that protrude into the combustion chamber, or sharp edges on the piston top. These parts may become hot enough to ignite the mixture.
- Carbon deposits in the combustion chamber that become red hot.
- A spark plug with a heat range too hot for the engine. Its electrode does not dissipate heat rapidly enough and may become hot enough to fire the mixture, even without a spark.
- Cross firing — electrical induction between spark plug wires. This may happen when the plug wires are cracked, or it can occur with perfectly good wires that are routed too close together.

Preignition usually causes heat-related damage. Prolonged preignition can create enough heat to melt spark plug electrodes, valve heads, and combustion chamber surfaces — or burn holes in piston crowns, figure 2-15.

Detonation: An unwanted explosion of a small portion of the air-fuel mixture caused by a sharp rise in combustion chamber pressure. The noise it makes is commonly called "knocking" or "pinging."

Preignition: An unwanted, early ignition of the air-fuel mixture.

■ Compression Ratio — What is the limit?

You know that a higher compression ratio increases horsepower and torque, but how high can a compression ratio be? Simply put, a compression ratio can be as high as the engine's fuel will tolerate without causing detonation. Racing engines fueled on conventionally blended "racing" gasoline are restricted to compression ratios of about 13:1, because of the relatively limited octane of that fuel. Methanol alcohol has a much higher octane, or detonation resistance, than conventionally blended gasoline. Racing engines fueled on methanol successfully use compression ratios of 15:1. Racing gasoline blended with a special anti-detonation agent such as toluene may tolerate compression ratios of 16:1 or even higher.

Racing engine builders do not always prepare their engines with compression ratios at these theoretical maximums. An engine built with the highest usable compression ratio runs on a critical edge just short of detonation. Any small loss of ring seal may allow oil into the combustion chamber and cause rampant detonation, destroying the engine. If the racing engine must live for 500 miles or 24 hours, the builder usually strives for an engine with a built-in margin of safety.

There are other practical limits to how high a compression ratio can be. When fuel is blended to tolerate it, static compression ratios above 15:1 theoretically increase power. Turbocharged and supercharged engines are another story, but to provide a compression ratio this high in a naturally aspirated engine requires the top of the piston to push deeply into the combustion chamber. The piston then does not allow optimum combustion chamber space for flame travel after ignition, reducing power. Ultra-high compression ratios may also cause valve-to-piston interference and not allow the engine builder to use the ideal valve timing for that engine. Finally, compression ratios over 15:1 are very hard on pistons. Current aluminum alloys may not always provide adequate piston life.

Octane Rating

Regular and premium grade gasolines contain the same amount of thermal energy. However, the premium grade is manufactured to better resist detonation and allow an engine to be tuned to release a greater amount of energy from the fuel. A measure of a gasoline's resistance to detonation is called its **octane rating**. In simple terms, it is a measure of the gasoline's ability to provide stable combustion at high temperatures and pressures. A higher octane rating permits higher pressures in the combustion chamber without detonation. The higher pressure allows more heat energy to be extracted from the gasoline to produce the most power. At lower pressures, some of the heat energy will go unused.

Several standards are used to rate gasoline octane. The two most common are the Research Octane Number (RON) and the Motor Octane Number (MON). The tests used to obtain these ratings employ a single-cylinder laboratory engine with a variable compression ratio. The gasoline to be rated is burned in the engine, and its anti-detonation characteristics are compared to a reference fuel whose RON and MON characteristics are known.

The RON and MON tests are similar, except that the laboratory engine is operated at a higher speed and higher inlet mixture temperature during the MON test. In general, this makes the RON a better indicator of gasoline anti-detonation characteristics when an engine is operated at full throttle and low rpm. In contrast, the MON is a better anti-knock indicator when an engine is operated at full throttle and high rpm, or at part throttle throughout the full rpm range. Today's typical pump gasoline has a RON rating approximately eight numbers higher than its MON rating.

As a result of the test differences, neither the RON nor MON rating can perfectly predict the performance of a gasoline in a road vehicle; the factors creating detonation are just too complex. After extensive testing, it was discovered that averaging RON and MON (adding the numbers and dividing by two), provided a good indicator of gasoline road performance. The resulting octane rating is called the Anti-Knock Index (AKI), and it is the number usually shown on the gas pump.

Anti-detonation compounds

Certain compounds are added to gasoline to increase its octane. High-octane gasoline does not contain any more thermal energy nor does its use increase power in an engine that does not require it. In fact, in an engine with low cylinder pressures, its high-octane feature is simply wasted. However, high-octane gasoline does allow engine designers to specify more spark advance and higher compression ratios for greater power without causing engine damage.

Figure 2-13. The extreme pressure of detonation hammered this piston.

For many years, **tetraethyl lead** was the most common anti-detonation compound added to gasoline to raise its octane rating. Tetraethyl lead slows flame propagation, reducing the chance of detonation.

With the advent of catalytic converters, lead had to be removed from gasoline. The lead coats the catalyst particles like paint, preventing the exhaust from touching it. Without lead, the octane rating of the gasoline is reduced. To keep the octane rating up, the gasoline requires further refining and blending. This results in less gasoline at higher cost.

In place of lead, gasoline manufacturers now use octane boosters made from alcohols and ethers such as tertiary butyl alcohol (TBA) and methyl tertiary butyl ether (MTBE). Gasolines with these octane improvers should not be confused with gasohol, a mixture of 90-percent unleaded gasoline and 10-percent ethyl alcohol.

Another gasoline additive used to reduce knock is methycyclopentadienyl manganese tricaronyl (MMT), a manganese compound. MMT is used in most low-lead gasolines, but its use in unleaded is restricted because manganese also coats catalytic converter particles, reducing the effectiveness of the catalyst.

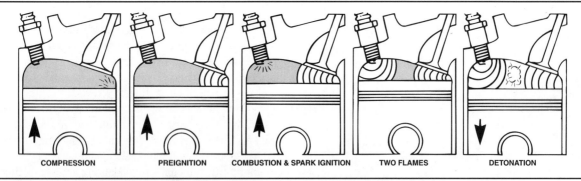

| COMPRESSION | PREIGNITION | COMBUSTION & SPARK IGNITION | TWO FLAMES | DETONATION |

Figure 2-14. Preignition is an early ignition of the air-fuel mixture. It can easily lead to detonation.

Figure 2-15. Heat caused by preignition melted through this piston.

ENGINE OPERATION AND AIR PRESSURE

Air is a substance with weight. The weight of the air exerts pressure on everything it touches. The greater the weight, the greater the pressure, but the weight of air is not always the same. It changes with temperature and with height above sea level. As a standard of reference, we use **Atmospheric pressure**, 14.7 pounds per square inch (101 kiloPascals), measured at sea level and at 32°F (0°C), figure 2-16.

You can think of an internal combustion engine as a big air pump. As the pistons move up and down in the cylinders, they pump in air and fuel for combustion and pump out exhaust gases. They do this by creating a difference in air pressure. The air outside an engine has weight and exerts pressure. So does the air inside an engine.

As a piston moves down on an intake stroke with the intake valve open, it creates a larger volume inside the cylinder for the air to fill. This lowers the air pressure within the engine. Because the pressure inside the engine is lower than the pressure outside, air flows into the engine to fill the low-pressure area and equalize the pressure.

The low pressure within the engine is called **vacuum**. You can think of the vacuum as sucking air into the engine, but it is really the higher pressure on the outside that forces air into the low-pressure volume inside. The difference in pressure between the two areas is called a **pressure differential**. The pressure differential principle has many applications in automotive fuel and emission systems.

An engine pumps exhaust out of its cylinders by creating pressure as a piston moves upward on the exhaust stroke. This creates high pressure in the cylinder, which forces the exhaust toward the lower-pressure volume outside the engine.

Pressure differential can be applied to liquids as well as to air. Fuel pumps work on this same principle. The pump creates a low-pressure area in the fuel system that allows the higher pressure of the air and fuel in the tank to force the fuel through the lines to the carburetor or the injection system.

Octane Rating: A measure of a gasoline's resistance to detonation.

Tetraethyl Lead: A gasoline additive used to help prevent detonation.

Atmospheric Pressure: The pressure on the earth's surface caused by the weight of air in the atmosphere. At sea level, this pressure is 14.7 psi (101 kiloPascals) at 32°F (0°C).

Vacuum: Air pressure lower than atmospheric pressure.

Pressure Differential: The difference in pressure between two points.

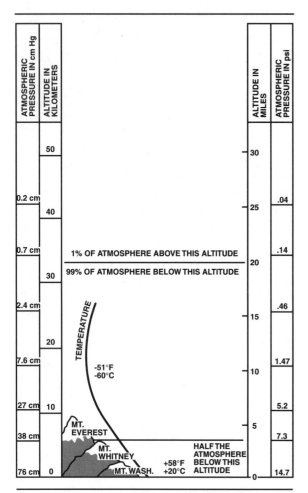

Figure 2-16. The blanket of air surrounding the earth extends many miles above the surface of the earth. Atmospheric pressure decreases at higher altitudes because the blanket is thinner.

AIRFLOW REQUIREMENTS

All automobile gasoline engines share certain air-fuel requirements. For example, a four-stroke engine can take in only so much air at any one time. How much fuel it consumes depends upon how much air the engine can take in. An engine's airflow requirement depends upon three factors:

- Engine displacement
- Engine revolutions per minute (rpm)
- Volumetric efficiency.

Engineers calculate an engine's airflow requirement, using the three factors listed above. This number is represented in cubic feet per minute (cfm) or cubic meters per minute (cmm). The engineer or designer then sizes the carburetor or fuel-injection intake airflow capacity to match the engine's maximum requirement. The following paragraphs describe volumetric efficiency and how it is related to engine airflow.

Volumetric Efficiency

Volumetric efficiency is a comparison of the actual volume of air-fuel mixture drawn into an engine to the theoretical maximum volume that could be drawn in. Volumetric efficiency is expressed as a percentage, and it changes with engine speed. For example, an engine might have 75 percent volumetric efficiency at 1,000 rpm. The same engine might be rated at 85 percent at 2,000 rpm and 60 percent at 3,000 rpm.

If the airflow volume is taken in slowly, a cylinder might be filled to capacity. A definite amount of time is required for the airflow to pass through all the curves of the intake manifold and valve port. So, manifold and port design directly relate to the engine's breathing, or volumetric efficiency. Camshaft timing and exhaust tuning also are important.

If the engine is running fast, the intake valve is not open long enough for a full volume to enter the cylinder. At 1,000 rpm, the intake valve might be open for $1/10$ of a second. As engine speed increases, this time is greatly reduced to a point where only a small airflow volume can enter the cylinder. Therefore, volumetric efficiency decreases as engine speed increases. At high speed, it may drop to as low as 50 percent.

A volumetric efficiency of 100 percent is never reached by a street engine. With a street engine, you can expect a volumetric efficiency of about 75 percent at maximum speed, or 80 percent at the torque peak. A high-performance street engine will be about 85 percent or a bit more efficient at peak torque. A race engine will usually have 95 percent or better volumetric efficiency. These figures apply only to naturally aspirated engines. Turbocharged and supercharged engines can easily achieve more than 100 percent volumetric efficiency.

AIR-FUEL RATIOS

Fuel burns best when it is turned into a fine spray and mixed with air before it is sucked into the cylinders. In carbureted engines, the fuel becomes a spray and is mixed with air in the carburetor. In fuel-injected engines, the fuel becomes a fine spray as is leaves the tip of the injectors, and the mixing with air takes place in the intake manifold. In both cases, there is a direct relationship between an engine's airflow and its fuel requirements. This relationship is called the **air-fuel ratio**.

The air-fuel ratio is the proportion by weight of air and gasoline mixed by the carburetor or injection system as required for combustion by the engine. This ratio is important, because there are limits to how rich (with more fuel) or how lean (with less fuel) it can be, and still remain combustible for burning. The mixtures with which an engine can operate without stalling range from 8 to 18.5:1, figure 2-17. These ratios are usually stated this way: 8 parts of air by weight combined with 1 part of gasoline by weight (8:1) is the richest mixture which an engine can tolerate and still fire regularly; 18.5 parts of air mixed with 1 part of gasoline (18.5:1) is the leanest. Richer or leaner air-fuel ratios will cause the engine to misfire badly or not run at all.

Weight rather than volume is used to calculate air-fuel ratios. Therefore, 14.7 pounds or kilograms of air are required to burn one pound or kilogram of gasoline, so our air-fuel ratio is 14.7:1.

Stoichiometric Air-Fuel Ratio

The ideal mixture or ratio at which all the fuel combines with all of the oxygen in the air and both are *completely burned* is called the **stoichiometric ratio** — a chemically perfect combination. In theory, an air-fuel mixture of about 14.7:1 will produce this ratio, but the exact ratio at which perfect mixture and

Volumetric Efficiency: The comparison of the actual volume of air-fuel mixture drawn into an engine to the theoretical maximum volume that could be drawn in. Written as a percentage.

Air-Fuel Ratio: The ratio in weight of air to gasoline in the air-fuel mixture drawn into an engine.

Stoichiometric Ratio: The chemically correct air-fuel mixture for combustion in which all oxygen and all fuel will be completely burned.

■ A Compressed History

With few exceptions, the compression ratios of road automobiles have risen as fast as fuel quality, metallurgy, and combustion chamber technology would allow. It is always desirable to engineers to raise compression ratios because higher compression for any given engine size increases power, fuel mileage, and throttle response. Most of the successful automotive engines of the late 1800s had compression ratios of only about 2.5:1. Fuel quality was low and metallurgy and combustion chamber design were primitive — not surprising because automotive technology was in its infancy.

One of the first production automobiles, the 1909 Ford model T, began its life with a 4.5:1 compression ratio (although it was reduced to 3.98:1 during its service life because Henry Ford was adamant that the T run on the fuel available anywhere it was sold). Compression ratios in the 4:1 range were common in the automotive era just prior to and just after World War I. After the public introduction of leaded gasoline in 1923, most production car compression ratios rose steadily to the upper 6:1 level by the late 1930s. Even though fuel quality was better than ever, the most common cylinder head of the time was the flathead, and its combustion chamber design would not tolerate compression ratios that were much higher.

During World War II, fuel quality improved because of the requirements of high-performance fighter aircraft. After the war, civilian gasolines benefited from some of this fuel technology. In the late 1940s and early 1950s, the first of the modern, overhead-valve V8s from Cadillac, Oldsmobile, and Chrysler featured compression ratios of about 7.5:1. However, engineers built these engines with strong materials and combustion chamber designs that could handle much higher compression ratios, should the proper fuel become available.

In the mid-1950s, a horsepower race was on and readily available high-octane gasoline allowed compression ratios to rise dramatically. The Hemi-head engines that powered the Chrysler 300-series cars had compression ratios in the 9:1 to 10:1 range.

Improved and more powerful versions of the Cadillac and Oldsmobile V8s from 1949 also featured compression ratios at this level.

By the 1960s, a few street-legal limited-production engines such as Chevrolet's 427 cubic-inch L-88 engine had a compression ratio of 12.5:1. The Dodge and Plymouth "Max Wedge" big-block engines could be ordered with a 13.5:1 compression ratio, the highest ever offered in a production car, although production was very limited. Many more high-performance, road-going engines, such as the fuel-injected Chevrolet Corvette engine and Ford's side-oiler 427 cubic-inch engine, used in the Shelby Cobra and optional in the Thunderbird and others, had compression ratios of 11:1 to 11.5:1. Although, during this era, ratios of 10:1 to 10.5:1 were far more common. Ratios this high, however, created high combustion chamber temperatures. This, in turn, created oxides of nitrogen (NO_x), an air pollutant.

In the early 1970s, two factors led to lower compression ratios and kept them low for more than a decade. First, ratios were lowered to around 8:1 to reduce NO_x formation and meet federal exhaust emission standards. Second, tetraethyl lead — a vital octane improver as well as a serious pollutant — was reduced in gasoline. With the resulting low-octane gasoline then available, ratios over 8:1 or 9:1 were impossible to use without detonation being a significant problem.

By the mid-1980s, technology improved to the point that compression ratios could again rise. Better combustion chamber shapes promoted rapid, but controlled burning of the air-fuel mixture, resisting detonation. Sophisticated computer controls for fuel injection kept air-fuel mixtures in the correct zone more accurately than ever before. Advanced emission controls components more effectively reduces pollutants. Knock sensors retarded ignition timing at the first hint of detonation. Complex, high-performance cars such as the Chevrolet Corvette ZR-1 successfully run 11.5:1 compression ratios on 92 AKI, unleaded gasoline. Ratios as high as 10:1 to 10.5:1 are not uncommon for many other road-going cars.

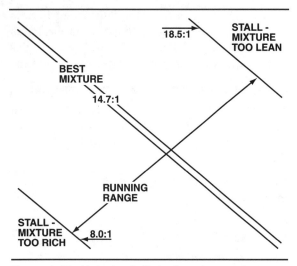

Figure 2-17. Air-fuel ratio limits for a four-stroke gasoline engine.

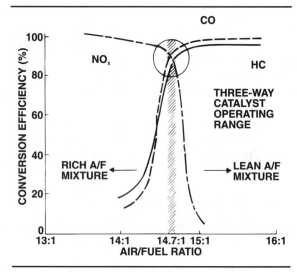

Figure 2-18. All three pollutants are best controlled with an air-fuel ratio of 14.7:1. A leaner mixture increases NO_x. A richer mixture increases HC and CO.

combustion occur depends upon the molecular structure of gasoline, which varies somewhat. The stoichiometric ratio is somewhat of a compromise between maximum power and maximum economy, but both are good at 14.7:1.

Emission control is also optimum at this ratio, if a 3-way oxidation-reduction catalytic converter is used. As the mixture richens, HC and CO conversion efficiency falls off. With leaner mixtures, NO_x conversion efficiency also falls off. As figure 2-18 shows, the conversion efficiency range is very narrow — between 14.65 and 14.75:1. A fuel system without feedback control cannot maintain this narrow range.

Engine Air-Fuel Requirements

An automobile engine will work with the air-fuel mixture ranging from 8 to 18.5:1. But the ideal ratio would be one that provides both the most power and the most economy, while producing the fewest emissions. But such a ratio does not exist because the fuel requirements of an engine vary widely depending upon temperature, load, and speed conditions.

Research has proven that the best fuel economy is obtained with a 15 to 16:1 ratio, while maximum power output is achieved with a 12.5 to 13.5:1 ratio. A rich mixture is required for idle, heavy load, and high-speed conditions; a leaner mixture is required for normal cruising and light-load conditions. No single air-fuel ratio provides the best fuel economy and the maximum power output at the same time.

Just as outside conditions such as speed, load, temperature, and atmospheric pressure change the engine's fuel requirements, other forces at work inside the engine cause additional variations. Here are two examples:

- The mixture is imperfect because complete vaporization of the fuel may not occur.
- Mixture distribution from a carburetor or a throttle body through the intake manifold to each cylinder is not equal; some cylinders get a richer or leaner mixture than others.

If an engine is to run well under such a wide variety of outside and inside conditions, the carburetor or the injection system must be able to vary the air-fuel mixture quickly, and to give the best mixture possible for the engine's requirements at a given moment.

Power Versus Economy

If the goal is to get the most power from an engine, all of the oxygen in the mixture must be burned, because the power output of any engine is limited by the amount of air it can pull in. To be sure that the oxygen combines properly with the available fuel, extra fuel must be provided. This increases the air-fuel ratio (makes it richer in fuel), resulting in some fuel which remains unburned.

To get the best fuel economy and the lowest emissions, the gasoline must be burned as completely as possible in the combustion chamber. This means that the greatest amount of economy will be produced with the least amount of leftover waste material, or emissions. If enough oxygen is to be available to combine with the gasoline, then more air must be provided. This results in a leaner air-fuel mixture (less gasoline) than the ideal ratio.

The air-fuel ratio required to provide maximum power will change very little, except at low speeds, figure 2-19. Reducing speed reduces the airflow into the engine. The result is a poorer mixing of the air and

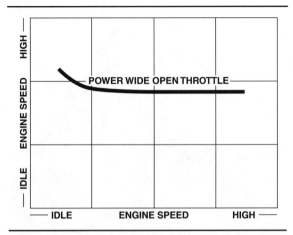

Figure 2-19. The air-fuel ratio needed for maximum power is relatively constant, except at very low speed, where it must be slightly richer.

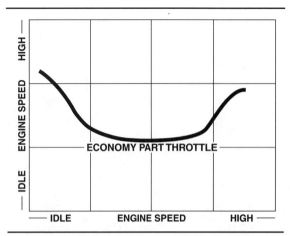

Figure 2-20. The air-fuel ratio for best economy is lean in the middle of the speed range but requires enrichment at high and low speeds.

fuel, and less efficiency in its distribution to the cylinders. Thus, at low speeds, a slight enrichment of the mixture is required to make up for this.

The same is true for maximum fuel economy: The leaner air-fuel ratio used will remain virtually the same throughout most of the operating range, figure 2-20. But enrichment will be required during idle and low speeds, as well as during higher speeds and under load — two conditions which require more power.

Enrichment also can occur when it is not required or wanted, as in the case of high-altitude driving. As altitude increases, atmospheric pressure drops and the air becomes thinner than it is at sea level. The same volume of air weighs less and contains less oxygen at higher altitudes. This means that an engine will take in fewer pounds or kilograms of air and less oxygen. The result is a richer air-fuel ratio, which must be corrected for efficient high-altitude engine operation. Altitude-compensating carburetors and fuel-injection air sensors solve this problem. The corrected air-fuel mixture allows the engine to burn its fuel efficiently, but the total horsepower is less, corrected or not. With less fuel being injected into the cylinder to mix with the lower oxygen content of the high-altitude air, fewer calories of energy are available to do work.

For these reasons, the carburetor or injection system must deliver fuel so that the best mileage is provided during normal cruising, with maximum power available whenever the engine is under load, accelerating, or at high-speed, figure 2-21.

INTRODUCTION TO EMISSION CONTROL

An internal combustion engine emits three major pollutants into the air: **hydrocarbons** (HC), **carbon monoxide** (CO), and **oxides of nitrogen** (NO_x),

figure 2-22. An engine also gives off many small liquid or solid particles, such as lead, carbon, sulfur, and other **particulates**, which contribute to pollution. All of these emissions are byproducts of combustion.

Unburned Hydrocarbons (HC)

Gasoline is a hydrocarbon, a compound of hydrogen and carbon. Unburned hydrocarbons given off by an automobile are largely unburned portions of fuel. However, hydrocarbons are also the only major automotive air pollutant that come from sources other than the engine's exhaust. Over 200 different varieties of hydrocarbon pollutants come from automotive sources. While most come from the fuel system and the engine exhaust, others are oil and gasoline fumes from the crankcase. Even a vehicle's tires, paint, and upholstery emit tiny amounts of hydrocarbons. Figure 2-23 shows the three major sources of hydrocarbon

Hydrocarbon: Any chemical compound made up of hydrogen and carbon. A major pollutant given off by an internal combustion engine. Gasoline is a hydrocarbon.

Carbon Monoxide: An odorless, colorless, poisonous gas. A major pollutant given off by an internal combustion engine.

Oxides of Nitrogen: Chemical compounds of nitrogen given off by an internal combustion engine. They combine with hydrocarbons to produce smog.

Particulates: Liquid or solid particles such as lead and carbon that are given off by an internal combustion engine as pollution.

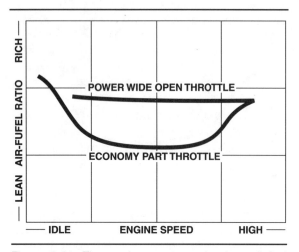

Figure 2-21. The engine must receive lean air-fuel ratios for best economy or rich ratios for maximum power at any given speed.

emissions from an automobile not equipped with emission controls:

• Fuel system evaporation — 20 percent
• Crankcase vapors — 20 percent
• Engine exhaust — 60 percent.

Hydrocarbons of all types are destroyed by combustion. If an automobile engine burned air and gasoline completely, there would be no hydrocarbons in the exhaust, only water and carbon dioxide. But when the vaporized and compressed air-fuel mixture is ignited, not all is burned. The relatively cold metal of the combustion chamber surfaces, cylinder walls, and piston crowns absorbs heat so rapidly that the fuel mixture in contact with these areas never gets hot enough to ignite. This unburned fuel then passes out with the exhaust gases. The problem is worse with engines that misfire or are not properly tuned.

Carbon Monoxide (CO)

Carbon monoxide is also found in automobile exhaust in large amounts. A deadly poison, carbon monoxide is both odorless and colorless. Carbon monoxide is absorbed by the red corpuscles in the body, displacing the oxygen. In small quantities, it causes headaches and vision difficulties. In larger quantities, it is fatal.

Because it is a product of incomplete combustion, the amount of carbon monoxide produced depends on the way in which hydrocarbons burn. When the air-fuel mixture burns, its hydrocarbons combine with oxygen. If the air-fuel mixture contains too much fuel, there is not enough oxygen to complete this process, so carbon monoxide is formed. To make combustion more complete, an air-fuel mixture with less fuel is used. This increases the ratio of oxygen, which reduces the formation of CO by producing relatively

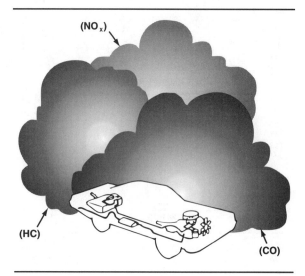

Figure 2-22. Hydrocarbons (HC), carbon monoxide (CO), and oxides of nitrogen (NO_x) are the three major pollutants emitted by an automobile.

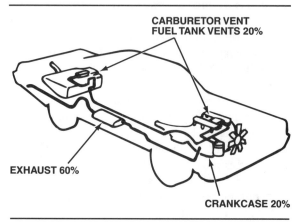

Figure 2-23. Sources of hydrocarbon emissions on a non-pollution controlled car.

harmless carbon dioxide (CO_2) instead. It is, however, one of the "greenhouse gases" that are being blamed for global warming.

Oxides of Nitrogen (NO_x)

Air is about 78 percent nitrogen, 21 percent oxygen, and 1 percent other gases. When the combustion chamber temperature reaches 2,500°F (1,400°C) or greater, the nitrogen and oxygen in the air-fuel mixture combine to form large quantities of nitric oxide (NO) and nitrogen dioxide NO_2, generically referred to as oxides of nitrogen (NO_x). NO_x is also formed at lower temperatures, but in far smaller amounts. When the amount of hydrocarbons in the atmosphere reaches a certain level, and the ratio of NO_x to HC is critical, the two pollutants will combine chemically to form photochemical smog.

The amount of NO_x formed can be reduced by lowering the temperature of combustion in the engine, but this causes a problem which is difficult to solve. Lowering the combustion chamber temperature to reduce NO_x results in less-efficient burning of the air-fuel mixture. This automatically means an increase in hydrocarbons and carbon monoxide, both of which are emitted in large quantities at lower combustion chamber temperatures.

NO_x is created by an oxidation reaction at high temperatures, in which oxygen is combined with nitrogen. If all high temperature oxidation could be prevented, all NO_x formation would be prevented. However, this would also increase HC and CO emissions. To get rid of NO_x once it is formed, designers create a **reduction reaction** in the catalytic converter. In the reduction reaction, oxygen is removed from the nitrogen.

Particulates

Particulates are microscopic solid particles, such as dust and soot. Because these fragments remain in the atmosphere for a long time, particulates are a prime cause of secondary pollution. For example, particulates such as lead and carbon tend to collect in the atmosphere. These are all harmful substances, large amounts of which can injure our health.

Particulates produced by automobiles are a small percentage of the total particulates in the atmosphere. Most come from fixed sources, such as factories. While automobiles do produce particulates, the amount can be reduced considerably. This is accomplished by eliminating certain additives such as lead from gasoline, and by changing other characteristics of the fuel. The amount of additives used in gasoline has been reduced and the types of additives are now carefully controlled.

Sulfur Oxides

Sulfur in gasoline and other fossil fuels (coal and oil) enters the atmosphere in the form of **sulfur oxides**. As these oxides break down, they combine with water in the air to form corrosive sulfuric acid, commonly referred to as acid rain.

Automotive Emission Controls

Early researchers dealing with automotive pollution and smog began work with the idea that all pollutants were carried into the atmosphere by the car's exhaust pipe. But auto manufacturers doing their own research soon discovered that pollutants were also given off from the fuel tank and the engine crankcase. The total automotive emission system, figure 2-24, contains three different types of controls. This picture illustrates that the emission controls on a modern automobile are not a separate system, but part of an engine's fuel, ignition, and exhaust systems. Modern automotive emission controls are very effective. Compared to a car without emission controls, a current car emits about 96 percent fewer unburned hydrocarbons, 96 percent less carbon monoxide, and 76 percent fewer oxides of nitrogen.

Automotive emission controls can be grouped into major families, as follows:

- **Crankcase emission controls**
  Positive crankcase ventilation (PCV) systems control HC emissions from the engine crankcase.
- **Evaporative emission controls**
  Evaporative emission control (EEC or EVAP) systems control the evaporation of HC vapors from the fuel tank, pump, and carburetor or fuel injection system.
- **Exhaust emission controls**
  Various systems and devices are used to control HC, CO, and NO_x emissions from the engine exhaust. These controls can be subdivided into the following general groups:

— Air injection systems — These systems provide additional air to the exhaust system to help burn up HC and CO in the exhaust and to aid catalytic conversion.

— Engine modifications — Various changes in the design of engines and in the operation of fuel and ignition system components help eliminate all three major pollutants.

— Spark timing controls — Vehicle manufacturers have used various systems to delay or retard ignition spark timing to control HC and NO_x emissions. Older systems modified the distributor vacuum advance. Late-model cars with electronic engine control systems have eliminated the need for mechanical or vacuum timing devices.

— Exhaust gas recirculation — An effective way to control NO_x emissions is to recirculate a small amount of exhaust gas back to the intake manifold to dilute the incoming air-fuel mixture.

— Catalytic converters — The first catalytic converters installed in the exhaust systems of 1975-76 cars

---

Reduction Reaction: A chemical reaction in which electrons are removed from a compound, such as when oxygen is removed from a compound.

Sulfur Oxides: Chemical compounds given off by processing and burning gasoline and other fossil fuels. As they decompose, they combine with water to form sulfuric acid.

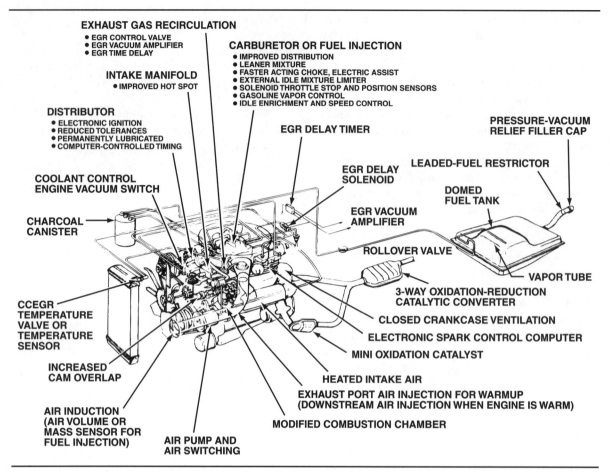

EXHAUST GAS RECIRCULATION
- EGR CONTROL VALVE
- EGR VACUUM AMPLIFIER
- EGR TIME DELAY

INTAKE MANIFOLD
- IMPROVED HOT SPOT

DISTRIBUTOR
- ELECTRONIC IGNITION
- REDUCED TOLERANCES
- PERMANENTLY LUBRICATED
- COMPUTER-CONTROLLED TIMING

CARBURETOR OR FUEL INJECTION
- IMPROVED DISTRIBUTION
- LEANER MIXTURE
- FASTER ACTING CHOKE, ELECTRIC ASSIST
- EXTERNAL IDLE MIXTURE LIMITER
- SOLENOID THROTTLE STOP AND POSITION SENSORS
- GASOLINE VAPOR CONTROL
- IDLE ENRICHMENT AND SPEED CONTROL

EGR DELAY TIMER

PRESSURE-VACUUM RELIEF FILLER CAP

LEADED-FUEL RESTRICTOR

EGR DELAY SOLENOID

DOMED FUEL TANK

COOLANT CONTROL ENGINE VACUUM SWITCH

EGR VACUUM AMPLIFIER

CHARCOAL CANISTER

ROLLOVER VALVE

VAPOR TUBE

3-WAY OXIDATION-REDUCTION CATALYTIC CONVERTER

CCEGR TEMPERATURE VALVE OR TEMPERATURE SENSOR

CLOSED CRANKCASE VENTILATION

ELECTRONIC SPARK CONTROL COMPUTER

MINI OXIDATION CATALYST

INCREASED CAM OVERLAP

HEATED INTAKE AIR

EXHAUST PORT AIR INJECTION FOR WARMUP (DOWNSTREAM AIR INJECTION WHEN ENGINE IS WARM)

AIR INDUCTION (AIR VOLUME OR MASS SENSOR FOR FUEL INJECTION)

MODIFIED COMBUSTION CHAMBER

AIR PUMP AND AIR SWITCHING

Figure 2-24. The complex emission controls on this modern Chrysler automobile engine are an integral part of the fuel, ignition, and exhaust systems.

helped the chemical oxidation or burning of HC and CO in the exhaust. Later catalytic converters, which began to appear on 1977-78 vehicles, also promoted the chemical reduction of NO_x emissions.

ENGINE DESIGN EFFECTS ON EMISSIONS

Of the three major pollutants, CO is affected the least by internal engine design. CO comes from rich mixtures, and the richness or leanness of a mixture is controlled by the carburetor and air intake. But the distribution of the mixture depends on the intake manifold. If the manifold and the passages through the cylinder head vary greatly from cylinder to cylinder, poor distribution is the result. With poor distribution, some cylinders get rich mixtures, while others get lean mixtures. Before the days of emission controls, engineers simply could let the overall mixture at the carburetor be a little rich so that no cylinder would run too lean. They did not have to worry about the amount of CO left over in the exhaust. If engineers simply tried to lean out the mixture at the carburetor to reduce

CO emissions, some cylinders would run too lean, which would cause stalling and poor performance.

Another solution to the problem of uneven fuel distribution would be to redesign the intake manifold. In fact, changes have been made in manifold design to improve fuel distribution and vaporization. These changes have been slight, however, when viewed as part of overall engine design. A perfect manifold cannot be designed unless all of the cylinders are an equal distance from the carburetor or throttle body. Short of arranging the cylinders in a circle, this would be hard to do. This is why many engines now use port fuel injection. Fuel distribution is much improved over carbureted and throttle-body systems and CO emissions are easier to control.

In contrast to CO emissions, HC emissions are greatly affected by internal engine design. Hydrocarbons will cling to the surface of the combustion chamber, which is cooler than the center of combustion in the chamber. These hydrocarbons that are cooled on the chamber surfaces may go unburned and be emitted in the exhaust. Therefore, modern engines are designed to keep the chamber surface as small as practical.

A sphere has the smallest surface area for its volume. A cube of the same volume will have a greater surface area. The more angles and curves there are in a combustion chamber wall, the greater the surface area will be. Engineers call this the **surface-to-volume ratio**, and they want to keep it as small as possible to reduce HC emissions.

Engineers also discovered that the small area above the top piston ring, between the piston and the cylinder wall, is a gathering place for unburned gasoline. Therefore, some years ago, changes were made in many piston and piston ring designs to reduce this area.

The compression ratio affects HC and NO_x emissions. With a high compression ratio, an engine can develop more combustion heat and produce more power. However, if domed pistons are used to achieve the higher compression ratio, the surface-to-volume ratio also goes up, which can produce more HC emissions.

For some years, engineers were puzzled about how to reduce NO_x emissions. Everything they did to reduce HC and CO either had little effect on NO_x or increased it. Remember that large amounts of NO_x are produced when combustion temperatures go over 2,500°F (1,400°C). Design features that get the most power from an engine, such as high compression ratios and advanced ignition timing, also increase NO_x. When engine compression ratios were lowered in the early 1970s, it helped to reduce NO_x formation. However, the most effective way to reduce NO_x is through exhaust gas recirculation (EGR). Exhaust gases are fairly nonreactive. They do not support combustion because they contain too little oxygen. When exhaust is recirculated to the intake mixture, it decreases the volume of combustible air-fuel mixture in the intake charge. This, in turn, lowers combustion temperatures and cuts NO_x formation. EGR also decreases horsepower, which is why such systems usually are equipped with extensive electronic and vacuum controls to help driveability and economy.

ENGINE MECHANICAL CONDITION EFFECTS ON EMISSIONS

The engine mechanical wear that affects exhaust emissions includes items such as valve train wear, which will change the valve timing and affect combustion. Increased clearances from mechanical wear will also allow excessive amounts of oil into the cylinders, contributing to HC emissions. Poor rings can also overload the PCV valve with deposits.

Surface-to-Volume Ratio: The ratio of the surface area of a three dimensional space to its volume.

SUMMARY

All substances are composed of matter. Many physical properties govern how an engine works: Mass is the measure of matter in an object. Weight is the measure of the earth's gravitational pull (force) on that object. Energy is the ability to do work. Force is the push or pull exerted by energy to do mechanical work (move something). Work is the application of force to move something a given distance. Torque is a twisting or turning force. Power is the rate at which work is done. Inertia is the tendency of a body to stay at rest or in motion.

Engine output is measured in torque and horsepower. Engine power measured on a dynamometer is called brake horsepower. The actual, usable brake horsepower is the indicated horsepower minus the mechanical, heat, and frictional losses.

Gasoline must be properly blended to prevent detonation. Tetraethyl lead was once the major octane booster, but its use was severely limited and eventually prohibited by the EPA.

Piston movement creates a pressure differential between the air inside the engine and the air outside the engine. This causes airflow into and out of the engine.

The engine air-fuel ratio is commonly measured in terms of the weight of the air taken in versus the weight of the gasoline taken in. Because an engine must operate with the best possible combination of power, economy, and clean exhaust, the desired air-fuel ratio changes as operating conditions change. The air-fuel ratio that provides the best compromise between power, economy, and low exhaust emissions is 14.7:1.

The automobile is a major source of air pollution resulting from the gasoline burned in the engine and the vapors escaping from the crankcase, the fuel tank, and the rest of the fuel system. The major pollutants are unburned hydrocarbons (HC), carbon monoxide (CO), and oxides of nitrogen (NO_x). The most visible and irritating form of air pollution is photochemical smog, which is formed when HC and NOx emissions combine in the presence of sunlight. The use of emission controls in the past two decades has reduced automotive pollutants by 65 to 98 percent.

Emission controls began as separate "add-on" components and systems, but are now completely integrated into engine and vehicle design. The major emission control systems used are PCV systems, evaporative control systems, air injection, spark timing controls, exhaust gas recirculation, catalytic converters, and electronic engine control systems.

Review Questions

Choose the single most correct answer.
Compare your answers with the correct answers on Page 328.

1. All substances are composed of:
   a. Inertia
   b. Gas
   c. Solids
   d. Matter

2. Mass is a measure of an object's:
   a. Matter
   b. Torque
   c. Weight
   d. Force

3. A calorie is a measure of:
   a. Force
   b. Mass
   c. Energy
   d. Torque

4. Power can be measured in:
   a. Foot-pounds
   b. Kilowatts
   c. Newton-meters
   d. Horses

5. Torque is:
   a. A measure of heat energy
   b. A kind of wrench
   c. The force applied to a lever being turned, multiplied by the length of the lever
   d. The rate of doing work

6. In a gasoline engine:
   a. A detonation occurs once in every cycle
   b. Engine output depends on the brand of fuel used
   c. All the fuel is burned and produces useful power
   d. None of the above

7. The engine output specification that most accurately predicts how the engine will perform when installed in a car is:
   a. The indicated horsepower
   b. Stated in gross brake horse-power
   c. Measured at the flywheel
   d. Stated in net brake horse-power

8. One horsepower is equivalent to:
   a. The output of a horse pulling a 100-lb weight a distance of 100 feet in 1 minute
   b. 0.746 kilowatt
   c. 3,300 foot-pounds per minute
   d. None of the above

9. How is engine output in horse-power calculated?
   a. Torque multiplied by rpm, divided by 5252
   b. Foot-pounds multiplied by rpm, divided by the diameter of the flywheel
   c. Crankshaft rotation speed multiplied by average piston thrust
   d. None of the above

10. The octane rating of a fuel indi-cates:
    a. Which grade of fuel is most powerful
    b. The proportion of pure gaso-line to additives
    c. The air-fuel mixture required to burn it
    d. The ability of a fuel to burn under high pressure without detonation

11. What is the stoichiometric ratio?
    a. The ratio of gallons of fuel to pounds of air
    b. A scientific term for volumet-ric efficiency
    c. The air-fuel ratio at which all of the liquid fuel present will blend with all the oxygen available and be completely burned
    d. The ratio of exhaust gas vol-ume to inlet mixture volume

12. Volumetric efficiency:
    a. Is the ratio of the actual vol-ume of air-fuel mixture drawn into an engine to the theoret-ical maximum volume that could be drawn in displace-ment during one cycle
    b. Decreases as engine speed increases
    c. Is expressed as a percentage
    d. All of the above

13. At maximum speed, the volu-metric efficiency of a stock engine is approximately:
    a. 75%
    b. 50%
    c. 10%
    d. None of the above

14. Obtaining maximum power results in:
    a. No change in air-fuel ratios
    b. Leaner mixtures
    c. Unburned oxygen
    d. Excess unburned fuel

15. Maximum fuel economy requires:
    a. Less air
    b. Leaner air-fuel mixtures
    c. Richer air-fuel mixtures
    d. Higher temperatures

16. For maximum power, the air-fuel ratio:
    a. Must be leaner at low speeds
    b. Must be richer at low speeds
    c. Must be richer at high speeds
    d. None of the above

17. The purpose of emission control is to minimize the release of:
    a. Hydrocarbons into the atmosphere
    b. Nitrogen oxides into the atmosphere
    c. Carbon monoxide into the atmosphere
    d. All of the above

18. The smallest possible surface-to-volume ratio of a combustion chamber produces:
    a. Minimum fuel consumption
    b. Maximum net brake horse-power
    c. Maximum combustion efficiency
    d. Minimum unburned hydro-carbons

19. Which of the following is *not* an automotive emission control system?
    a. Exhaust gas recirculation
    b. Air injection
    c. Fuel injection
    d. Catalytic converter

20. Automotive emission controls can be grouped into which of the following categories:
    a. Crankcase emission controls
    b. Evaporative emission controls
    c. Exhaust emission controls
    d. All of the above

3

Engine Materials and Manufacture

ENGINE MATERIALS

Many different materials are used in an automotive engine. Not long ago metals, such as iron, steel, and aluminum, were the only substances auto manufacturers had at their disposal. While metals are still the dominant materials, man-made plastics, ceramics, and composites are beginning what is sure to be an ever increasing use in the construction of automotive engines.

In order to service an engine properly, you do not need to know how to select the raw material to make an engine part. However, you should know why different metals and man-made materials are used for different parts of an engine. In this chapter we compare the strengths, weaknesses, and compatibilities of these materials. This knowledge will help you do a better job in the various machining and assembly operations that you will use in the repairing or rebuilding of an engine.

METALLURGY

Metallurgy is the science of metals. Metal is divided into two basic categories. The first is ferrous metals, those metals that contain iron, such as cast iron and steel. The second is non-ferrous metals, those that contain no iron, such as aluminum, magnesium, and titanium. You can use a magnet to make a quick check to determine whether a metal is ferrous or not. The magnet will stick to the ferrous metal because it is attracted to its iron content, but will not stick to the non-ferrous metal.

When two or more different metals are added together, the result is called an **alloy**. The resulting alloy may be significantly different than either of the original metals. For example, the alloy of copper and zinc makes brass, and the alloy of copper and tin makes bronze.

To understand how one metal is stronger than another, you should realize that all metals in the solid state have a crystalline or granular structure. Metal from raw ore must be formed into different shapes for storage, handling, and use. It would not be practical to keep metal in a molten state from the time it is extracted from the ore until a finished product is made. Therefore, the molten metal is cooled and formed into bars, sheets, **billets**, and other shapes. As this forming is done, the metal develops a grain because of the way the molecules of the material align with each other. The grain structure is not visible on the surface of rolled or cast metal. You must fracture the metal to reveal the grain structure inside.

Figure 3-1. Pig iron is made by combining iron and carbon during the smelting process, and pouring the liquid into channels dug into the sand in a sand-casting box.

This grain structure is made up of millions of identical unit cells that can change size or position through the application of temperature changes and external forces. You do not need to know about the chemistry involved, but you should know that grain size and position are important factors in determining the strength and properties of metal. Grain size may vary from coarse to fine.

Each metal is developed or selected to perform a job because of the particular properties that metal processes. The following paragraphs will acquaint you will the properties of various metals.

Iron

Iron, the most common metal used in the industry, comes from iron ore mined from the earth. Blast furnaces heat the iron ore to burn off impurities. Coke, a hot-burning fuel made from coal, burns in the furnace, and some of the carbon from the coke combines with the iron. When the liquid iron drains from the furnace, it contains about 3- to 5-percent carbon and several

impurities. The iron flows into channels dug in sand that look somewhat like a mother pig feeding piglets, hence the name **pig iron**. The high percentage of carbon makes the iron **brittle**. Pig iron is used to make steel and several kinds of cast iron, figure 3-1.

Slag is the name for the mixture of rock and stone impurities in the iron ore and by-products released during the production of pig iron. Slag is lighter than iron. It floats to the top of the molten iron in the blast furnace and drains off through a separate hole and is allowed to solidify. Slag is used as building blocks or crushed to make gravel.

Cast iron

Foundries remelt pig iron and pour it into a mold to form a product shape called **cast iron**. When cast iron is produced, metallurgists control its properties by regulating the amount of carbon in its makeup, adding alloying elements, or specifying heat treatment.

The kind of cast iron commonly used in automotive engines is gray cast iron. Cylinder blocks and heads of gray cast iron usually contain about 3 percent carbon, which is scattered in the form of flakes throughout the grain structure of the metal. Gray cast iron is easy to cast and machine, absorbs vibrations, and resists corrosion well. To improve the strength and hardness of the castings, manufacturers sometimes add small amounts of nickel, molybdenum, and chromium.

A stronger form of cast iron is ductile cast iron, also known as nodular iron. It contains about 2- to 2.65-percent carbon and small amounts of magnesium and other elements. These additives, together with heat treatment, cause the carbon in the molten iron to form small balls or nodules, giving the iron more **ductility** and greatly reducing brittleness.

Designers often specify ductile cast iron for strong automotive castings, such as crankshafts, camshafts, and differential gears.

Steel

Steel is iron that has a carbon content of as little as 0.05 percent to a maximum of 1.7 percent. Steelmakers heat pig iron in a furnace so that most of the carbon burns off. Other impurities such as phosphorus, sulfur, silicon, and manganese also burn off, except for small quantities. During the refining process, the correct amount of carbon is added back in. The term plain carbon steel is used to differentiate them from the special steel alloys detailed in the next section.

Low-carbon steel contains about 0.05 to 0.30 percent carbon. Medium-carbon steel contains about 0.30 to 0.60 percent carbon. High-carbon steel has a carbon content of about 0.70 percent to 1.70 percent. It may seem peculiar that the best so-called "high-carbon" steels do not have as much carbon in them as pig iron. The high-carbon steels are not being compared to iron, but to lower carbon steels.

Low-carbon steel, also called mild steel, is the most common form of steel, and is easily bent, formed, or welded. Vehicle manufacturers use it for automobile bodies, chassis, valve covers, and some gears and shafts. Medium-carbon steel has greater **hardness** than mild steel, meaning that it better resists being dented or penetrated. Medium-carbon steel is also more difficult to weld or bend. It is used for connecting rods, some crankshafts, and axles. High-carbon steel is the hardest, but most brittle, form of plain carbon steel. Manufacturers often call it tool steel because they use it to make drills, files, taps, dies, hammers, and other tools.

Steel Alloys

Metallurgists often add small amounts of other metals to steel to strengthen and improve it, figure 3-2. Most steel alloys used in automotive engines or machine work contain one or more of the following metals: chromium, nickel, vanadium, molybdenum, and tungsten. The addition of chromium hardens steel and helps it resist rust. Steels with high amounts of chromium, 11 to 26 percent, are often called stainless steels. Chromium steels are used for fine measuring tools and ball bearings. Adding nickel to make a chromium-nickel steel improves **toughness**, the ability to withstand sudden shock loadings. Auto manufacturers therefore use chromium-nickel steels to make gears, springs, and axles. Small amounts of vanadium, usually 0.03 to 0.20 percent, are added to make chromium-vanadium steels where improved **tensile strength** (great resistance to being pulled apart) is important. Springs, axles, connecting rods, and steering knuckles are sometimes made of chromium-vanadium steels.

Adding molybdenum in small amounts to steel improves toughness, resistance to softening at high working temperatures, and the ability to be hardened through heat treatment. Manufacturers use molybdenum steels for connecting rod bolts, coil and leaf springs, differential gears, and transmission shafts.

Alloy: A mixture of two or more metals. Brass is an alloy of copper and zinc. Stainless steel is an alloy of steel and chromium.

Billet: An unfinished block or bar of metal that is ready for machining.

Pig Iron: Small solidified blocks of iron poured from the blast furnace into channels dug in sand. Used to make steel and cast iron.

Brittle: A metal is brittle if a blow can easily crack or break it.

Cast Iron: Pig iron remelted and poured into a product shape.

Ductility: The ability to be drawn or stretched without breaking. Ductile metals are not brittle.

Hardness: The resistance to being dented or penetrated. Hard metals are usually brittle.

Toughness: An ability of a metal to withstand a sudden blow. The test for toughness uses a weight falling onto the test metal.

Tensile Strength: A metal's resistance to a force pulling it apart. The force is measured in pounds per square inch (psi) or millions of pascals (MPa).

THE EFFECT OF ALLOYING ELEMENTS ON STEEL

Effect	Element											
	Carbon	Chromium	Cobalt	Lead	Manganese	Molybdenum	Nickel	Phosphorus	Silicon	Sulphur	Tungsten	Vanadium
Increases tensile strength	X	X			X	X	X					X
Increases hardness	X	X										
Increases wear resistance	X	X			X		X				X	
Increases hardenability	X	X			X	X	X					X
Increases ductility					X							
Increases elastic limit		X				X						
Increases rust resistance		X					X					
Increases abrasion resistance		X			X							
Increases toughness		X			X	X	X					X
Decreases ductility	X	X										
Decreases toughness			X									
Imparts fine grain structure					X							X
Creates soundness in casting									X			
Facilitates rolling and forging					X				X			
Improves machinability				X						X		

Figure 3-2. This table shows the effects of adding other elements and metals to steel.

Adding chromium and molybdenum to steel makes a very strong and useful alloy. Machinists usually call this alloy "chrome-moly" and it is a good material for custom-made racing crankshafts and frame tubes for race cars.

Alloy tool steel

Alloy tool steels are special steel alloys used to make high-quality cutting tool bits for lathes or boring bars, milling cutters, drills, punches, and dies. There are hundreds of different kinds of alloy tool steels, but they generally contain one or more of the following elements: chromium, vanadium, molybdenum, cobalt, and tungsten. Machinists also call steels with these elements high-speed steels, because machines that use them can run at faster speeds without wearing out the tool. Alloy tool steels are stronger and harder than plain carbon tool steels and they also maintain a sharp cutting edge for longer working periods and at higher temperatures.

The highest grade alloy tool steel contains tungsten carbide. It is the hardest man-made substance, nearly as hard as a diamond. Tungsten carbide is expensive, so cutting tools made from it are usually very small.

Aluminum

Refineries obtain aluminum from bauxite, an ore generally taken from open pit mines. The element aluminum is very common in the earth's crust, but it is expensive to extract the pure metal because it requires a great deal of electrical power in the smelting process.

Aluminum is not as strong as steel, but it is much lighter, about one-third the weight of steel. To improve the strength of aluminum, metallurgists add copper, manganese, silicon, magnesium, or zinc to produce many types of aluminum alloys. Aluminum and its alloys are much easier to machine and cast than steel, but require specialized equipment to weld.

With few exceptions, pistons in gasoline engines have been made of aluminum for more than half a century. Before that, iron or steel were popular materials because the aluminum alloys then available could not withstand the heat and pressure in the combustion chamber.

Aluminum is a superior material for pistons because its light weight reduces the reciprocating mass, and increases the rate at which an engine can pick up speed. Also, since a piston must change direction violently many times a second, light pistons improve engine durability by reducing loads on the connecting rods and crankshaft.

Since aluminum is more expensive than steel, automobile designers did not commonly specify it where a thin, lightweight piece of steel would do the job. An example is an engine rocker arm cover, traditionally stamped of thin sheet steel. Today, designers specify aluminum parts more frequently because of the weight advantage, which translates into better fuel mileage. Many engines now have aluminum cylinder heads, engine front covers, or water pump housings.

Aluminum is also a good material for engine blocks. However, an aluminum block requires either steel cylinder liners or special casting and cylinder bore preparation to make the cylinder durable for long-term use in a car engine. Either of these alternatives increases costs and complicates high-volume engine production. Because of this extra cost, aluminum blocks were once reserved for engines in high-priced, limited-production, or high-performance cars. Aluminum blocks are more common now because they reduce vehicle weight but are still not as common as aluminum cylinder heads.

Magnesium

Magnesium is most notable for its extremely light weight, about two-thirds the weight of aluminum. The first light-weight, "mag" racing wheels were made of magnesium. Pure magnesium burns readily, is expensive, and has low tensile strength, so it is typically alloyed with other metals, usually aluminum. Automotive street use of magnesium is fairly rare, but some Corvette valve covers and the engine block, or cases, of the Porsche 911 are magnesium alloy.

Titanium

Titanium is a nonferrous metal with some of the properties of steel. The main advantages of titanium alloys are their light weight and high strength. Titanium alloys are almost as strong as common steel alloys, but about half the weight. However, titanium alloys are difficult to weld and difficult to machine. Titanium alloys are also very expensive, so only race engines usually benefit from titanium-alloy connecting rods, valves, and valve retainers. One notable exception is

■ SAE-AISI Steel Numbering System

The Society of Automotive Engineers (SAE) and the American Iron and Steel Institute (AISI) developed a numbering system that shows machinists the chemical composition of different kinds of simple alloy steel. Each steel is given a number with four (or sometimes five) digits. The first digit frequently, but not always, indicates what kind of steel it is: 1 is a plain carbon steel, 3 is a nickel-chromium steel, 4 is a molybdenum steel, etc. For most alloy steels, the second digit tells the approximate percentage of the major alloying element that is represented by the first digit. The last two or three digits represent the average percentage of carbon the steel contains in "points," or hundredths of one percent: 1-point carbon is 0.01 percent, 45-point carbon is 0.45 percent, and 100-point carbon is 1.00 percent.

By this numbering system, we can look at the accompanying chart and tell that a SAE-AISI 4130 steel is a chromium-molybdenum steel with approximately 1 percent chromium and 0.30 percent carbon.

Steels with five digits have one or more percent of carbon. For example, SAE-AISI 52100 is a chromium steel that contains approximately 2 percent chromium and 1 percent carbon.

Kind of Steel	Series Number
Carbon steels	1xxx
Non-resulfurized	10xx
Resulfurized	11xx
Resulfurized and rephosphorized	12xx
Nickel steels	2xxx
Nickel-Chromium steels	31xx
	33xx
Molybdenum steels	4xxx
	44xx
	45xx
Chromium-Molybdenum steels	41xx
Nickel-Molybdenum	46xx
	48xx
Chromium steels	50xx
	51xx
	52xx
Chromium-vanadium steels	61xx
Nickel-Chromium-Molybdenum steels	43xx
	47xx
	81xx
	86xx
	87xx
	88xx
	93xx
	94xx
	98xx
Silicon-Manganese steels	92xx

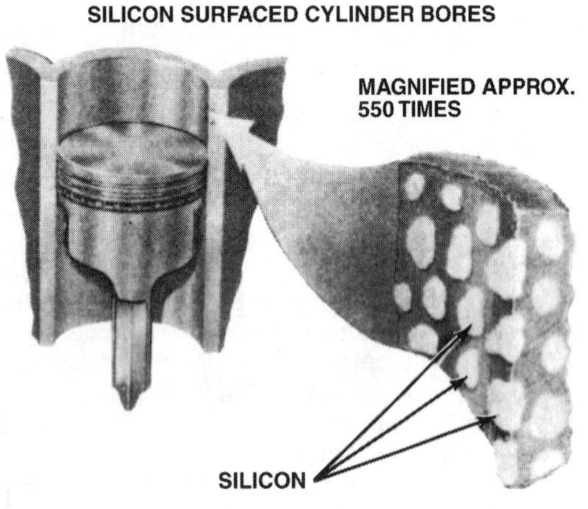

SILICON SURFACED CYLINDER BORES

MAGNIFIED APPROX.
550 TIMES

SILICON

Figure 3-3. Small particles of silicon are a part of the surface of the cylinder in this aluminum-silicon Chevrolet engine block.

Acura's NSX, which was introduced in 1990 with titanium connecting rods.

Aluminum-silicon

Race engines pioneered the process, but the 1970 Chevrolet Vega 4-cylinder engine had the first aluminum cylinder block without cylinder liners. To make this arrangement work, Chevrolet die-cast the engine out of an aluminum alloy containing 17-percent silicon. The silicon is in the form of small particles, each measuring about 0.001 inch (0.025 mm). After casting, the surface of the metal is a mix of small particles of silicon and aluminum. The cylinder bores are honed to a 7-microinch finish. Then an electrochemical etching process removes about 0.00015-inch (0.0038 mm) of aluminum. The resulting finish is pure silicon, with minute gaps between the silicon particles, figure 3-3. However, the silicon particles are so close together that they provided a good wearing surface for the pistons and rings.

To make the aluminum pistons operate in the hard silicon bores without seizing, the pistons are electroplated with a hard iron coating. This gives approximately the same durability as aluminum pistons in a cast iron bore. The Vega engine was used for seven years, and then discontinued. Its main problem was not the cylinder bore or pistons, but damage occurring when the cooling system was neglected and the engine overheated.

The aluminum-silicon process did not disappear with the death of the Vega. The 1978 Porsche V8 and 1981 Mercedes-Benz V8 engines introduced a similar process. These cars have all-aluminum engines with aluminum cylinders, figure 3-4. Like the Vega, the aluminum blocks of these engines have a very high

Figure 3-4. Mercedes-Benz V8s are all-aluminum engines with aluminum cylinders, such as this 380 SL engine.

silicon content. After casting, the foundry etches the inside walls of the cylinders with an acid to remove most of the aluminum, allowing the pistons to ride on silicon particles. Both Porsche and Mercedes-Benz used liner-less aluminum blocks for many years.

MAN-MADE MATERIALS

Engineers are always striving to develop superior materials. There are some man-made materials that are better than metals in some critical areas, or materials with properties that are acceptable but are cheaper to manufacture than the metal they replace. The man-made materials used in automotive engines include:

- Plastics
- Ceramics
- Other composites.

Plastics

One of the earliest uses of plastic composites on a production car was in 1981 when American Motors introduced plastic or glass-filled nylon valve covers that reduced engine weight, figure 3-5. General Motors introduced a composite pulley in its V6 engines in 1988, and added plastic valve covers in 1990. These parts reduce overall weight and cost, transmit less noise, and, in the case of the pulley, reduce reciprocating mass.

Plastic intake manifolds are becoming a common engine part because they are lightweight and have dampening characteristics that reduce noise and vibration. The first was introduced by BMW on its 1991 2.5-liter 6-cylinder engine. The manifold is cast from a recyclable fiberglass-reinforced polyamide. It

STEEL
VALVE COVER

PLASTIC
VALVE COVER

Figure 3-5. American Motors reduced the weight of the valve covers through the use of plastics.

weighs only 6 pounds (2.7 kg), about half the weight of a comparable aluminum manifold. BMW claims that the pastic material makes it easier to create contours for favorable airflow.

The raw plastic composite is molded around a metal core. After the mold is complete, the foundry melts out the core, leaving a smooth, hollow inner space.

Ceramics

A number of materials are used to make automotive ceramics, including: zirconium, alumina, silicon carbide, and silicon nitride.

Ceramic parts are lighter than steel, have better frictional properties, require less lubrication, are much less sensitive to heat, and are good insulators. On the negative side, ceramic parts are expensive and brittle, requiring careful handling.

Some ceramic parts are already used in engines. A ceramic rotor fills the turbocharger of the Buick Grand National GNX, produced in 1987. The ceramic rotors are lighter than their steel counterparts, allowing the rotor to accelerate and build boost quicker and reduce turbo lag. Porsche fitted ceramic liners in the 1986-89 944 Turbo's exhaust ports. The liners insulate the engine block and head from exhaust heat, thus allowing a smaller radiator and sending more heat to run the turbocharger to improve its efficiency. Mitsubishi used ceramic rocker arm pads on some of their engines.

Ceramic intake and exhaust valves, valve seats, valve spring retainers, rollers for roller lifters, solid lifters, and wrist pins are already available in the aftermarket, figure 3-6.

Other Composites

Composite materials consist of two or more physically or chemically different components tightly bound together. The composite material then has properties that neither component possesses alone. Some experimental connecting rods and pushrods have been built from graphite-reinforced fiber materials. Ford Motor Company has built a series of production engines with accessory mounting brackets made from this material.

The 1991 Honda Prelude's cylinder block is made of aluminum, but has composite rather than iron or steel cylinder liners, figure 3-7. Honda casts the block around composite cylinder cores of carbon and alumina fibers. The cores bind to the block during the casting process. Honda then bores the cylinders slightly smaller than the core to produce composite liners. Honda specifies iron-plated pistons for these cylinders — like the pistons in the aluminum-silicon, liner-less blocks.

Composite cylinders have better sliding properties for lower friction, and they add rigidity, so the block itself requires less material for strength and consequently is lighter. Furthermore, the new block shows better heat transfer from the cylinders to the block and water jackets than cast-iron liners.

Weight reduction is a major advantage of engine parts built with composite materials. Less engine weight means better fuel mileage. While these materials are not very common in production engines at present, engineers will continue to develop them for use in the near future.

MANUFACTURING OPERATIONS

You have learned that all steel alloys in an engine are not the same, nor are all iron or all aluminum alloys. The steel used in a connecting rod may differ in its alloy composition from that used in a crankshaft or a camshaft. Equally important, a vehicle manufacturer builds different parts with various processes, giving the part distinct properties. As a skilled engine machinist, you should know the differences between:

- Casting
- Forging
- Machining
- Stamping.

Composite: A man-made material that consists of two or more physically or chemically different components tightly bound together. The composite material has properties that neither of the components possess alone.

Figure 3-6. Ceramic materials are excellent for parts subjected to great heat and wear, such as these valves, valve seats, and lifter rollers.

Casting

A foundry produces a metal casting by pouring molten metal into a mold and allowing it to cool to a solid state. An iron casting usually has a coarse, grainy appearance because sand forms the mold. Engine blocks and cylinder heads are made this way. After the casting cools, the foundry breaks the sand mold away from the metal. In mass production, machines load the parts onto a vibrating conveyor where the sand falls away due to the vibration. If the interior of the casting is hollow, the sand that fills the cavities is poured out. The round holes in the sides of the engine block are there to make sand removal easier. These holes are closed with core plugs during engine assembly.

Normally, a foundry packs sand around a pattern of the desired shape to be cast, and then removes the pattern before the metal is poured in. In the "full-mold" or "lost-foam" process, a polystyrene foam pattern is left within the sand mold. When the foundry pours in the molten metal, the intense heat vaporizes the foam, figure 3-8. This is a very accurate, quick, and economical casting method because the pattern removal step is eliminated. General Motors casts the aluminum cylinder heads for the 2.5-liter, 4-cylinder engines this way.

Although sand casting is the oldest and the most basic casting method, there are many other ways to cast metals. Aluminum is often cast in permanent molds that are reused for thousands of castings.

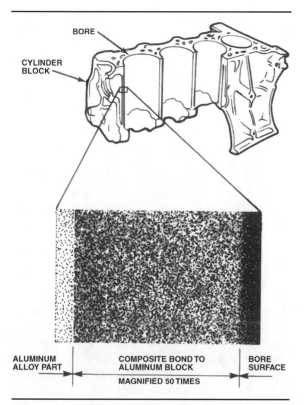

Figure 3-7. The Honda Prelude aluminum engine block uses cylinder liners made of carbon and alumina fibers.

Theoretically, permanent molds produce a part to a more accurate size and with a better finish.

In die casting, the foundry forces liquid metal into a permanent mold under pressure. In centrifugal casting, a machine spins the mold as the metal is poured in. Centrifugal action forces the metal into the outer parts of the mold. Casting operations that force the metal into the mold do not significantly strengthen the finished part, but they do help eliminate air holes or other imperfections.

Forging

A metal to be forged is heated to its plastic state, a temperature where the metal holds a shape, molds easily, but is not hot enough to be a liquid. The two-piece mold, or forging die, is forced onto the hot metal with tremendous force. The metal is forced into the cavities of the mold and assumes its shape, figure 3-9. It may require two or more dies and forging operations to get the metal into the desired, finished shape. A small amount of metal, called flash, squeezes out between the halves of the mold and is trimmed off. Because of the tremendous pressure, forged parts are densely packed and immensely strong, without air holes or imperfections.

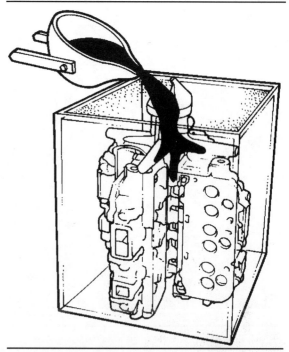

Figure 3-8. In the full-mold casting process, casting sand is packed around polystyrene foam molds. As the molten metal is poured in, it vaporizes the foam and fills the cavities made by the foam molds.

■ The Engine That Could Not Be Made

Before 1932, there were few V8 engines. When a V8 did appear, it usually consisted of two 4-cylinder engines bolted together. The crankcase was separate, and a 4-cylinder block was bolted on each side to form a V8.

To mass-produce such a V8 engine was too expensive. The only way that low-priced cars would ever get V8 power was by casting the V8 block in one piece. But engineers and foundry workers said it couldn't be done. There were just too many core pieces that might shift and ruin the casting. Nobody wanted to try it, but Henry Ford insisted.

Ford had been up against negative thinkers before. Every time he wanted to do something different, everybody told him it couldn't be done. So he just developed his own methods, and went ahead and did it.

The famous flathead Ford V8 of 1932 was the first mass-produced V8 engine with a block cast in one piece. It was the first time that low-priced cars had the advantage of the smoothness and power of a V8. But it wasn't easy. The cores for the casting used more than 50 pieces. And all of them had to be held in precise position, or the resulting block had to be scrapped. After the engine came out, everybody hailed it as a tremendous accomplishment. Mr. Ford just repeated his oft-quoted statement, "Anything that can be drawn up can be cast."

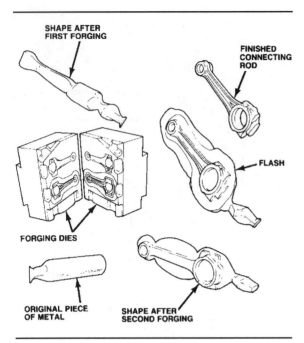

SHAPE AFTER
FIRST FORGING

FINISHED
CONNECTING
ROD

FLASH

FORGING DIES

ORIGINAL PIECE
OF METAL

SHAPE AFTER
SECOND FORGING

Figure 3-9. Two-piece dies are used to forge connecting rods.

An important difference between a machined part and a forging is that a forging allows the metal's grain structure to follow the shape of the finished product. On a machined part, a large portion of the billet is cut away to make the desired shape. This disrupts the natural flow of the grain structure and produces a weaker part than a forging.

A forging is far stronger than a casting when comparing identical parts made from the same material. However, modern alloying processes and casting methods have made it possible to cast some parts that formerly could only be strong enough when forged. Years ago, crankshafts and camshafts were usually forged because of their strength requirements. Today, cast crankshafts and camshafts are strong enough for hundreds of thousands of miles of use in passenger car engines. Cast parts are cheaper to make because the tooling required is cheaper than forging equipment.

In spite of the advances in metallurgy and casting technology, some applications still require the strength of forged parts. High-performance street engines and race engines usually have forged connecting rods, crankshafts, and pistons. Because of the high heat and cylinder pressures in turbocharged engines, they almost always require forged pistons.

Machining

Machining a part from a billet is the most time-consuming and expensive way to make a part, but it produces very strong and precise engine parts. A steel billet is placed in a lathe, or similar machine tool, and the part is cut with extreme accuracy to the size and shape wanted. A skilled machinist with modern machine tools can produce parts without design compromises.

Parts machined from high-quality billets are stronger than castings, but forged parts are the strongest of the three when made with equal-quality design and material. Crankshafts and camshafts for race engines are often machined from steel billets when a cast part is too weak, a forged part is not available, or when the forging process requires compromises in the design that make the part less desirable for racing.

Stamping

If a part is thin and does not have too many complicated angles and surfaces, it can be made by stamping it out of a large sheet of cold metal. Manufacturers usually stamp engine oil pans, valve covers, and timing covers. This is the same basic method used to make most sheet metal car body parts.

If the finished part is to be flat, it is stamped out on a punch press. This works much like a cookie cutter stamping out cookies from a sheet of dough. If the finished part is not flat, it is stamped, or drawn, into shape and then trimmed. For a part that is rather shallow, the stamping can be done in one draw, followed by trimming. For deeper parts, two or three draws may be necessary to form the part into its final shape.

MATERIAL TREATMENTS

After vehicle manufacturers form or machine metal parts, the parts usually need further treatment before they are ready for use in an engine. Manufacturers may treat the part by applying heat, blasting the part with metal shot, or applying one of several chemical processes. Also, they may create the desired surface finish (smoothness or roughness) by machining or by chemical processes.

Metal is one of the few substances whose properties can be changed by heat treating. All heat treatment processes involve heating and cooling metal to a temperature-time cycle, which changes the grain structure, altering the metal's properties. The temperature and time involved vary depending upon the type of metal used and the desired results. Heat treating includes these three steps:

- Heating the metal to the desired temperature.
- Holding the metal at this temperature for a period of time.
- Cooling the metal at a certain, controlled rate.

Hardening

Heating a carbon steel to about 1,400°F (760°C) makes the grain structure finer. Quickly cooling the steel at that temperature freezes the grain structure in that form, making the metal very hard. Depending upon how hard the steel must be, it is cooled in water, brine, or oil. Of the three liquids, water cools metal the quickest and oil the slowest, with brine between the two. The more carbon the steel contains and the more quickly it is cooled, the harder it becomes. The steel may become so hard that it is difficult to cut or work with and it may become very brittle.

To make a steel workable and less brittle after hardening, metal manufacturers use the process of **tempering**. In tempering, the steel is heated and allowed to cool slowly. The result is a steel that is slightly softer and more easily worked, but with improved toughness. Tempering temperatures range from about 300°F to 1,100°F (150°C to 600°C). Tempering temperatures vary, depending upon the metal, figure 3-10. Generally, the higher the tempering temperature used, the more hardness is lost, but the greater the toughness produced in the metal.

Annealing is a heating process for softening hard metal to make it easier to machine. To anneal steel, the heat-treating facility heats the metal to above 1,000°F (550°C) and then allows it to cool very slowly. It may be packed in ashes or lime for slow cooling, or the furnace shut off and the part allowed to cool in the slowly cooling furnace. After machining, the same steel can be hardened and tempered, if desired. When aluminum is annealed, it is heated to 650°F (350°C) and allowed to air-cool slowly.

To relieve internal stresses created in the metal due to forging, machining, or bending, a metal is **normalized**. It is usually done prior to final machining to ensure dimensional stability. Normalizing can also remove the effects of other heat treatment process. The metal to be normalized is heated to its normalizing temperature, held there for about one hour for each inch of thickness, and allowed to cool in the open air. Normalized steel is softer and easier to machine, but not as soft as when annealed.

Casehardening

Casehardening, also called surface hardening, is a process that hardens the surface of steel down to a relatively shallow depth. It does not alter the core properties of the part. Casehardening provides a protective surface for a completed part. A manufacturer only performs casehardening after all machining operations are completed. Crankshafts are especially good candidates for casehardening because the process

TEMPERING COLORS AND APPROXIMATE TEMPERATURES FOR CARBON STEEL		
Color	Celsius temperature	Use
Pale yellow	220°	Lathe tools, shaper tools.
Light straw	230°	Milling cutters, drills, reamers.
Dark straw	245°	Taps and dies.
Brown	255°	Scissors, shear blades.
Brownish purple	265°	Axes and wood chisels.
Purple	275°	Cold chisels, center punches.
Bright blue	295°	Screwdrivers, wrenches.
Dark blue	315°	Wood saws.

Figure 3-10. Steel glows different colors as it is heated. This chart indicates the colors and uses for carbon steel when annealed.

protects the main and rod bearing journals from cracks and retards wear.

A common form of casehardening for crankshafts is called **nitriding**. To perform nitriding, a manufacturer heats the part to about 900°F to 1,050°F (475°C to 575°C) in a furnace filled with ammonia gas, then allows the part to gradually cool. The process adds nitrogen into the surface of the metal down to approximately 0.007 inch, and requires about a day or more to complete. Extremely hard nitrides form in the surface of the metal and provide a harder surface than if the metal were simply hardened by heat treating.

Another form of nitriding introduces nitrogen into the metal in a molten salt bath. The temperature used is about the same as for gas nitriding. Several salt mixtures, which include cyanide salts, are available for this process. All are extremely poisonous and must be handled with care.

Tempering: The process of reheating steel to a temperature between 300°F to 1,100°F (150°C to 600°C), then allowing it to cool slowly. Tempering makes the steel more workable and improves toughness.

Annealing: The process of heating a metal to a high temperature and allowing it to cool very slowly. Annealing makes metal softer and easier to machine.

Normalizing: The process of heating metal to its normalizing temperature, holding it there for about one hour for each inch of thickness, and allowing it to cool in the open air. Normalizing removes internal stresses in the metal.

Nitriding: A form of casehardening where nitrogen is placed into the surface of a metal.

A tradename for a cyanide-salt nitriding process favored by General Motors is **tuftriding**. GM uses tuftriding because it is easier to manage and less expensive to perform on a large scale.

Another form of casehardening is ion nitriding. The part is placed in a pressurized chamber filled with hydrogen and nitrogen gas. An electrical current is applied through the part to introduce the nitrogen into the surface of the metal.

When a machinist grinds a crankshaft to refinish it, the grinding usually removes all of the casehardening. It is not common to reapply the casehardening when a crankshaft is refinished. However, some machine shops that rebuild race or exotic engines have small nitriding operations to reapply the casehardening.

Shot-Peening

Vehicle manufacturers shot-peen metal parts to strengthen them by reducing the chance for surface cracks to develop. Most metal fatigue failures originate as surface cracks.

In the **shot-peening** process, the manufacturer blasts the surface of a metal part with round, steel shot in a controlled manner, figure 3-11. The bombarding shot makes thousands of overlapping, tiny dents that compress the metal's surface. Tension forces cause surface cracks. By pre-stressing the surface in compression, the tension forces must overcome the compressive force before the part experiences any cracks on its surface.

Many parts in an engine could benefit from shot-peening. The only common use is the shot-peening of valve springs and high-performance connecting rods.

Plating

When a coating of copper, nickel, chromium, or other metal is applied to the surface of a part, it is called plating, figure 3-12. The plating facility applies the coating by a process called electroplating. The part is immersed in a tank of chemicals that contain the plating metal in solution. Direct current is then applied to the solution through an electrode. The part to be plated is the cathode, or negative electrode. The plating metal is the anode, or positive electrode. When the current is flowing, metal from the anode is deposited on the cathode.

Plating can increase wear resistance or improve the appearance of the part, or both. For example, decorative chrome plating helps keep bumpers from rusting and improves their appearance. Manufacturers sometimes apply hard chrome plating to intake or exhaust valve stems or gearbox internals to reduce wear.

Figure 3-11. A metal part is shot-peened to strengthen its surface and reduce the occurrence of surface cracks.

Anodizing

In **anodizing**, the plating process is reversed. The anode does not take part in the chemical process, it only supplies the electric current. When current is flowing, the metal leaves the anode and is deposited at the cathode. The metal to be anodized is the anode. The metal is part of the solution, and it flows to the cathode when the current is flowing. A special solution is used that results in oxygen combining with the anode as the metal leaves. The result is a coating of oxide on the anode.

In the case of aluminum, this oxide coating is very durable. It does not conduct electricity, and it reduces corrosion of the aluminum. Various processes have been developed to give the anodizing a color tint. Although it is most often used to prevent corrosion, it can also be used for decoration. Magnesium and titanium are two other metals that are anodized.

Surface Finishes

Every surface has minute grooves or imperfections. A highly polished metal surface may appear to be perfectly smooth, but if it is examined under a microscope, many hills and valleys will appear. The most widely used instrument for measuring surface finish is a **surface roughness indicator**. The indicator moves a diamond stylus over the surface of the metal. As the stylus rises and falls over the uneven surface, its movement is magnified and measured by the machine. The result is a measurement in millionths of an inch (or meter) of how rough the surface is.

A millionth of an inch is called a **microinch**, abbreviated μin. In the metric system, the standard measurement for surface finish is the **micro-meter**, abbreviated μm. A surface roughness indicator usually gives its readings in the form of a graph. The actual hills and valleys, magnified many times, can be

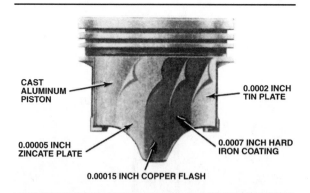

Figure 3-12. When a part is coated with a thin surface of metal it is called plating. Chevrolet iron-coated the Vega's pistons to give them long life in the aluminum-silicon cylinder bore. Mercedes-Benz used either chrome plating or iron coating for the same purpose.

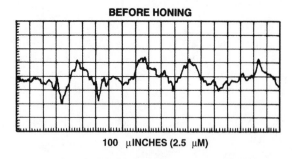

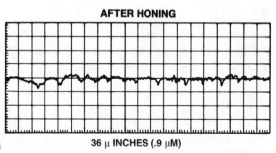

Figure 3-13. A surface roughness indicator gives a graphic printout of surface quality. Comparing these two charts, you can see that the honed surface, shown in the bottom chart, has far smaller peaks and valleys than the surface without honing, shown in the top chart.

seen on the graph. To express the graph picture in numbers, the surface roughness indicator calculates the average of the height of the ridges and gives the reading in microinches or micro-meters.

There are two ways to calculate the average. If it is done by simple arithmetic, it is called the arithmetic average, or aa method. A more complicated calculation is the root mean square, or rms method. Microinch readings are often listed with aa or rms after the reading, to indicate the method of calculation. The difference between the two readings is about 10 percent, the aa being smaller. Root mean square measurements are more common.

A rough cut from a saw gives a finish of about 800 µin (20 µm). A casting surface is about 700 µin (18 µm), but there is great variation. A drilled hole gives a reading of 250 to 60 µin (6 to 1.5 µm). It is only when surfaces are ground, honed, or polished that they become very smooth, figure 3-13. Grinding can produce finishes from 63 to as low as 4 µin (1.6 to 0.10 µm). Boring will give a surface finish between 250 and 16 µin (6 to 0.4 µm), depending on the depth of cut and feed of the boring bar. Honing with fine stones can reduce that to below 5 µin (0.13 µm). The smoothest finish is the superfinish. It is a honed finish, but the hones are scrubbed back and forth until they polish the surface. Superfinishing can produce a surface as smooth as 1 microinch (0.025 µm).

Tuftriding: A tradename for a form of salt-bath nitriding.

Shot-Peening: A surface treatment process in which the surface of a metal part is blasted with round, steel shot in a controlled manner. Shot-peening reduces the chances of surface cracks developing.

Anodizing: The process of oxidizing a metal, such as aluminum or magnesium, by running electric current through a special solution containing the metal anode.

Surface Roughness Indicator: An electrical instrument that measures surface roughness by running a diamond stylus over the surface of a metal.

Microinch: One millionth of an inch, abbreviated µin. The standard measurement for surface finish in the American Customary System.

Micro-meter: One millionth of a meter, abbreviated µm. The standard measurement for surface finish in the metric system.

MACHINING OPERATIONS

When rebuilding an engine, you will use various machine tools and perform many different machining operations. The two primary categories are internal and external machining. Some machine operations actually fall into both groups. Understanding the terms for these operations, as well as their differences and similarities, will help you do a better job of engine service.

Internal Machining

Internal machining involves working previously machined holes or inside the engine cavities as well as making new holes or cavities. It includes:

* Boring
* Honing
* Drilling
* Reaming
* Tapping
* Broaching.

Boring

In automotive machine work, the **boring** process is performed in a hole that has already been drilled or cast, figure 3-14. Boring provides a smooth-surface, accurately placed hole that is much more precise than drilling, but not as smooth as honing.

An automotive machinist typically bores a hole with a machine that holds a bar with a small cutting tool on the side. The part to be bored is held securely while the machinist inserts the rotating boring bar in the hole to make the cut. As the bar and cutting tool turn, a feed mechanism slowly feeds the bar through the hole. Boring is commonly used on automotive engines for enlarging the cylinders to the next available piston size.

Another boring process aligns main bearing seats and camshaft bearings to ensure that they are concentric with or have the same center as each other. This process is called **align boring** or line boring because a series of holes is bored in a straight line.

The machine used to bore cylinders can only be used for that purpose. Similarly, a lineboring machine can only be used for its specific purpose. Other boring tools are made to bore connecting rods and pistons in order to fit piston pins.

Honing

A hone is a tool with rotating abrasive stones used to finish an opening, such as a cylinder, to an exact dimension and a specific finish, figure 3-15. Honing produces a finish that is much smoother than boring. In cylinder honing the finish differs depending on the piston rings used.

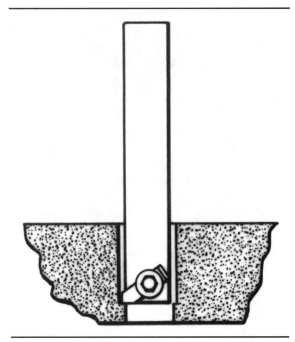

Figure 3-14. Boring is more precise than drilling, but requires a hole that has already been cut.

Honing is done by rigid or flexibly mounted stones. The rigid stones are usually several inches long, but not as long as the hole being honed. Before honing, the machinist expands the rigid hone against the walls of the hole or cylinder. The pressure against the walls can be increased or decreased depending on the amount of material to be removed and the type of finish desired. As the hone rotates in the hole, the machinist moves the hone in and out so the stones touch the entire surface.

Stones mounted flexibly cannot be adjusted to different diameters. They are used only for getting the correct surface finish. Flexible hones are available in many sizes and the machinist selects the correct diameter hone for the cylinder bore size.

Drilling

A drill is an implement with a cutting edge for making holes in various materials, figure 3-16. A drill has a point that a machinist forces against the material. The point penetrates the material slightly, and the cone-shaped cutting lips of the drill carve out the material. A drill works like a rotating knife blade or chisel. All of the cutting is done with the end of the drill. Drilling produces a relatively accurate hole size with a moderately good surface finish.

Where close tolerances are needed, a drill may not produce a hole diameter that is accurate enough. If the tip of the drill is slightly off center, the diameter of the hole will not be the same as the diameter of the drill, and the hole will not be perfectly round.

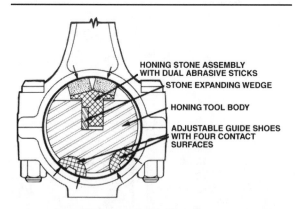

Figure 3-15. A rigid hone finishes a hole to the exact diameter needed and with the proper surface finish.

To improve the accuracy of the drilled hole, a machinist first drills a small pilot hole to properly locate where the finished hole will be. Then, the machinist drills with one or two undersized drills before drilling the hole to its final size.

Reaming

When a hole of a very precise size is needed, but the hole is too small to easily bore, it is usually drilled first and then reamed to the desired size. The reamer has cutting edges ground into a rod several inches long, figure 3-17. These cutting edges can be straight, or they can spiral around the rod. Unlike a drill, a reamer cuts with its sides, not its tip. Reaming removes only a few thousandths of an inch of material.

The machinist first drills the hole accurately, just slightly under the size of the reamer. The machinist then forces the reamer into the hole and turns it at the same time. Reaming smoothes the hole and straightens the sides, but is not as accurate a method for sizing a large hole as honing.

Tapping

Cutting threads in a hole with a tap is called tapping, figure 3-18. The tap is a solid piece of metal with external threads around it, similar to a bolt. Flutes are ground into the sides of the tap so that there are either three or four cutting edges. Taps are made of tool steel that is hardened and tempered and they will cut threads in any material that is softer than the tap.

The end of the tap is usually tapered for easy entry into the hole. The machinist turns the tap until it cuts into the metal. After a tap starts, it is self-feeding while it turns.

The hole must be drilled accurately before the tap is used. If the hole is too large, the threads will be too shallow and weak. If the hole is too small, the tap may seize or break trying to cut the threads. The usual depth of the tapped threads is 75 percent of the

Figure 3-16. The drill uses an edge that cuts through the surface to make an opening.

threads on the tap. Under 75 percent results in weak threads. Over 75 percent makes it too difficult to turn the tap.

In any specific size of tap there are three types. A taper, or starter tap, has a long taper so it can be easily started in an untapped opening. A plug tap has a taper, but only for the first few threads on the tap. It cuts a little deeper into the hole. The bottom tap cuts threads all the way to the bottom of the hole. It has little or no taper on the end, just enough to allow the tap to enter the hole.

Broaching

A broach is a cutting tool used in a press, most often used to change the shape of openings, figure 3-19. If a round hole is going to be made square, a machinist forces the broach through to cut out the four corners. Broaches may have cutting teeth all around their sides. When changing the round hole to a square shape, the broach might cut all four corners at once.

Boring: A process by which a machinist enlarges a hole or cylinder.

Align Boring: A boring process that aligns main bearing seats or camshaft bearings to ensure that they are concentric. Also called line boring.

Honing: A process by which a machinist enlarges or finishes the interior surface of a cylinder.

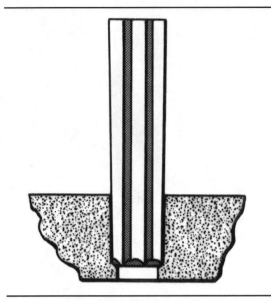

Figure 3-17. The cutting edges of the reamer finish a hole to the exact size.

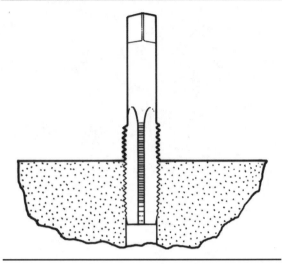

Figure 3-18. The threads around the piece of metal called a tap are capable of cutting corresponding threads in a hole.

Broaches cut in one direction, and the metal to be removed is taken in one pass through the work. Each tooth on a broach removes about 0.001 inch (0.025 mm), figure 3-20. The broach is moved through the work with considerable force. The longer the broach, the more material will be taken out, because there are more teeth. Machinists also use a broach to cut an external keyway or a slot in a shaft.

External Machining

When working on the outside surface of a part to change its shape or the texture of the surface, you are doing external machining. External machining includes:

- Milling
- Grinding
- Polishing.

Milling
A milling machine cuts metal with a multiple-tooth cutting tool called a milling cutter. A machinist mounts the part on a table which can be moved up, down or sideways. Milling machines can also cut slots or keyways on the outside of parts, figure 3-21.

Machinists sometimes use a single-edge cutting tool in a milling machine to remove material from a flat surface. If a surface is warped, grooved, or pitted, it can be mounted on a milling table for resurfacing. A cylinder head is a good example. The surface of the head is mounted horizontally, so that the milling tool moves across it when the table is moved from side to side. One pass removes only a few thousandths of an

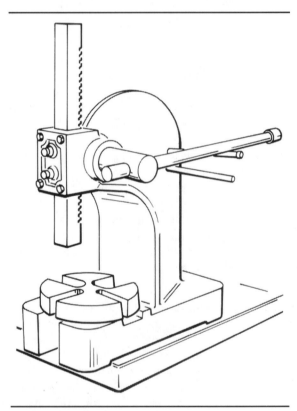

Figure 3-19. One way to change the shape of an opening is to use a broaching tool in a press.

inch. The machinist makes more passes until the entire surface is flat.

Grinding
Grinding is the removal of metal with an abrasive wheel made of aluminum oxide or silicon carbide,

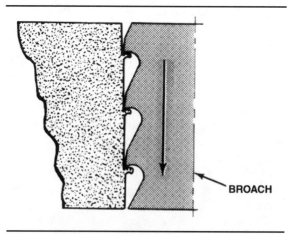

Figure 3-20. The broach cuts in only one direction and removes all the metal required in one pass.

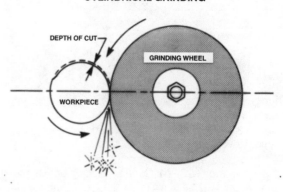

Figure 3-22. When you grind, you remove metal with an abrasive wheel or stones. Two common types of grinding are surface and cylindrical grinding.

Figure 3-21. Milling is often used to cut slots or keyways.

figure 3-22. The applications of grinding are almost unlimited. Machinists use machines that will grind just about any type of internal or external surface or shape to a precise dimension. About the only thing grinding is not used for is drilling holes. It could be done, but a drill is much faster. However, after the drilling, a grinder may be used to finish the hole to an exact size.

Automotive machinists use grinding extensively for resurfacing. When cylinder head and block gasket surfaces are warped or otherwise damaged, machinists resurface them on a grinding machine. Crankshaft journals are also resurfaced by grinding.

Machine grinding is highly accurate. Hand grinding is less accurate. Hand grinding is limited to areas where parts do not have to be joined.

Polishing

Polishing makes a surface smooth and usually bright. It does not remove enough metal to change the size of

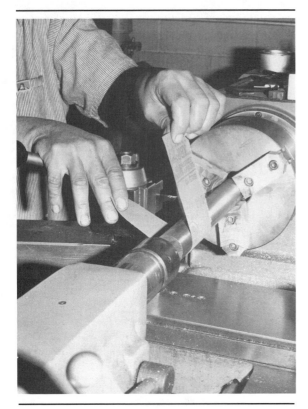

Figure 3-23. Polishing makes a surface smooth, but removes very little metal.

a part, figure 3-23. Machinists usually polish parts with cloth buffing wheels or a narrow belt sander. Automotive crankshaft journals are polished after they are ground. The object in polishing is to make a surface that rubs against a bearing as smooth as possible to reduce wear.

MATERIALS AND MACHINING RELATIONSHIPS

Most flat metal surfaces appear smooth to the naked eye, but under microscopic examination the true roughness of the surface is revealed. When using a surface roughness indicator, the small diamond tip moves over the surface while the machine graphs the peaks and valleys. The amount of movement of the diamond is measured in millionths of an inch or millimeter. A perfect surface would measure zero.

From reading about the different machines discussed in this chapter you have learned that it is possible to obtain surfaces ranging from rough to very smooth. Different operating conditions for different parts of the engine require different surface finishes.

Engineers have established that there is an ideal surface finish for every application. For example, if two surfaces are going to be joined together, with a gasket in between, they should be slightly rough. A smooth or polished surface has a tendency to let the gasket slide, or squeeze out between the pieces. The ideal gasket surface should have what is known as a small amount of "tooth." It is flat and straight, but there is enough surface roughness to make the gasket adhere.

On the other hand, bearing surfaces should be as smooth as possible. If crankshaft journals were rough, they would act as a grinding wheel and remove metal from the bearings. A ball bearing is another good example. The balls and the race must be finely finished so that there is no grinding action between them.

Cylinder walls are a special case. It is impossible to make piston rings that will fit against a cylinder wall perfectly. There are small irregularities in both the rings and the walls that must be worn off in the first few hundred miles. This is known as seating the rings. If the cylinder walls are just the right surface, the rings will wear in quickly and seat perfectly. Once the rings are seated, they will last many thousands of miles.

If the cylinder wall is too rough, the rings will grind away. This can happen in only a few hundred miles. The surface produced by a boring bar is about 100 µin (2.54 µm). The boring bar actually plows a furrow in the metal. For the best ring seating, the cylinder wall surface should have approximately a 25-µin (0.64-µm) finish. The only way to get this finish is with a hone, after boring.

The type of finish is also important. We can analyze a surface and measure the peaks and valleys in microinches or micro-meters. But this measurement does not indicate how sharp the peaks are. A "peaky" finish on a cylinder wall, with lots of sharp points, is not as desirable as a finish with those peaks blunted. The ideal finish for ring seating is the one in which the peaks have been cut off, forming small plateaus instead of a point. This is known as a **plateaued finish**. The plateaued finish is usually achieved on cylinder walls at the end of the honing process with a ball hone.

When using a machine tool on a part, the material of the part affects the machining process. Aluminum is relatively soft, and therefore machines more easily than other metals or alloys. Some machinists sharpen the tool bit differently for each type of material on which they work.

Plateaued Finish: A surface finish in which the highest parts of a surface have been honed to flattened peaks. A plateaued finish is the best finish possible for new cylinder bores.

Machining is usually done before heat treating or hardening. Before hardening, the metal is more easily machined. In some situations, such as restoring or rebuilding a part, it is necessary to machine a used part. The part may have been hardened before it was put into service. This makes machining very difficult, if not impossible. A good example is the resurfacing of hydraulic lifters or tappets. If the tappet was surface hardened, and the wear goes through the thin hardening, resurfacing is a waste of time. Resurfacing by grinding will certainly restore the surface and remove the wear, but it does not restore the hardening. For this reason, many engine rebuilders refuse to resurface a hardened part, such as a tappet. They prefer to use new ones.

SUMMARY

Many different materials are used in an engine. These include iron, steel, and aluminum, as well as man-made plastics, ceramics, and other composites.

Metallurgy is the science of metals. Metal is divided into two basic categories, ferrous and non-ferrous. When two or more different metals are added together, the result is called an alloy. Metal has a crystalline grain structure that causes different metals to have different properties. Properties of metals include hardness, brittleness, toughness, ductility, and tensile strength.

Iron is the most common metal. It is made from iron ore mined from the earth. Pig iron remelted and poured into a mold to form a product shape is called cast iron. Steel is iron that has a carbon content of as little as 0.05 percent to a maximum of 1.7 percent. Steel alloys may include chromium, nickel, vanadium, molybdenum, and tungsten. Non-ferrous metals used on engines include aluminum, magnesium, and titanium. Man-made materials include: ceramics, plastics, and composites. They may be lighter, more heat resistant, less expensive, or have properties that exceed metals in some other area.

Metal parts are made by casting, forging, machining, or stamping. Parts machined from high-quality billets are stronger than castings, but forged parts are the strongest of the three when made with equal-quality design and material.

A metal's properties may be changed by heat treating. Heat treating includes these three steps: heating the metal to the desired temperature, holding the metal at this temperature for a period of time, and cooling the metal at a certain, controlled rate. Heat-treating processes include: hardening, annealing, tempering, and casehardening. Steel's strength may also be improved by shot-peening, a process of blasting the part with metal shot.

Machining or working of metals is done to make the part the correct size, or to achieve the correct surface finish. Machining operations include boring, honing, drilling, reaming, tapping, broaching, milling, grinding, and polishing.

Review Questions
Choose the single most correct answer.
Compare your answers with the correct answers on Page 328.

1. Aluminum is often used for:
   a. Camshafts
   b. Valves
   c. Cylinder liners
   d. Cylinder heads

2. Which of the following is most often used to produce a smooth, accurate hole?
   a. Broaching
   b. Drilling
   c. Boring
   d. Milling

3. Plain carbon steel is made of:
   a. Carbon and vanadium
   b. Carbon and iron
   c. Carbon and nickel
   d. Carbon and chromium

4. Pig iron contains how much carbon?
   a. 3 to 5 percent
   b. 0.7 to 1.7 percent
   c. Almost none
   d. About 5 percent

5. High carbon steel has a carbon content of approximately:
   a. 2 to 6 percent
   b. 0.7 to 1.70 percent
   c. 0.1 to 3.7 percent
   d. 2 percent

6. The metal characteristic of "hardness" is:
   a. Resistance to cracking
   b. Resistance to rust
   c. Resistance to being dented or penetrated
   d. All of the above

7. Alloy tool steels may contain which metal besides steel?
   a. Vanadium
   b. Tungsten
   c. Chromium
   d. All of the above

8. Which of the following make hard steel easier to machine?
   a. Annealing
   b. Plating
   c. Casehardening
   d. None of the above

9. When a metal is forged:
   a. It is poured into a mold.
   b. It is cut out on a lathe or milling machine.
   c. A mold is forced over the metal.
   d. It is heated until it is soft and easily machined.

10. Heat treating can make a metal:
    a. Harder
    b. Softer
    c. Neither a or b
    d. Both a and b

11. A metal is shot-peened to:
    a. Improve its machinability
    b. Improve its resistance to surface cracks
    c. Reduce its hardness
    d. Normalize it

12. In the process of_____, steel is heated to 300°F to 1,100°F (148°C to 593°C) and allowed to cool slowly.
    a. Annealing
    b. Forging
    c. Hardening
    d. Tempering

13. When made from equal-quality material, which process produces the strongest parts?
    a. Casting
    b. Machining
    c. Forging
    d. Stamping

14. Which of the following is used in electroplating?
    a. Direct current
    b. Alternating current
    c. Electromagnetism
    d. High voltage

15. When a precision sized hole is needed, drilling is usually followed by:
    a. Boring
    b. Honing
    c. Reaming
    d. Broaching

16. Each of the following is an example of external machining *except*:
    a. Polishing
    b. Milling
    c. Reaming
    d. Grinding

17. Surface roughness is measured in:
    a. Millimeters
    b. Angstroms
    c. Multimeters
    d. None of the above

18. Plastic engine parts offer:
    a. Lower weight
    b. Reduced cost
    c. Better noise control
    d. All of the above

19. Technicians use align boring to:
    a. Enlarge a series of holes in a straight line
    b. Enlarge cylinders
    c. Restore keyways
    d. Surface cylinder heads

20. A technician may use milling to:
    a. Enlarge cylinders
    b. Surface cylinder heads
    c. Restore keyways
    d. Enlarge a series of holes in a straight line

4

Cooling Systems

All engines use a cooling system to maintain the correct engine temperature. Without such a system, the engine would quickly fail. The well-being of the cooling system has a direct effect on the life and performance of any engine. Engine damage is often the result of cooling system neglect or failure.

This chapter explains the purpose of the cooling system and how it functions. Next, the development of engine coolant is discussed. Finally, the cooling system components and the relationship between engine temperature and mechanical wear is described.

COOLING SYSTEM FUNCTION

When the air-fuel mixture in a combustion chamber burns, the cylinder temperature can soar to 6,000°F (3,300°C). During the complete four-stroke cycle, the average temperature is about 1,500°F (800°C). Only about one quarter to one third of this heat energy is turned into mechanical energy by the engine. The remaining heat must be removed by other means to maintain engine efficiency and prevent overheating.

About half of this waste heat remains in the combustion gases, and is removed through the exhaust system. The other half of this waste heat energy is absorbed by the metal of the engine block, which is then cooled by the cooling system. The engine cooling system, in turn, can recycle much of this waste heat through the passenger compartment heating system.

Proper operation of the cooling system is essential. If the excess heat is allowed to build up in the engine:

- Engine oil temperature will rise, reducing viscosity and engine lubrication, and decomposing the oil.
- The incoming air-fuel mixture will become too hot and reduce engine efficiency.
- Excessive cylinder head temperatures can cause damaging preignition or detonation.
- Metal engine parts can expand to the point of damage, seizure, or total engine failure.

An engine that is too cool will have poor fuel vaporization, poor lubrication, excessive acids in the blowby gases, and high hydrocarbon (HC) emissions. An engine that is too hot will have poor volumetric efficiency, poor lubrication, high oxides of nitrogen (NO_x) emissions, and in extreme cases the fuel may detonate or preignite.

Obviously, there must be an optimum engine operating temperature that will minimize these problems. This temperature varies for each engine, depending on the design. Before the extensive use of emission controls, engine coolant temperatures averaged about 180°F (80°C). Late-model, emission-controlled engines average about 10 to 15 degrees hotter to reduce exhaust emissions and improve fuel economy. These average temperatures can vary depending on driving conditions. Engines use either an air or liquid cooling system to maintain the proper operating temperature.

Air-Cooled Systems

For many years, vehicle manufacturers produced air-cooled engines. These engines rely on airflow to reduce and stabilize engine temperature, figure 4-1. These air-cooled engines have deeply finned heads and cylinders. The fins provide more surface area to absorb heat and draw it away from the cylinders and combustion chambers. The fins also expose more surface area to the air to help dissipate the heat. The oil crankcases of air-cooled engines have internal as well as external fins that reduce oil temperature to help keep overall engine temperature under control, figure 4-1. The internal fins speed heat absorption from the oil to the crankcase and the external fins dissipate the heat to the air. However, when an engine is installed in a vehicle and surrounded by the body, the fins alone are not sufficient to keep engine temperature in the correct range.

Air-cooled engines require an auxiliary fan to furnish an adequate supply of air for cooling the engine, figure 4-1. A shroud directs the air flow so that it circulates around the heads and cylinders efficiently. A thermostat opens and closes either the fan's air intake or outlet (depending on design) so that the engine warms up quickly and the operating temperature does not vary over a wide range.

Most automotive air-cooled engines require an oil cooler for additional engine heat reduction. These oil coolers are small radiators through which the oil circulates after being pumped through the engine, before returning to the crankcase. Air-cooled engines depend on their oil supply to maintain the correct engine temperature.

Vehicles built with air-cooled engines are no longer being sold in the United States due to their inability to meet stricter emissions standards.

Liquid-Cooled Systems

A liquid-cooled system is the most common automotive cooling system. The coolant is circulated outside of the engine and exposed indirectly to the air by a radiator. The air absorbs heat from the coolant, so that the coolant can flow back into the engine and absorb more heat. The greater the difference in temperature between the coolant and the air, the more heat will be absorbed by the air.

Uniform engine temperatures reduce thermal stress. The consistent temperatures also allow very precise fuel metering which helps keep exhaust emissions low. These are the main reasons why most engines are liquid-cooled.

THE DEVELOPMENT OF ENGINE COOLANT

Early cooling systems used plain water as engine coolant. Since plain water has a high specific heat and thermal conductivity, it can transfer heat quite well. However, it has disadvantages when used in a cooling system:

- Iron, steel, and aluminum engine parts react with water to form rust and other types of corrosion, which weaken parts, clog water jackets, and slow heat transfer from the coolant to the engine castings.
- Using hard water introduces minerals to the cooling system, which clogs water jackets with thick scale deposits, and also reduces heat transfer.
- When the car is not running and the system is not pressurized, water will freeze at its normal freezing point of 32°F (0°C). The expansion of water as it freezes can crack radiators, fittings, and even engine blocks.

Alcohol coolants worked much better. A 50/50 mixture of either methanol or ethanol and water would protect against freezing to -20°F (-29°C). However, alcohols boiled at lower temperatures than water and therefore evaporated from the engine fairly quickly. To maintain freeze protection, the alcohol concentration had to be tested periodically and additional alcohol had to be added to bring the protection up to the necessary level.

Most cooling systems use a mixture of water and **ethylene glycol** for coolant. Ethylene glycol coolant is about 96 percent pure ethylene glycol, 2 percent added water, and 2 percent additives. Although ethylene glycol does not transfer heat as well as water, the greater temperature range in which an ethylene glycol-water mixture remains a liquid makes up for this disadvantage.

Concentrated ethylene glycol freezes at about -8°F (-22°C), figure 4-2, but a 50/50 mix of ethylene glycol and water resists freezing down to a temperature of -34°F (-37°C). Maximum protection against freezing is achieved by a solution of approximately 68 percent coolant. Concentrations higher than this cause the coolant freezing point to rise, instead of fall. The freezing protection of concentrated, 100 percent coolant is no better than that of a 35-percent solution in water.

Since ethylene glycol vaporizes at such a high temperature, mixing it with water also protects against coolant boilover. Boiling is harmful because the gases

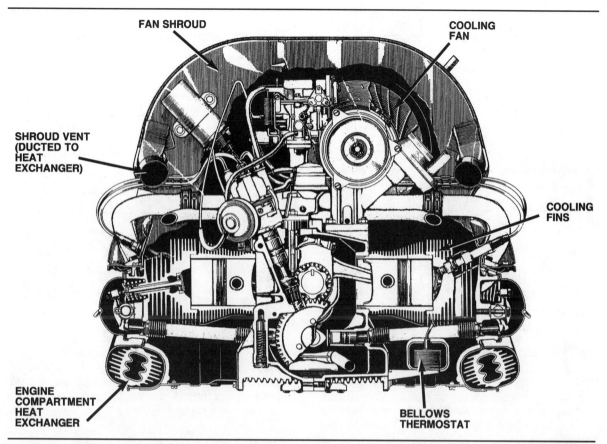

FAN SHROUD

COOLING FAN

SHROUD VENT (DUCTED TO HEAT EXCHANGER)

COOLING FINS

ENGINE COMPARTMENT HEAT EXCHANGER

BELLOWS THERMOSTAT

Figure 4-1. A typical air-cooled engine relies on a fan and deep fins to maintain the proper operating temperature.

form steam pockets in the water jackets. Steam in the cooling system absorbs much less heat than liquid, so localized hot spots may form that can warp or crack castings. In addition, steam pressure forces coolant out of the radiator, reducing efficiency even further.

Ethylene glycol: A chemical compound that increases the coolant's resistance to both freezing and boilover.

■ Nucleate Boiling

When most technicians think about an automotive cooling system, they think of it as a means of preventing engine coolant from boiling. After all, the purpose of radiator pressure caps and coolant is to raise the temperature a cooling system can withstand before the coolant boils.

In fact, things are not quite so simple. Some of the coolant in a normally operating engine is actually boiling all of the time. Were the process to stop, the engine would rapidly overheat, and either seize or detonate into scrap metal.

Nucleate boiling is the key to low engine temperature. Ordinary ethylene glycol and water coolant boils at around 263°F (128°C). In the cylinder head, however, the temperature of the combustion chamber or exhaust port walls may approach 500°F (250°C). As individual droplets of coolant flow through the water jackets against the other sides of these hot surfaces, they boil into a gas and are instantly washed away, to

be replaced by fresh liquid coolant. The superheated bubbles are swept into the lower-temperature coolant flowing through the middle of the passage, where they cool off and condense back to liquid.

Boiling the coolant into a gas removes heat from the metal surface much more efficiently than would merely warming the coolant. With nucleate boiling, the overall temperature of the coolant still stays below its boiling point — only the small nuclei of coolant actually touching the metal flash into a gas. The result is much better heat transfer between the metal and the coolant, and lower operating temperatures.

Nucleate boiling requires a strong, turbulent coolant flow to perform properly. If the coolant were to stop moving, steam pockets and steam-blanketed areas would form very quickly. This would lead to local overheating, detonation, boilover, and warped or cracked castings. Many engine blocks and cylinder heads have special passages called steam holes that permit bubbles from boiling coolant to escape when the engine is operated at low speeds or turned off.

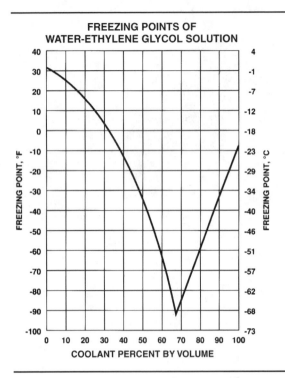

Figure 4-2. Up to a concentration of 68 percent, adding ethylene glycol extends the lower temperature limit of engine coolant by protecting against freezing.

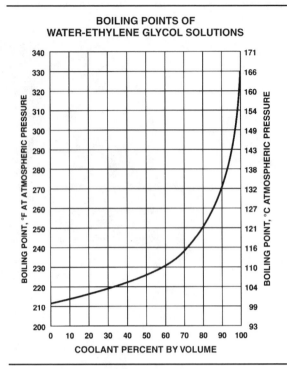

Figure 4-3. Adding ethylene glycol also extends the upper limit of engine coolant by protecting against boiling, even without a radiator pressure cap.

Concentrated ethylene glycol coolant boils at 330°F (165°C), and a 50/50 mix resists boiling up to 226°F (108°C), figure 4-3. Pressurizing the system with a sealed radiator cap raises the boiling point even higher.

Coolant usually contains rust inhibitors, antifoaming agents, and sometimes water-soluble lubricants. It may also contain small particles designed to seal minor leaks in the cooling system. Always follow the vehicle manufacturer's specifications for the coolant concentration. Too much or too little coolant in the mixture will affect the ability to cool the engine components. In certain weather conditions, the ambient temperatures will be a factor in determining the specified coolant concentration.

A 50-percent coolant mixture is the industry standard. However, under extreme weather conditions, the percentage may be different. Vehicle manufacturers may specify a higher concentration of coolant if temperatures below -37°F (-35°C) are expected. Some manufacturers recommend a lower concentration of coolant if extremely high temperatures are expected. Again, follow manufacturers' specifications. A coolant concentration that is too high will cause the coolant to gel and restrict the cooling system. A coolant concentration that is too low may protect the system from freezing; however, it does not give the corrosion protection available with a 50-percent concentration.

Other Engine Coolants

For years, most vehicle manufacturers specified ethylene glycol and water as the only acceptable engine coolant. In recent years, other coolants have been introduced. Always use the required coolant that meets manufacturers' specifications. Different types of coolants are not compatible, and cannot be mixed. Even topping off with a different type of coolant will affect the protection level. There is not an accurate way to determine the exact percentage of each type of coolant, or what effect topping off with a different fluid has on the overall protection level. Incompatible fluids break down the additives that protect and cool the engine. Some manufacturers do allow top-off with a different type of fluid in an emergency, when the regular type of coolant is not available. (In these circumstances, the coolant life is shorter, and must be replaced at a sooner service interval.) Other fluids are incompatible under any circumstance. You must learn and follow the exact specifications of each manufacturer.

Any liquid, whether ethylene glycol, propylene glycol, or pure water, that has circulated in an automotive cooling system will pick up traces of heavy metals and other contaminants. Lead that leaches out of the solder used in radiator construction is the most harmful of these contaminants. For these reasons, all used coolants must be properly recycled or disposed of following EPA guidelines.

Propylene glycol

Propylene glycol is similar to ethylene glycol, since both are glycol compounds. As with ethylene glycol, the highly concentrated blend of propylene glycol used in coolant is extremely toxic.

The freezing and boiling points of propylene glycol are different from other coolants, so propylene glycol cannot be mixed with any other type of coolant. Propylene glycol coolant is not approved by vehicle manufacturers for use in all systems. When it is approved, propylene glycol must be used by itself, and the system must be flushed of any other coolant before propylene glycol is introduced.

DEX-COOL™

DEX-COOL™ is an ethylene glycol-based coolant that is factory-installed in late-model General Motors vehicles. DEX-COOL™ is silicate-free and has an orange color to distinguish it from other coolants. The extended-life DEX-COOL™ has a service interval that is longer than the green, ethylene glycol-based coolants. DEX-COOL™ manufacturers caution that you should not mix DEX-COOL™ with other coolants, since this compromises many of the DEX-COOL™ benefits.

VW/Audi coolant

Volkswagen and Audi use an ethylene glycol-based, extended-life coolant in their water-cooled engines. This phosphate and silicate-free coolant is red in color and designed to last the service life of the engine. There is no scheduled replacement. The new coolant, which is specifically formulated for aluminum engines, offers improved corrosion protection, thermal stability, and heat transfer. The coolant also improves hard water tolerance.

Due to the different chemical additives, the red coolant is not compatible with any other coolant formulation. Coolant that has turned brown or purple, or has foamy deposits in the radiator, indicates contamination caused by other coolants. If this occurs, the cooling system must be drained and flushed immediately, or engine damage may occur.

COOLING SYSTEM COMPONENTS

Most engines use a liquid-cooled system, figure 4-4. Coolant circulates through the engine, absorbing excess heat as it does so. A typical cooling system consists of the following components:

- Water pump
- Thermostat
- Radiator
- Radiator cap
- Coolant recovery tank
- Radiator fan

- Hoses
- Core plugs
- Heater core.

Water Pump

The water pump is located between the engine block and the cooling fan. The water pump, figure 4-5, uses centrifugal force to circulate the coolant, and consists of a fan-shaped impeller set in a round chamber with curved inlet and outlet passages. The chamber is called a scroll because of these curved areas. The impeller is driven by the crankshaft pulley and spins within the scroll. Coolant from the radiator or bypass enters the water pump inlet and is picked up by the impeller blades. Centrifugal force moves the coolant outward, and the scroll walls direct it to the outlet passage and into the engine. Since the water pump is driven by the engine, coolant circulates whenever the engine is running.

Water pumps require so little attention that they are easy to neglect. Centrifugal pumps are subject to worn or loose drive belts, belt pulleys, or bearings. Early water pumps required periodic lubrication, but modern pumps are lubricated for life at the factory. The pump shaft is sealed to prevent coolant from leaking into and past the bearings.

Thermostat

The thermostat is a temperature-sensitive valve that regulates the flow of coolant between the engine and the radiator. Since the engine creates so much heat, it is designed to operate at high temperatures. When the engine is cold, it must be warmed quickly to its ideal operating temperature. To speed the warmup, the thermostat stays closed when the coolant is cold, figure 4-6, position **A**. This prevents the coolant from circulating through the radiator. Instead, the water pump pushes it through a passage (thermostat bypass) that bypasses the radiator hose inlet, and returns the coolant to the engine. The cold coolant continuously circulates through the engine as it warms. Until the engine warms up and starts giving off excess heat, there is no heat for the coolant to carry away. Since the coolant is already cold, it does not need to travel to the radiator to cool.

Once the engine has warmed up, the coolant begins absorbing the excess heat. When the coolant is warmed up, the thermostat opens, and the coolant circulates through the radiator, figure 4-6, position **B**. This removes much of the heat from the coolant before it circulates back through the engine. Some systems are designed so the thermostat also seals the bypass when it opens, forcing all of the coolant to flow into the radiator.

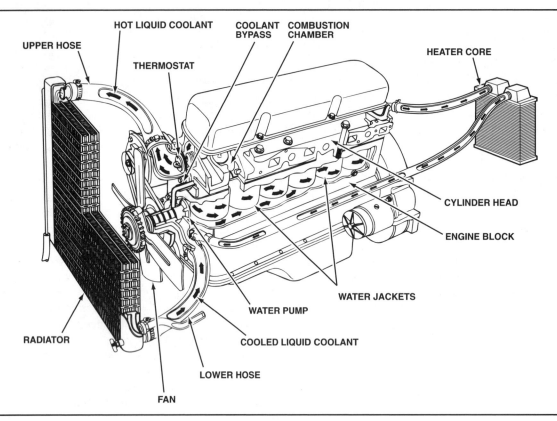

UPPER HOSE
HOT LIQUID COOLANT
COOLANT BYPASS
COMBUSTION CHAMBER
THERMOSTAT
HEATER CORE
CYLINDER HEAD
ENGINE BLOCK
WATER JACKETS
WATER PUMP
COOLED LIQUID COOLANT
RADIATOR
LOWER HOSE
FAN

Figure 4-4. A liquid cooling system uses the circulation of coolant to maintain the proper engine operating temperature.

Once the temperatures reach normal operating range, the cooling system maintains the appropriate temperature range to prevent excessive heat and pressure buildup. The coolant temperature rises and falls as it circulates through the system. The coolant is hot as it leaves the engine, and is cold as it leaves the radiator. When the coolant is cold, the thermostat is completely closed. As the coolant temperature rises, the thermostat opens and the coolant is directed into the radiator to be cooled. When the coolant temperature falls, the thermostat closes slightly, and decreases the flow of coolant through the radiator. The thermostat's position varies during normal operation, but it is rarely open completely. Operating conditions that can open the thermostat completely include long uphill climbs, idling in heavy traffic, extremely hot days, or the selection of a thermostat with a temperature rating that is too low for the engine.

Thermostats are activated by an integral wax-pellet, figure 4-7. A sealed chamber in the thermostat is filled with wax that is solid when cold. The wax melts and expands when heated. When the wax is cold, a spring provides enough tension to hold the thermostat closed. As the coolant temperature nears the thermostat rating, the wax melts and expands, overcoming spring tension to open the thermostat.

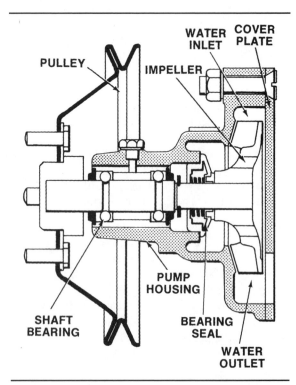

PULLEY
WATER INLET
COVER PLATE
IMPELLER
SHAFT BEARING
PUMP HOUSING
BEARING SEAL
WATER OUTLET

Figure 4-5. Cutaway of a typical water pump, showing the impeller, seals, and bearings.

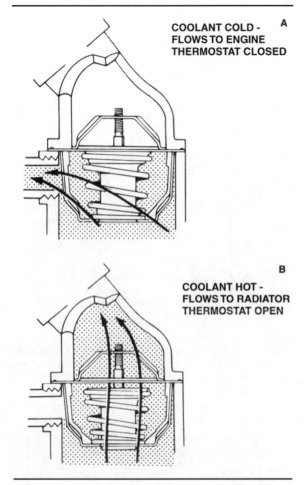

Figure 4-6. With a cold engine, the thermostat stays closed (A). As the engine is warmed, the thermostat opens (B).

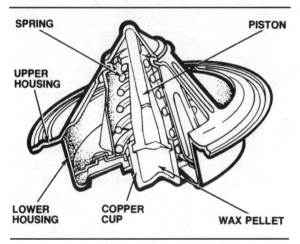

Figure 4-7. A cross section of a typical, wax-actuated thermostat made by Stant, showing the position of the wax pellet and the spring.

This type of thermostat will work consistently under different system pressures.

Most thermostats are mounted in a metal housing at the top of the engine, figure 4- 8. Thermostats are available with different opening temperatures. Common settings are in the range of 180° to 195°F (82° to 90°C). Thermostats should be fully open about 20°F (11°C) higher than their rated temperature.

Thermostats have various temperatures at which they open. The opening temperature is the key factor when selecting a replacement thermostat. The thermostat must open at the specified temperature for the particular system being serviced. If the opening temperature is too low, the thermostat opens too soon, and the coolant circulates to the radiator before it is hot enough to cool. If the warm coolant is circulated out of the engine prematurely, it prevents the engine from warming up completely, and can cause poor gas mileage, reduced engine performance, sludge formation in the crankcase, and severe engine wear. This will also prevent normal operation of the heater and defroster.

If the opening temperature is too high, the thermostat does not open fast enough, and the hot coolant remains in the hot engine instead of being circulated to the radiator to be cooled. An engine that is operated at too high of a temperature can cause various problems including engine ping, detonation, increased oil consumption, or reduced engine life.

Thermostats can be damaged by the corrosion and scale that build up in the cooling system. This material

■ **Engines Without Water Pumps**

All modern liquid-cooled engines use a pump to circulate the coolant. But there was a time when the water pump was something that only the more expensive cars had. The 1909 Model T Ford had a water pump, but Ford deleted it in the interest of simplicity and low cost. In addition to the Model T, many automotive engines had cooling systems operated by the thermo-syphon principle. This means, simply, that hot water rises and cold water sinks. As the water in the block was heated, it rose out to the top of the radiator. Once in the radiator, it started to cool, and slowly sank. This heating and cooling created enough circulation to keep the engine from overheating, as long as the weather was cool. In hot weather, or when pulling hard up a hill, the engines would often overheat. Because overheating was common, motorists accepted it as just one more hazard of motoring, like flat tires or getting stuck in the mud. By the late 1920s, car buyers were beginning to expect more sophistication and trouble-free operation even in low-price automobiles. In 1928, Ford brought out its first all-new model in almost 20 years, the Model A. It had — and kept — a water pump.

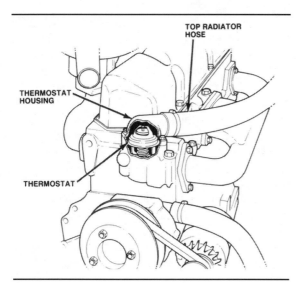

Figure 4-8. A typical thermostat location at the front of the cylinder head.

will cling to the central plunger of the thermostat, causing erratic and sticky valve movement. A thermostat that sticks open or shut can cause the same problems as a thermostat with an incorrect opening temperature.

Engines are equipped with a **thermostat bypass** circuit. Some designs are a separate bypass circuit, and some are integral. The separate circuit design has an additional temperature-sensitive valve that permits coolant to flow through the water pump and back into the engine when the main thermostat is closed. The bypass itself may be an external hose or a passageway within the engine or water pump body.

With an integral bypass, the main thermostat closes the bypass circuit as it opens the passage to the radiator, forcing all the coolant through the radiator when the engine is warm. Other systems leave the bypass open, and split the flow between the bypass and the radiator circuits when the engine is running.

Some engines have a spring-loaded bypass circuit, figure 4-9. When the cold engine is first started, the thermostat remains closed. Coolant pressure builds enough to open a spring-loaded bypass valve installed just below the thermostat. This allows the coolant to circulate through the bypass channel and back into the engine block. The passage can be an internal passage in the block, figure 4-10, or an external hose or tube, figure 4-11. When the operating temperature is reached, the thermostat opens, and coolant circulates through the radiator.

Radiator

The radiator is located between the cooling fan and the vehicle frame. The radiator is a heat exchanger. It absorbs heat from the hot coolant exiting the engine

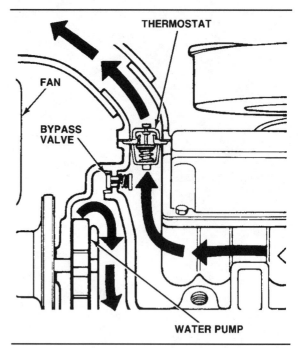

Figure 4-9. Some engines use a spring-loaded bypass valve that opens under coolant pressure when the thermostat is closed.

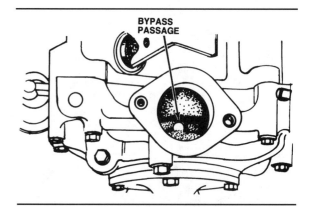

Figure 4-10. This internal passage in the thermostat housing directs cold coolant to the water pump.

and transfers it to the cooling air stream passing through the radiator.

Radiator cores are made of thin metal tubes, usually copper, brass, or aluminum, with cooling fins attached, figure 4-12. Coolant flows through the tubes and transfers its heat to the radiator fins. The air flowing past the radiator tubes and fins absorbs this heat. The number of tubes and fins in a radiator, and their condition, determine the unit's heat-transferring capacity.

Two large reservoirs called header tanks are connected to opposite ends of the tubes in order to keep a steady stream of coolant flowing into and out of the core. Header tanks are usually constructed from copper, brass, aluminum, or plastic.

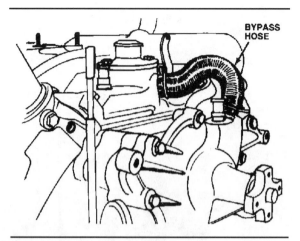

Figure 4-11. This external bypass hose connects the thermostat housing to the water pump.

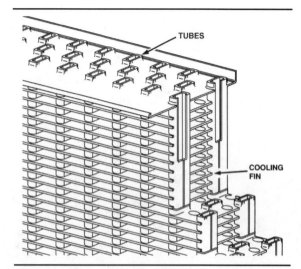

Figure 4-12. The tubes and cooling fins of the radiator core.

There are two types of radiator construction. Figure 4-13 illustrates a downflow radiator which has vertical tubes and tanks at the top and bottom of the core. Figure 4-13 illustrates a crossflow radiator which has horizontal tubes and a tank on either side of the core. These radiators are generally wider and shorter than the downflow type.

Most radiators are fitted with a draincock in the lower tank. This fitting, generally installed with a wing-screw closure, is used to drain the cooling system for periodic maintenance. Some radiators simply have a hex- or square-head bolt as the draincock fitting. Radiators without a draincock must be drained by removing the lower radiator hose.

On vehicles with automatic transmissions, the lower header tanks may contain an integral transmission oil cooler, figure 4-14. Although it is called an "oil" cooler, it actually cools automatic transmission fluid. The pump in the transmission forces hot transmission fluid through the lines and into the coils in the tank. Waste heat from the transmission is absorbed by the coolant surrounding the coil, and the cooled fluid is then returned to the transmission.

Since radiators are generally mounted directly in the front of the engine compartment, behind the grille, or air intake openings, they are subject to a lot of bugs, rocks, mud, and other debris. All of these can damage the radiator fins and restrict coolant flow, air flow, and heat transfer. Check the radiator surface on a regular basis, wash or brush off any collected debris, and, if necessary, use a fin straightening tool to restore any bent fins to their proper condition.

Radiator Cap

As you know, the temperature at which a liquid boils depends on the nature of the liquid and the pressure and heat applied to it. To raise the boiling point of the

■ Early Coolant

To prevent freezing, early motorists added many substances to the water in their cooling systems. Many of these did work, but had various drawbacks. Salt, calcium chloride, and soda were used as coolants, but formed acids in the coolant and corroded the engine and radiator. Adding sugar or honey to the cooling system also prevented freezing, but only at concentrations so high that the syrup was difficult to circulate through the engine. Kerosene and engine oil were sometimes used because of their low freezing points, but were flammable and caused the rubber hoses to deteriorate.

Methanol (wood alcohol) or ethanol (grain alcohol) were much more successful. Alcohols work well as coolants; a 50/50 mixture of either methanol or ethanol and water will protect against freezing to -20°F (-30°C). However, alcohols boil at lower temperatures than water and will therefore evaporate from the engine fairly quickly. To maintain freeze protection, the alcohol concentration must be tested periodically and enough fresh alcohol added to bring the protection up to the necessary level.

At one time, glycerine was also popular for use in cooling systems, as the first "permanent coolant" (permanent in that it did not evaporate like the alcohols). Concentrated glycerine boils at 227°F (110°C), and a 50/50 mixture protected against freezing as well as alcohol. Glycerine coolants were marketed partially diluted with water and with added corrosion inhibitors.

Thermostat bypass: An additional outlet port from the thermostat housing that allows cold coolant to bypass the thermostat while it remains closed. The coolant also bypasses the radiator, which assists the engine in warming up faster.

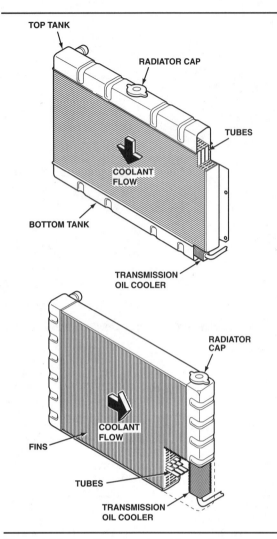

Figure 4-13. A radiator may be either a downflow type or a crossflow type.

liquid in the cooling system, and to obtain greater heat-transfer efficiency from the cooling system, the system is kept sealed and pressurized. The pressure rises inside the cooling system as the coolant is warmed.

The radiator cap is designed with a pressure-relief mechanism to keep the appropriate amount of pressure in the system. The pressure-relief mechanism performs three functions:

- It allows pressure to build as the temperature increases. This raises the boiling point of the coolant, which creates a greater differential between internal and external temperatures for better heat exchange.
- It allows the system to vent any excess pressure, so the system components do not rupture from internal pressure when operating temperatures go beyond their peak, such as after extended hot weather driving with the air conditioner on.

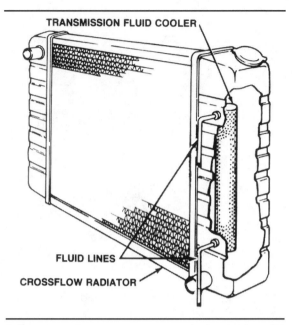

Figure 4-14. This automatic transmission fluid cooler is installed in one of the radiator tanks.

- On early radiator caps, it allows atmospheric pressure to re-enter the system as the coolant cools and contracts after the engine is shut off. On later systems with coolant recovery tanks, the vacuum produced when this valve opens draws excess coolant from the recovery tank back into the radiator.

The cap, figure 4-15, contains two valves. The pressure relief valve allows the pressure in the system to rise 7 to 18 psi (88 to 124 kPa) above normal atmospheric pressure before the coolant is allowed to escape, usually back to the overflow reservoir. This pressurizing action can raise the temperature at which the coolant boils to 260°F (127°C) or more. The vacuum valve allows outside air or coolant from the reservoir to enter the system when the engine cools.

Radiator caps are usually on the top or side tank, figure 4-16. Occasionally, when there is restricted clearance under the hood, the radiator cap is in an unusual location, such as on the top radiator hose, figure 4-17.

System pressure
There are two reasons for pressurizing a cooling system: to increase water pump efficiency, and to raise the boiling point of the coolant.

Pump efficiency is affected by pressure. Without a radiator cap installed in the system, a water pump is only about 85 percent efficient. With a 14-psi (97 kPa) pressure cap, the pump becomes almost 100 percent efficient, because the pressure decreases **cavitation**. Cavitation is the forming of low-pressure bubbles by

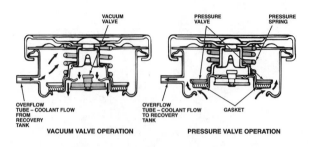

Figure 4-15. A cross section of a radiator pressure cap used in a recovery system.

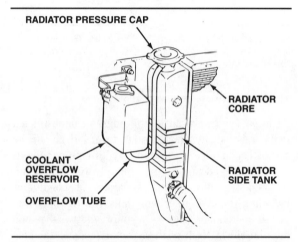

Figure 4-16. Most radiator caps are mounted on one of the radiator tanks, similar to this Ford system.

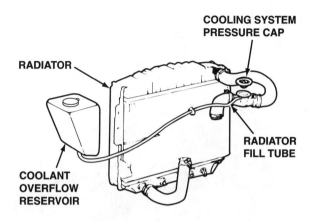

Figure 4-17. Some radiator caps are mounted away from the radiator to lower the radiator profile.

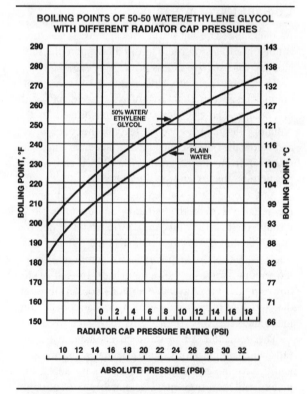

Figure 4-18. Pressurizing the cooling system provides protection against boilover for both water and ethylene glycol-based engine coolant.

the water pump blades. A pressurized system makes it difficult for these bubbles to form. Cavitation is undesirable because the bubbles collapse with enough power to blast small cavities in any metal surface they are near. Unchecked, cavitation will erode the water pump blades and housing.

Coolant boiling-point control is equally important. Water boils at 212°F (100°C) under atmospheric pressure (14.7 psi or 100 kPa) at sea level. A 50 percent solution of ethylene glycol and water boils at around 223°F (110°C) under the same conditions. Even this higher figure may be dangerously close to, or even below, the operating temperature of late-model engines. Pressurizing the system raises the boiling point of the coolant, figure 4-18. Each 1-psi (7 kPa) increase in pressure raises the boiling point about 3°F (2°C). Thus, coolant under 15-psi (105 kPa) pressure boils at approximately 268°F (130°C), 45°F (24°C) higher than without the cap. Most cooling systems are pressurized at 12 to 17 psi (83 to 117 kPa). Remember that when removing the radiator cap on a hot engine, as the pressure is released, the coolant will boil instantly.

Cavitation: An undesirable condition in the cooling system caused by the water pump blades. As the blades turn, they form low-pressure (vacuum) bubbles behind them.

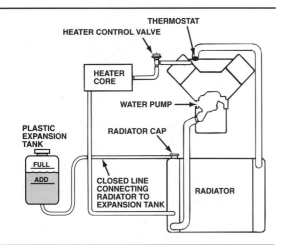

Figure 4-19. The level in the coolant recovery system rises and lowers with engine temperature.

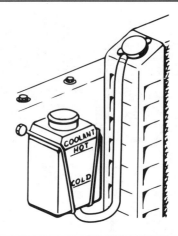

Figure 4-20. When adding coolant to the recovery system resevoir, be sure to take note of the "hot" and "cold" fill marks and fill accordingly.

Coolant Recovery Tank

As the engine coolant warms, it expands. When there is not any expansion room left in the system, the radiator cap releases the excess pressure and coolant.

The coolant recovery tank, figure 4-19, is a simple plastic reservoir that is always about half full of coolant. The overflow tube is positioned in the radiator filler neck, above the level of the relief valve, and below the outer sealing gasket. The tube enters the reservoir from the bottom, or through a cap with a plastic tube that extends to the bottom.

As the coolant in the radiator expands and pushes the relief valve in the radiator cap open, it flows through the overflow tube into the reservoir. When the engine is shut down, the coolant in the radiator contracts as it cools, which creates a vacuum and opens the vacuum valve in the radiator cap. Instead of air, however, the vacuum valve siphons coolant through the overflow tube from the recovery tank, back into the radiator.

The coolant level in the recovery tank rises and falls with engine temperature, however, the radiator and the engine cooling system are always kept completely full. By preventing as much air as possible from entering the cooling system, rust formation and corrosion are greatly reduced. Keeping the air out of the cooling system also raises the heat-exchange efficiency of the system by ensuring that air and foam are not circulated with the coolant. This also prevents trapped air bubbles from eroding the water jackets.

When more coolant must be added to a recovery system, it is added to the overflow tank, not directly to the radiator. The tank is marked with "hot" and "cold" fill levels, figure 4-20.

Radiator Fan

At highway speeds, airflow through an unobstructed radiator is usually great enough to absorb all of the excess coolant heat. Unfortunately, driving conditions and vehicle designs are not always so ideal. Automobiles must endure a variety of conditions, including stop-and-go driving, hills, towing, operating the air conditioner, extremely hot or cold weather, and high humidity. Each of these factors adds to the heat load that the radiator is expected to dissipate.

A fan is mounted in front of or behind the radiator to increase the airflow through the radiator core, figure 4-21. The typical design of most cooling fans in the past had a fan blade on a pulley, mounted on the same shaft as the water pump impeller, and driven by a belt. However, this design is not practical on front-wheel drive vehicles equipped with a transverse engine. To accommodate the transverse engine, engineers designed an electrical fan-and-motor assembly which is mounted in a shroud attached to the radiator and air-conditioning condenser.

The electric fan offers more flexibility in automobile design, figure 4-22. Another advantage is that an electric fan can be controlled directly by a thermostatic device, such as a coolant temperature sensor mounted in the radiator or engine block. When the sensor detects heat above a certain level in the radiator, it switches the fan on. The fan runs only until the coolant temperature is reduced to the desired level.

Generally, electric fans are connected to their own power source and not affected by the ignition system. If the sensor detects a high enough temperature, the fan will switch on any time, whether the engine is on or off. Always temporarily disconnect the power to

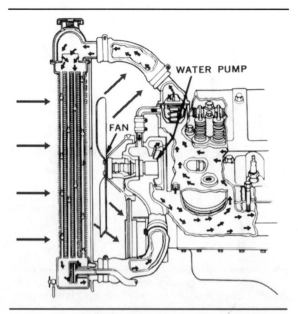

Figure 4-21. The engine-driven fan blows air into the engine compartment through the radiator core.

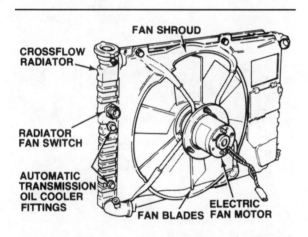

Figure 4-22. An electric motor drives the radiator fan on many late-model vehicles.

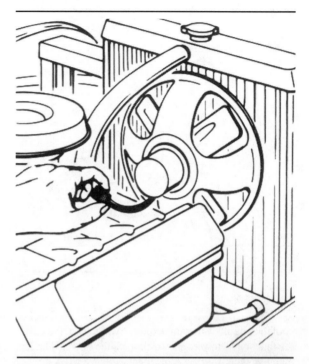

Figure 4-23. Always disconnect the fan power lead before working near an electric fan.

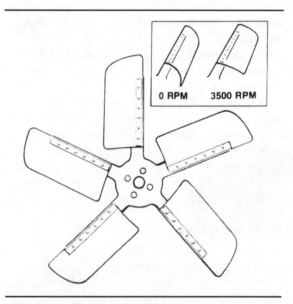

Figure 4-24. The flexible blades of this fan change shape as engine speed changes.

the fan when working in the area, and reconnect it when your work is complete, figure 4-23.

The simplest fans have rigid blades. These work quite well when used with the thermostatic controls; however, on conventional belt-driven fans, they are noisy and absorb a measurable amount of engine horsepower at higher speeds. Manufacturers have developed several alternative designs to increase fuel economy and decrease noise.

A flex-blade fan, figure 4-24, changes shape to reduce operating drag. At low speeds, the curve of the fan blades pulls air through the radiator core. As the speed of the fan and engine increases, the resistance of the air flattens the blades slightly, and the fan

requires less horsepower to turn than one with rigid blades. Although this also reduces the amount of air the fan can move with each rotation, both the fan and the vehicle itself are moving fast enough to maintain a strong airstream through the radiator.

Another way to avoid power loss and excess noise on fan assemblies is the clutch fan. When additional airstream is needed, the clutch fan rotates enough to

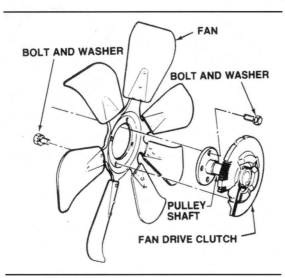

Figure 4-25. This Ford fan has a fluid clutch that controls the speed of rotation.

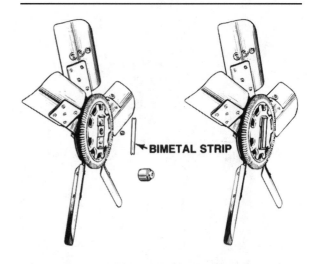

Figure 4-26. The bimetal temperature sensor spring controls the amount of silicone that is allowed into the drive and that, in turn, controls the speed of the fan.

cool the engine. When maximum rotation is not needed, the fan "slips" to prevent wasting engine power. The fan clutch design is either centrifugal or thermostatic.

Both types use a silicone-fluid coupling, figure 4-25. Above a certain speed, the torque required to rotate the fan exceeds the ability of the viscous fluid coupling to transfer without slipping. The fan then slips harmlessly.

The centrifugal fan clutch uses centrifugal force to engage and disengage the fluid coupling. The thermostatic fan clutch has a bimetallic temperature sensor spring attached to a valve that controls the flow of silicone, figure 4-26. The bimetallic spring reacts to changes in the temperature. When the air flowing through the radiator is warm, the spring relaxes and permits more fluid to enter the coupling, which increases fan speed. When the airflow is cool, the spring retracts the valve and allows greater slippage of the fan.

Radiator Hoses and Drive Belts

Radiator hoses connect the cooling system passages to the radiator tanks, figure 4-27. The hoses are flexible and will absorb the motion between the vibrating engine and the stationary radiator. The hoses are made of synthetic rubber, and are often reinforced with a wire coil. Typically, hoses are preformed to fit the specific vehicle application.

The hose ends fit over necks on the engine and radiator and are held in place with hose clamps, figure 4-28. Hose clamps can be held in place by spring tension or by a screw.

For many years, the drive belt that operated the radiator fan and water pump was a reinforced rubber V-belt, figure 4-29. It fit tightly over the crankshaft

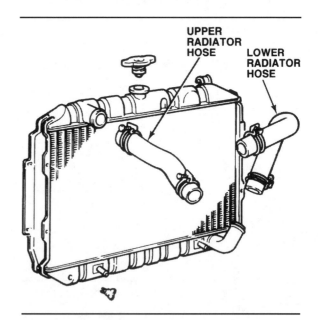

Figure 4-27. Radiator hoses must be large enough to carry the coolant when the thermostat is fully open.

pulley, the fan pulley, and usually, the alternator pulley. Belt tension was usually adjusted by moving the alternator in its mounting.

Today, most vehicles use a single serpentine belt to drive all accessories, figure 4-30. The belt is made of reinforced rubber, and has several small V-shaped grooves that fit corresponding grooves in the accessory drive pulleys. On some serpentine belt systems, belt tension is adjustable. Other serpentine systems have a spring-loaded automatic tensioner assembly that maintains spring tension at all times.

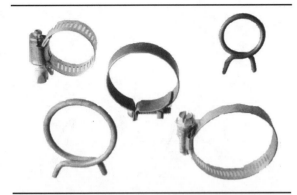

Figure 4-28. Hose clamps are available in different sizes and designs for different applications.

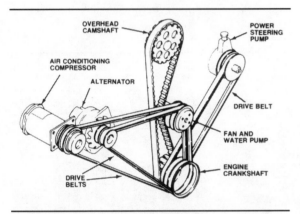

Figure 4-29. On V-belt drive systems, most accessories have separate drive belts.

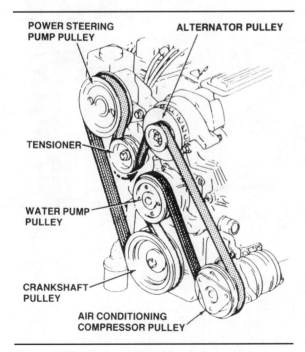

Figure 4-30. A serpentine drive belt system uses a single belt to drive all accessories.

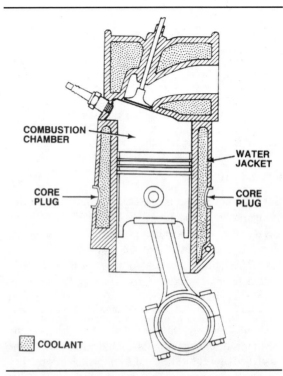

COOLANT

Figure 4-31. Coolant circulates through the water jackets in the engine block and head.

Water Jackets

Nearly all of the heat produced in an engine is the product of the combustion, or extreme revolutions of internal engine components. The hottest parts of an engine are those parts exposed to the burning fuel. Coolant passages are designed into the engine block and cylinder head castings to allow coolant to pass through areas near the exhaust valves, and combustion chambers. These coolant passages are collectively called the **water jackets**, figure 4-31.

Coolant enters and leaves the engine at the points where the radiator hoses attach. In some engines, water distribution tubes receive coolant at the front of the engine and distribute it near each valve seat. Varying hole sizes in this tube allow all areas to receive equal amounts of coolant. The water jacket holes punched in the head gaskets are usually sized to restrict coolant transfer at the front of the engine and force it toward the rear, figure 4-32.

Water Jackets: Passages in the head and block that allow coolant to circulate throughout the engine.

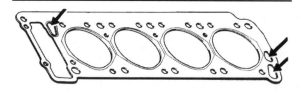

Figure 4-32. These head gasket holes show the placement of coolant passages that carry engine coolant the full length of the cylinder head.

Core Plugs

Core plugs seal the holes used during the manufacturing process. The plugs are round discs of sheet metal pressed into openings on the sides of the engine block that lead directly to the water jackets, figure 4-33. These plugs may leak or pop out if the cooling system freezes or is not properly maintained. Core plugs are also (incorrectly) known as freeze plugs. Core plugs tend to pop out if the block freezes. Originally, it was thought that these plugs were designed for this reason, to lessen the pressure on the block and prevent it from cracking, which is why they are sometimes referred to as freeze plugs.

The Heater Core

You may not normally think of the heater core as part of the cooling system. However, the heater core is directly connected to the cooling system by the heater water hoses. A heater core failure can lead to other cooling system malfunctions.

The heater core is a heat exchanger, figure 4-34, and is constructed like a radiator. It is mounted inside a housing assembly and is located either on the engine compartment bulkhead or on the passenger compartment cowl panel. As warmed coolant circulates through the heater core tubes, the heater core fins absorb the heat. At the same time, a blower motor blows air across the heater core fins. The air absorbs the heat from the fins and the warm air is vented into the passenger compartment.

The **heater control valve**, also known as the coolant control valve or water control valve, is used to control the rate of coolant flow to the heater core. The valve is located on the heater inlet hose. In some applications, a cable connects the valve to the temperature control setting knob on the heater control panel. On other systems, the valve is controlled by vacuum or electricity.

If the heater control valve, heater core, or heater hose become damaged or blocked, the coolant will not be able to flow at the proper rate. Improper coolant flow may result in a no-heat condition.

Some modern heater systems use the blend air door to control coolant entering the heater core. The blend

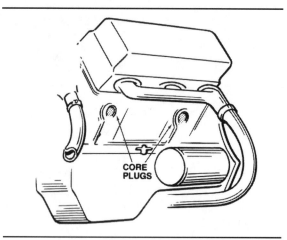

Figure 4-33. Core plugs are pressed into the cylinder block.

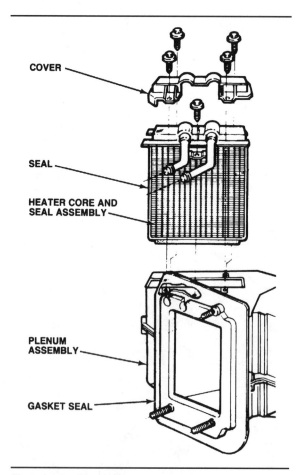

Figure 4-34. The heater core is similar to a radiator, but is mounted within the heater ductwork.

air door is a metal flap in the heater core ducts that controls the pattern of airflow across the heater core or to a bypass duct. When the blend air door is closed, there is no airflow across the heater core; therefore, no heated air outflow occurs. The door can be opened partially or

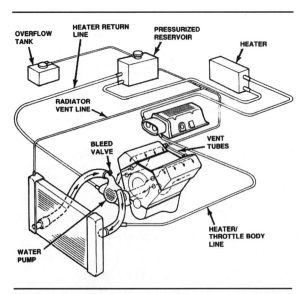

Figure 4-35. The reverse-flow coolant path of the LT1 V8 cools the cylinder head combustion chambers more efficiently than conventional-flow coolant paths.

completely to allow moderate or full heat. The door position is controlled by cable, vacuum, or electricity.

Reverse-flow Cooling System

Most cooling systems send coolant from the water pump to the engine block, up to the cylinder heads, and back to the radiator. The major disadvantage of conventional coolant flow is that the hottest parts of the engine, the combustion chambers in the cylinder heads, are cooled with the hottest coolant because they are the last destination in the cooling system circuit. The reverse-flow cooling system of the Chevrolet Corvette, first used on the LT-1 engine in 1992, and continued in the LS1 engine, changes this by pumping the lowest-temperature coolant directly from the radiator to the cylinder heads *first*, figure 4-35. Warmer coolant then travels to the block before returning to the radiator. This sounds simple, but the advantages are significant.

The ability of the coolant to absorb the greatest amount of heat as it reaches the head reduces the chances of detonation. The coolant is also at a more stable temperature when it reaches the heads, allowing better control of the combustion process, offering better fuel economy, and decreasing exhaust emissions.

ENGINE TEMPERATURE EFFECTS ON MECHANICAL WEAR

When the engine is cold, it suffers poor fuel vaporization. The liquid fuel entering the cylinders tends to wash oil off the cylinder walls, reducing lubrication. The liquid fuel runs past the piston rings and into the crankcase, where it dilutes the oil and further hinders

lubrication. A cold engine does not burn off moisture that condenses in the crankcase. This moisture mixes with the oil to form sludge, which increases engine friction, robs it of power, and increases mechanical wear.

An engine that is too hot forms carbon and varnish deposits that affect engine operation. Excessive heat thins the engine oil to the point that it no longer lubricates well, which increases mechanical wear. The thin oil will also be drawn into the combustion chamber in greater amounts, causing carbon buildup, raising the compression ratio, and increasing the chance of detonation.

SUMMARY

The cooling system must remove about one third of the heat produced by an engine. Air-cooled engines use a flow of air over their outer surfaces to reduce engine temperature. Most cooling systems use a liquid coolant to remove the excess heat.

Modern coolant is a mixture of water and chemical additives, such as ethylene or propylene glycol. The pressurized cooling system absorbs excess engine heat into the liquid coolant flowing through the water jackets in the engine castings. The water pump circulates the coolant. The thermostat regulates the coolant flow between the engine and the radiator. If the heater is in use, the hot coolant flows to the heater first, and then to the water pump. If the heater is not in use, the coolant bypasses the heater core and goes directly to the water pump. From there, the coolant travels to the radiator. As the hot coolant flows through the radiator, the heat is transferred to the airstream moving through the radiator. The cooled coolant then returns to the engine. Airflow is enhanced by an electric or engine-driven fan mounted between the engine and the radiator.

Coolant expands as it warms. Cooling systems include a recovery tank to contain the excess coolant when it expands. When there is no additional expansion room, the radiator cap opens and vents the excess coolant into the recovery system, which relieves the excess pressure. As the coolant cools, it contracts and vacuum draws the coolant back into the radiator from the recovery tank. Then the coolant circulates back to the engine and begins the cycle all over again. Engine operating temperatures that are either too hot or too cold will produce greater mechanical engine wear.

Core Plugs: Shallow metal cups inserted into the engine block to seal holes left during manufacturing.

Heater Control Valve: The valve that controls the rate of coolant flow to the heater core.

Review Questions

Choose the single most correct answer.
Compare your answers with the correct answers on Page 328.

1. Air-cooled engines require which of the following parts to reduce and stabilize engine temperature?
   a. A crankcase that holds at least six quarts of oil
   b. An auxiliary fan and shroud
   c. A rich air-fuel mixture
   d. A dual exhaust system

2. Which of the following terms describes a type of radiator?
   a. Downflow
   b. Backflow
   c. Throughflow
   d. Acrossflow

3. The engine converts about _____ of its total heat energy into mechanical energy.
   a. One fourth
   b. One third
   c. One half
   d. Two thirds

4. All of the following conserve power at high engine speeds except:
   a. Flexible blade fans
   b. Multiblade fans
   c. Clutch fans
   d. Electric fans

5. Which mixture of ethylene glycol and water gives protection down to the lowest temperatures?
   a. 100% ethylene glycol
   b. 73% ethylene glycol and 27% water
   c. 68% ethylene glycol and 32% water
   d. 50% ethylene glycol and 50% water

6. Core plugs are installed in the engine block to:
   a. Relieve pressure if the coolant should freeze
   b. Allow the block to be flushed during engine overhaul
   c. Provide inspection points for internal block condition
   d. None of the above

7. An engine that is too hot will have:
   a. Poor lubrication
   b. Poor cooling system circulation
   c. High oxides of carbon
   d. High thermal inefficiency

8. A 50% water and a 50% ethylene glycol mixture in a cooling system with a 14 psi pressure cap will boil at:
   a. 248°F (120°C)
   b. 255°F (124°C)
   c. 263°F (128°C)
   d. 270°F (132°C)

9. The water pump uses_____to circulate the coolant.
   a. Positive pressure
   b. Gravitational force
   c. Centripetal force
   d. Centrifugal force

10. Technician A says that coolant flows from the radiator to the block first in most engines. Technician B says that coolant flows from the radiator to the cylinder heads first in some engines. Who is right?
    a. A only
    b. B only
    c. Both A and B
    d. Neither A nor B

11. A pressurized cooling system will:
    a. Reduce engine temperature
    b. Increase coolant warmup rate
    c. Reduce the coolant boiling point
    d. Increase the coolant boiling point

12. If the cooling system makes the engine run too cool, the engine will:
    a. Make more torque
    b. Have less efficiency
    c. Preignite
    d. Produce higher HC emissions

13. A disadvantage of ethylene glycol when used as a coolant is that:
    a. Its freezing point is much lower than water
    b. It does not transfer heat as well as water
    c. Its boiling point is much higher than water
    d. None of the above

14. The thermostat regulates coolant flow between the engine and the:
    a. Cylinder head
    b. Heater
    c. Radiator
    d. Transmission cooler

15. DEX-COOL™ is a(n):
    a. Propylene glycol-based coolant
    b. New red-colored coolant
    c. Ethylene glycol-based coolant
    d. Coolant specially designed for diesel engines

16. For every one psi increase in pressure, the coolant boiling point is raised about:
    a. 3°F (1.7°C)
    b. 2°F (1.1°C)
    c. 1°F (0.6°C)
    d. 4°F (2.2°C)

17. Technician A says that thermostats have their opening temperature stamped on the outside. Technician B says that modern thermostats are wax actuated. Who is right?
    a. A only
    b. B only
    c. Both A and B
    d. Neither A nor B

18. Technician A says an engine that runs too cool will have poor lubrication. Technician B says an engine that runs too warm will have poor lubrication. Who is right?
    a. A only
    b. B only
    c. Both A and B
    d. Neither A nor B

19. Technician A says that water jackets are cast into the block and head of liquid-cooled engines. Technician B says the internal size of water jackets makes no difference to coolant distribution. Who is right?
    a. A only
    b. B only
    c. Both A and B
    d. Neither A nor B

20. Technician A says that during the complete four-stroke cycle, the average combustion chamber temperature is 750°F (400°C). Technician B says that the average temperature is about 1,500°F (800°C). Who is right?
    a. A only
    b. B only
    c. Both A and B
    d. Neither A nor B

5

Intake and Exhaust Systems

We have already covered engine air and fuel requirements, but to fully understand automotive engines, you must also study the intake and exhaust systems. In the first half of this chapter we will cover airflow, fuel atomization, vaporization, and the role fuel delivery and intake manifold design play in determining these functions. The second half of this chapter covers the parts and arrangement of the exhaust system and how the exhaust system design affects backpressure.

THE INTAKE SYSTEM

The primary parts of the intake system are the manifold and fuel delivery system, figure 5-1. The manifold is a casting with a series of enclosed passages that route the air-fuel mixture from the carburetor or throttle body to the intake ports in the engine's cylinder head. With port fuel-injection systems, the intake manifold routes only air and the fuel injectors deliver the fuel directly to the intake port.

Airflow

We pointed out in Chapter 2 that you can think of an automotive engine as a large air pump. As the pistons move up and down in the cylinders, they draw in air and fuel for combustion and pump out the burned exhaust. They do this by creating a difference in air pressure.

The downward movement of the piston on the intake stroke creates a partial vacuum, lowering the air pressure in the combustion chamber and intake manifold. Opening the carburetor or fuel-injection throttle valve, figure 5-2, causes air to move from the higher pressure area outside the engine, through the intake, to the lower pressure area of the manifold. The opening of the throttle valve determines how much and how fast the air travels. When the intake valve in the cylinder head opens, this vacuum then draws in the air-fuel mixture from the intake manifold into the combustion chamber, figure 5-3.

Air-fuel mixture control
Gasoline must be metered, atomized, vaporized, and distributed to each cylinder in the form of a burnable mixture. To do this, a metering device mixes the gasoline with air in the correct ratio and distributes the mixture as required by engine load, speed, throttle valve position, and operating temperature. On most late-model engines, a fuel-injection system does the job of air-fuel mixing done by a carburetor on older engines.

Cars and trucks are no longer being built with carburetors. However, we give a short explanation of carburetors first because this makes it easier to understand some aspects of fuel injection and why fuel

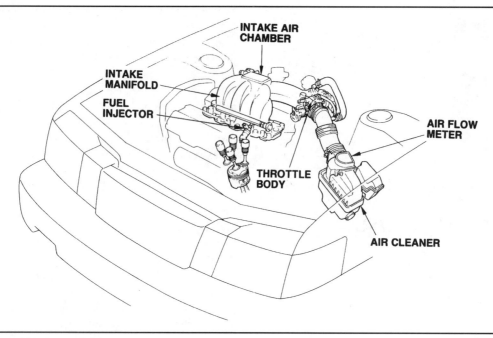

Figure 5-1. This Lexus ES 250 intake system features port fuel injectors.

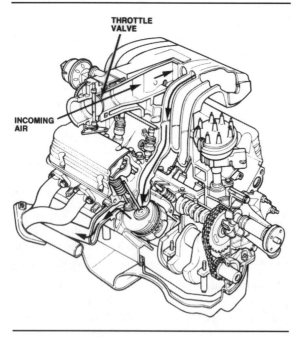

Figure 5-2. Opening the throttle valve allows air to enter the intake system.

injection is superior. For thorough coverage of carburetors see a companion book in our Chek-Chart Automotive Series, *Fuel Systems and Emission Controls*.

Carburetors

Carburetors consist of light-weight metal bodies that have one, two, three, or four barrels, figure 5-4.

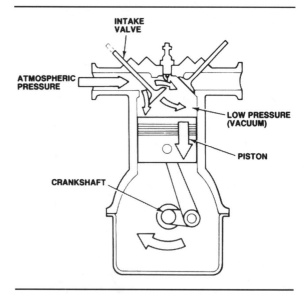

Figure 5-3. Downward movement of the piston lowers the air pressure inside the combustion chamber. The pressure differential between the atmosphere and inside the engine forces air into the engine.

Carburetors use airflow through their barrels to draw fuel out of a **float bowl**. As air rushes through the barrel, pressure inside the barrel drops. A passage or **jet** connects the barrel to the fuel-filled float bowl. The higher, atmospheric pressure in the float bowl forces the fuel through the jet and into the low-pressure stream of air rushing through the barrel, figure 5-5. The speed of the air flowing through the carburetor affects how much of fuel is forced through the jet.

Figure 5-4. Carburetors are complex devices, but they work on two simple principles: airflow and pressure differential.

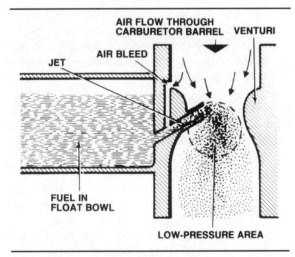

Figure 5-5. Airflow through the carburetor barrel creates a pressure differential. The higher, atmospheric pressure in the float bowl pushes fuel into the low pressure area in the barrel.

The higher the airflow velocity, the lower the pressure will be in the carburetor barrel, and the more fuel will flow through the jet and out of the float bowl.

Instead of having a carburetor barrel with straight sides, designers make carburetors work better by placing a constriction, or **venturi**, inside the barrel. When air flows through the venturi constriction, the air speeds up. This increase in speed lowers the air pressure inside the venturi and draws more liquid fuel into the airflow.

In addition to mixing liquid fuel with air, the carburetor must also atomize the liquid as much as possible. To help break up the liquid fuel for better atomization, designers put a small opening called an air bleed in the fuel inlet passage. This mixes air with

◼ Vapor Lock

Vapor lock occurs when fuel bubbles form in the fuel lines, reducing fuel flow to the engine. Partial vapor lock leans the air-fuel mixture and reduces engine power. Complete vapor lock causes an engine to stall, and makes restarting impossible until the bubbles in the fuel system disperse — usually when the fuel system cools. Any one or any combination of these four factors can cause fuel bubbles to form and vapor lock to occur:

- High gasoline temperature due to engine over-heating or high temperature weather
- Low fuel-system pressure
- Gasoline with too high a volatility
- Low air pressure due to driving at high altitude.

Vapor lock is much less of a problem today for several reasons. Oil companies are very good at reducing the vapor-locking tendencies of gasoline by adjusting its volatility to weather and geographic requirements. Generally, lower volatility fuels are used in summer or at high altitudes to reduce their chances of vapor lock. Higher volatility fuels are used in winter for easier starting. Engines are able to avoid vapor lock with the higher volatility fuels in winter because of the cooler weather.

These rules of thumb about fuel volatility still hold true, but the trend is toward relatively lower volatility fuel year round because it pollutes the atmosphere less. Low volatility fuels evaporate more slowly, putting fewer hydrocarbons into the air.

Vapor lock is theoretically possible in fuel-injected engines, but the high fuel-line pressures that are normal with fuel injection make vapor lock very unlikely. The higher the pressure, the higher the boiling point of the fuel, making vapor difficult to form. On today's vehicles an electric fuel pump is usually mounted at the fuel tank, pressurizing 98 percent of the fuel system. Older cars with mechanical fuel pumps mounted on the engine actually created a low pressure condition in the entire fuel line from the tank to the pump, increasing the likelihood of vapor lock.

Float Bowl: Gasoline is delivered by the fuel pump from the fuel tank to the carburetor float bowl, where it is ready for use.

Jet: A metal orifice in a carburetor that meters fuel flow. Larger jets allow more fuel to flow than smaller jets. Jets are usually replaceable and available in different sizes.

Venturi: A restriction in an airflow, such as in a carburetor, that speeds the airflow and helps create a pressure differential.

the liquid fuel as it is drawn into the main airflow. The air bleed opening is *above* the venturi, at high pressure, causing the air to flow in.

The carburetor must also change the air-fuel mixture automatically. It has circuits that deliver a richer mixture for starting, idle, and acceleration, and a leaner mixture for part-throttle operation. Even with all its various circuits, a carburetor is a mechanical device that is neither totally accurate nor particularly fast in responding to changing engine requirements. Adding electronic feedback mixture control improves a carburetor's fuel metering capabilities under some circumstances, but many mechanical jets, passages, and air bleeds still do most of the work. Adding feedback controls and other emission-related devices results in very complex carburetors that are expensive to repair or replace.

The intake manifold also is a device that, when teamed with a carburetor, results in less than ideal air-fuel control. If there is only one carburetor for several cylinders, it is difficult to position the carburetor an equal distance from all cylinders and remain within the space limitations under the hood. In the longer intake manifold passages, the fuel has a tendency to fall out of the airstream. Bends in the passageways tend to slow down the flow or cause puddles of fuel to collect. The manifold runners have to be kept as short as possible to minimize fuel delivery lag, and there cannot be any low points where fuel might puddle. These restrictions severely limit the amount of manifold tuning possible, and even the best designs still have problems with fuel condensing on cold manifold walls.

Electronic fuel injection

Multiple carburetors helped eliminate some of the compromises in carburetor and manifold design by improving mixture distribution. Few multiple carb systems are used today because fuel injection is a better solution. The first fuel-injection systems were mechanical. They relied on an injection pump, mechanically driven by the crankshaft and timed to engine rpm, to allow the injectors to fire at the proper time.

The development of reliable solid-state components during the 1970s made the best fuel-injection systems, **electronic fuel injection (EFI)**, practical. EFI provides precise mixture control over all speed ranges and under all operating conditions. Its fuel delivery components are simpler and often less expensive than a feedback carburetor. Most designs allow a wider range of manifold designs. Equally important is that EFI offers highly reliable and very precise electronic control. With few exceptions, modern engines use one of two types of electronically controlled fuel injection:

- Port injection
- Throttle-body injection.

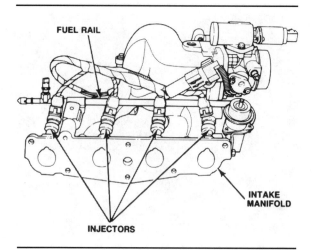

Figure 5-6. The injectors mount near the intake ports of a port fuel-injection system. A fuel rail carries the gasoline to the injectors.

With **port fuel injection**, (also called multi-point or multi-port injection by some vehicle manufacturers) individual injection nozzles for each cylinder are located in the ends of the intake manifold passages, near the intake valves, figure 5-6. Only air flows through the manifold up to the location of the injectors. At that point, the system injects a precise amount of atomized gasoline into the airflow. The fuel vaporizes immediately before it passes around the valve and into the combustion chamber, figure 5-7.

With **throttle body fuel injection** (TBI), a throttle body on top of the intake manifold houses one or two injection nozzles, figure 5-8. The throttle body looks like the lower part of a carburetor, but does not have a fuel bowl or any of the fuel metering circuits of a carburetor. The system injects fuel under pressure into the airflow as it passes through the throttle body.

Fuel atomization and vaporization

Proper fuel distribution depends on six factors:

- Correct fuel volatility
- Proper fuel atomization
- Complete fuel vaporization
- Intake manifold passage design
- Intake throttle valve angle
- Carburetor, throttle-body, or fuel injector location on the intake manifold.

Volatility is a measure of gasoline's ability to change from a liquid to a vapor and is affected by temperature and altitude. The more volatile it is, the more efficiently the gasoline will vaporize. Efficient vaporization promotes even fuel distribution to all of the engine's cylinders for complete combustion.

Volatility is controlled by blending different hydrocarbons that have different boiling points. In this way,

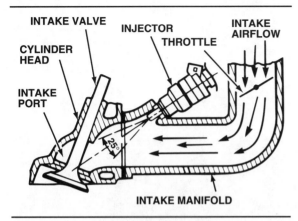

Figure 5-7. In a port fuel-injection system, air and fuel combine at the intake valve.

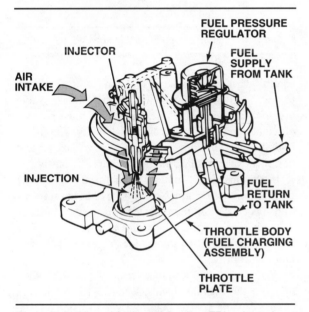

Figure 5-8. Air and fuel combine in a TBI unit at about the same point as in a carburetor.

it is possible to produce a fuel with a high boiling point for use in warm weather, and one with a lower boiling point for cold-weather driving. Such blending involves some guesswork about weather conditions, so severe and unexpected temperature changes can cause a number of temperature-related problems ranging from hard starting to vapor lock.

Several factors contribute to changing gasoline from a liquid into a combustible vapor. **Atomization** is an important factor. Gasoline must atomize, or break up into a fine mist, before it will properly vaporize. In a carbureted engine, the liquid fuel first enters the carburetor where it is sprayed into the incoming air and atomized, figure 5-9. The resulting atomized air-fuel mixture then moves into the intake manifold

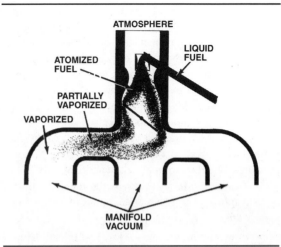

Figure 5-9. Changing liquid fuel into a combustible material is a two-stage process. First, it is atomized to a mist and mixed with air. The air-fuel mixture then vaporizes in the manifold.

where manifold heat vaporizes the many fine droplets of the atomized fuel.

Vaporization is another important factor. Rapid vaporization occurs only when the fuel is hot enough to boil. The boiling point is related to pressure: the higher the pressure, the higher the boiling point; the lower the pressure, the lower the boiling point.

Electronic Fuel Injection (EFI): A computer-controlled fuel-injection system that gives precise mixture control at all operating conditions at all speed ranges.

Port Fuel Injection: A fuel-injection system in which individual injectors are installed in the intake manifold at a point close to the intake valve. Air passing through the manifold mixes with the injector spray just as the intake valve opens.

Throttle-Body Fuel Injection (TBI): A fuel-injection system in which one or two injectors are installed in a carburetor-like throttle body mounted on a conventional intake manifold. Fuel is sprayed at a constant pressure above the throttle plate to mix with the incoming air charge.

Volatility: The ease with which a liquid changes from a liquid to a gas or vapor. Gasoline is more volatile than water because it evaporates quicker.

Atomization: Breaking a liquid down into small particles or a fine mist.

Vaporization: Changing a liquid, such as gasoline, into a vapor by evaporation or boiling.

Because intake manifold pressure is usually quite a bit less than atmospheric pressure, the boiling point of gasoline drops when it enters the manifold, and the fuel quickly begins to vaporize. Fuel injectors spray the fuel into the incoming air under much greater pressure than do carburetors, so the fuel atomizes more thoroughly and is able to quickly vaporize.

Heat from the intake manifold floor combines with heat absorbed from air particles surrounding the fuel particles to speed vaporization. The higher the temperature, the more complete the vaporization will be. Thus, raising the temperature of the intake manifold helps vaporization. Several factors can cause poor vaporization:

- Low mixture velocity or low fuel injection pressure
- Insufficient fuel volatility
- A cold manifold
- Poor manifold design
- Cold incoming air
- Low manifold vacuum.

When poor vaporization occurs, too much liquid fuel reaches the cylinders. Some of this additional fuel escapes through the exhaust as unburned hydrocarbons and some washes oil from the cylinder walls, causing engine wear. Blowby gases carry the rest past the piston rings.

Carburetors can be precise and consistent, but fuel injection can always provide the *correct* amount of fuel for conditions — cold start, wide open throttle, etc. — while carburetion is a compromise. Also, injecting the fuel under pressure provides good vaporization under all loads and speeds. The six causes of poor vaporization listed above all apply to carburetors, but only the first four have any effect on throttle-body fuel-injection systems and only the first two on port fuel-injection systems. With fuel injection, accurate metering, precise fuel control, and improved vaporization provide more power across the speed range of the engine, better fuel economy, and reduced exhaust emissions.

With TBI, manifold temperature still affects vaporization. As the mixture travels the length of the manifold, some of it separates and fuel condenses on the walls of the manifold. Since TBI atomizes the fuel more consistently under all conditions, separation is less likely than with a carburetor. However, with port fuel-injection, the manifold carries only air, so the air-fuel mixture does not separate as it travels through the manifold.

The Intake Manifold

The air-fuel mixture flowing from the carburetor or throttle body must be evenly distributed to each cylinder. The intake manifold does this with a series of

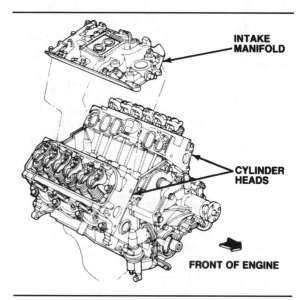

Figure 5-10. V engines, such as this Ford V6, have their intake manifold between the two banks of cylinders.

carefully designed passages that connect the carburetor or throttle body with the engine's intake valve ports. To do its job, the intake manifold must provide efficient vaporization and air-fuel delivery. The manifolds for port fuel-injected engines carry only air and designers size, or tune, them to do this efficiently.

Intake manifolds on older engines are usually cast iron, but modern engines often have aluminum manifolds because of aluminum's superior heat conductivity and light weight. Aluminum's heat conductivity helps transfer heat to the air-fuel mixture faster and more uniformly.

A V engine has its intake manifold in the valley between the cylinder banks, figure 5-10. An in-line engine has its intake manifold on the side of the engine. The average in-line engine has both intake and exhaust manifolds on the same side, figure 5-11. High-performance in-line engines have their two manifolds mounted on opposite sides of the cylinder head. This is called a crossflow design, figure 5-12.

Most intake manifolds are separate pieces that unbolt from the cylinder head, figure 5-13. However, Ford 200- and 250-cid (3.3- and 4.1-liter) engines had the intake manifold cast into the head to reduce cost and simplify overall engine manufacturing. The last of the in-line Chevrolet 6-cylinder engines in the late 1970s had intake manifolds that were integral with the heads. This design was an attempt to improve mixture temperature and distribution, which are hard to control uniformly for an in-line 6-cylinder engine.

Intake manifold design

The design of an intake manifold for a carburetor or fuel-injection system has a direct effect on mixture

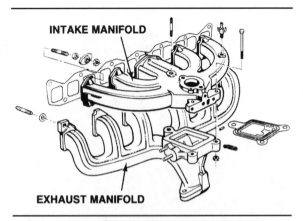

Figure 5-11. Most in-line engines have their intake and exhaust manifolds on the same side of the engine.

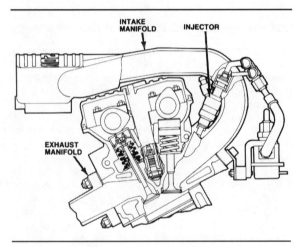

Figure 5-12. This Audi port fuel-injection system mounts to a crossflow cylinder head.

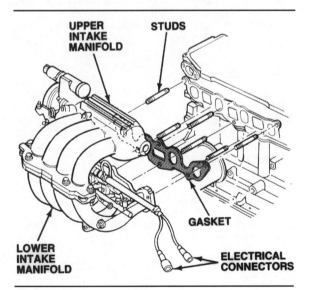

Figure 5-13. Most intake manifolds unbolt from the cylinder head.

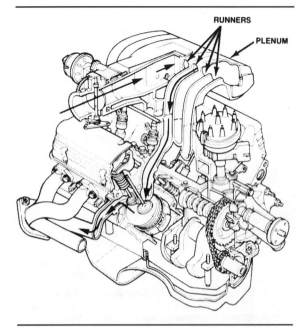

Figure 5-14. Intake manifold runners travel from the plenum to the intake ports of the cylinder head.

distribution and volumetric efficiency over the speed range of an engine. Both velocity and heating are also affected by the size of the manifold passages, or **runners**, through which the mixture must travel, figure 5-14. If the passages are large, the mixture will travel slowly at low rpm, allowing fuel particles to cling to the manifold wall and avoid vaporization. Small passages create a higher velocity but restrict the volume of the mixture that can pass through at high rpm. The angles at which internal manifold passages turn are also critical, figure 5-15. When they are too sharp, fuel tends to separate out of the mixture by puddling and condensing.

The efficiency of a manifold is determined by the shape, interior surface, and size of its runners. Manifold design is usually a compromise. Short, large diameter runners produce the best power at high rpm, but long, smaller diameter runners produce the best power at low to medium rpm. Runners must be just large enough in diameter to supply all cylinders with equal amounts of the air-fuel mixture at the necessary rpm. Passages that are too large or too small reduce efficiency. Manifold runners should be without sharp corners, bends or turns to interfere with mixture flow. In carburetor or TBI manifolds the ports may be

Runners: The passages or branches of an intake manifold that connect the manifold's plenum chamber to the engine's inlet ports.

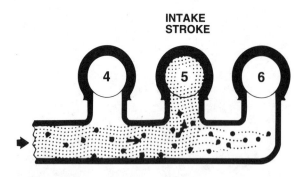

Figure 5-15. A manifold with large passages and sharp angles will cause the liquid fuel to separate out of the air-fuel mixture.

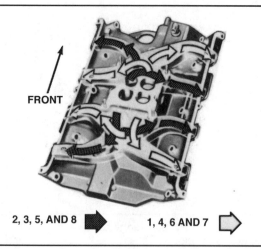

Figure 5-16. This V8 dual-plane manifold separates the air-fuel mixture for good distribution.

■ Sonic Tuning of Intake and Exhaust Systems

What exactly is a "tuned exhaust?" Why do racers spend so much time and money cutting and welding different header systems for their cars? And what about "tuned intakes?" Why are the velocity stacks on a racing fuel injection induction system sometimes short, and sometimes long, and sometimes even different lengths for different venturis in the same system? There's obviously horsepower to be found in changing with the intake and exhaust system, but where does it come from, and how does it work?

The answers to these questions lie in understanding resonance, or sonic tuning. Sonic tuning in the exhaust is a way to scavenge the exhaust gases from a cylinder more efficiently than with piston action alone. It also can help the induction system, like a sort of natural supercharging, which gets more mixture into the cylinder than with engine vacuum alone. Most street engines get around 80 to 85 percent volumetric efficiency at their torque peak (where VE is highest). With a tuned intake and exhaust system, on the other hand, a race engine can get better than *100 percent* volumetric efficiency over a narrow portion of its power band.

Here's how it works. As the exhaust valve opens, the pressurized gases burst out of the cylinder into the exhaust port, forming a high pressure pulse in the exhaust gases already in the header primary pipe. This pressure pulse runs through the header pipe at the speed of sound — about 1,700 feet per second at exhaust temperature. The speed of the pulse does *not* depend on the speed of the gases. In fact, the pulse reaches the end of the header pipe much sooner than the gases do. When it reaches the end, it inverts — becoming a negative pressure pulse, or vacuum — and rushes back up the pipe toward the engine.

During this time the rising piston has been pushing exhaust gases through the exhaust port. When the negative pulse reaches the exhaust port, the extra vacuum helps scavenge the gases left in the cylinder,

actually pulling them out past the valve. Because the intake valve has also been open for a little while (the valves are at overlap), the vacuum also helps pull fresh mixture into the cylinder before the piston even starts on its way down.

The end result is more horsepower, because the cylinder charge is both denser and less diluted with residual exhaust gases. Obviously, the way to optimize the system is to adjust the length of the header pipes so that the negative pressure pulse arrives at the exhaust valve at just the right time. And equally obviously, a tuned exhaust is only going to work at its best over a narrow range of engine rpm, perhaps several hundred rpm. Making the pipes shorter lets the pressure pulse reach the exhaust port sooner, tuning the pipes for higher rpm. Making the pipes longer does the opposite, tuning the pipes for lower rpm. Racers can find noticeable differences in horsepower when they change the length of their pipes by as little as three inches.

Pipes with megaphones at the ends make the pressure pulse longer and less distinct, spreading the effect over a wider rpm than a straight, cut-off pipe, but making it weaker. A reverse-cone megaphone (one that increases and then decreases in diameter) can produce an additional power boost by sending a positive pressure pulse up the exhaust pipe, chasing the vacuum pulse. At some engine speed, this pressure pulse will catch the fresh mixture escaping from the exhaust port and actually stuff it back into the combustion chamber.

Pressure pulses resonate back and forth in the induction tract, too, caused by opening and closing the intake valve repeatedly in the path of the moving column of air-fuel mixture. Racers sometimes use sonic tuning in the induction system, but this is much trickier than in the exhaust. Ideally, a positive pulse should arrive late in the intake stroke to reduce reverse pumping, and a negative pulse should arrive just as the intake valve closes to ease the disruption the closing valve has on the moving column of slightly rough to aid vaporization by reducing puddling and breaking up mixture flow.

The air-fuel mixture should be distributed as evenly as possible among the cylinders. Figure 5-16 shows a carbureted V8 intake manifold design that promotes good distribution to all cylinders. If one or more cylinders receives an overly lean mixture, a richer overall mixture will be necessary for that cylinder to fire properly. This will cause the other cylinders to receive a mixture that is too rich. Overly lean combustion produces oxides of nitrogen (NO_x), while overly rich combustion produces unburned hydrocarbons (HC) and carbon monoxide (CO). Neither condition is desirable, since they raise emissions and lower fuel economy.

Manifolds usually have an induction **plenum** chamber, a storage area for the air-fuel mixture that stabilizes

the incoming air charge and actually allows it to rise slightly in pressure, figure 5-17 and 5-18. The mixture accumulates within the plenum until it is drawn out by one of the cylinders. This provides a uniform mixture charge that can be distributed equally to each cylinder, unaffected by momentary variations in throttle position and intake turbulence. The plenum feeds the mixture into runners leading to the intake ports.

Many fuel-injected engines have tuned intake manifolds. The design of these manifolds varies considerably according to the type of fuel injection used, figures 5-19 and 5-20, but all share similar features.

Plenum: A chamber that stabilizes the air-fuel mixture and allows it to rise to a pressure slightly above atmospheric pressure.

---

mixture. And a third positive pulse can pressurize the port just ahead of the valve, letting the intake charge burst into the cylinder the instant the valve opens.

In practice, however, sonic tuning in the intake is very difficult to control. The temperature of the intake charge is hard to predict, because it is simultaneously heated by the hot manifold, and cooled by the evaporating gasoline. In addition, opening and closing the throttle valves raises and lowers the pressure in the intake manifold and port, which has a strong effect on pulse speed. And finally, a standard log-type or common plenum chamber manifold gives a designer little room for tuning, because the pulses mix together and cancel or reinforce each other wherever the passages meet.

Port fuel-injection systems *do* allow some intake tuning, though, because the long runners are separated between the intake plenum and the intake valve. The manufacturers can adjust the length of the runners to move the power boost up and down the rpm range of the engine, perhaps to fill a "hole" in the

power band, or to add a little more low or high rpm torque, depending on the expected use of the engine.

Aftermarket induction systems with a separate carburetor or injection nozzle for each cylinder can also be tuned by making the velocity stacks longer or shorter. Some tuners use different-length stacks on the same engine to widen the power band by making different cylinders resonate at different rpm. In general, long runners seem to help low-rpm horsepower, and short runners seem to help high-rpm horsepower, but in practice tuners usually find the best length for a given engine and application by trial and error.

Sonic tuning has the most effect on piston-port two-stroke engines, which require it for cylinder scavenging. Four-strokes with poppet valves gain much less from sonic tuning, because they have a camshaft to force the intake and exhaust gases to move at certain times. Nonetheless, resonance still has a profound effect on racing four-stroke engines. With an intake and exhaust system working in harmony, the designer can specify a camshaft with much more duration than the engine could otherwise tolerate. This means that higher valve lifts are possible while still keeping valve acceleration low for reliability. The result is greater volumetric efficiency from the camshaft, made possible by a tuned intake and exhaust system.

For those of you contemplating modifying a street engine, the benefits from chasing horsepower through sonic tuning are less than you might hope. Big power gains are only possible over a narrow range of engine speeds — useful at the track, but not in traffic. A better approach would be to plan a system concentrating on low restriction and high gas velocities in both intake and exhaust. Using small diameter intake runners and header pipes helps horsepower with small displacement engines and at low rpm, keeping gas velocities high enough so that induction and scavenging are more efficient. For high-rpm use or larger engines, larger intakes and header diameters keep the passages from restricting high gas flow.

Unequal length air intake stacks eliminate peaks and valleys in the torque curve.

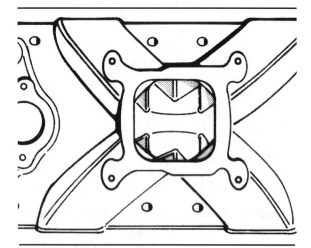

Figure 5-17. The plenum on this carbureted V8 single-plane manifold feeds the intake runners to all cylinders.

Individual "tuned" runners connect the plenum chamber to each intake port. These runners are specifically designed for the intake ports to which they connect and provide increased airflow at high speed for maximum power. In figure 5-20, a large-diameter passage located inside the plenum chamber behind the airflow entry point provides secondary tuning. This increases airflow at low speeds for better torque.

Unequal distribution and manifold design
The air-fuel mixture reaching the engine cylinders may vary between cylinders in amount and ratio for several reasons:

- The mixture flow is directed against one side of the manifold by the throttle valve in the carburetor or throttle body, figure 5-21. To some extent, the carburetor choke plate can influence flow in a similar manner.
- Smaller liquid particles of the air-fuel mixture will turn corners in the manifold more easily, while the larger particles tend to continue in one direction.
- Cylinders closer to the carburetor or throttle body will receive a richer mixture than those farther away from the carburetor or throttle body. However, this can be minimized by carburetor or throttle body placement and good manifold design. Depending on manifold design, using a 2- or 4-barrel carburetor, or 2-bore throttle body, can improve mixture distribution, because each barrel or bore supplies fewer cylinders.
- Charge robbing can result between two cylinders that are located close together in the block and whose intake strokes occur at about the same time in the firing order. When charge robbing

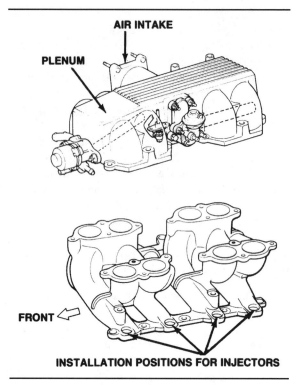

Figure 5-18. The plenum on this fuel-injected V8 manifold is located in the upper section of the intake manifold.

occurs, the cylinder whose intake stroke occurs first draws in a balanced mixture, but the intake stroke occurring later draws a weaker, leaner mixture because less fuel is available. Careful manifold design can minimize charge robbing, but the only real cure is port fuel-injection.

Manifold and intake charge heat
As you learned in Chapter 2, liquid gasoline is composed of various hydrocarbons which vaporize at different temperatures. If all of the hydrocarbons in gasoline vaporized at a fast rate, the job of the intake manifold would be simple. But they do not, so it is essential that the intake manifolds of engines with carburetors and throttle bodies are heated to keep the air-fuel mixture properly vaporized. Although it is not as critical, many port fuel-injection manifolds are heated as well.

Intake manifolds are heated with exhaust gas heat, engine coolant heat, exhaust manifold heat or all three, depending on the engine design, figure 5-22. On V engines an exhaust crossover passage inside the manifold carries hot exhaust gases near the base of the carburetor or throttle body, figure 5-23. In-line engines with both manifolds on the same side of the engine, position the intake manifold on top of the exhaust manifold, figure 5-24. A passage between the two manifolds fills with exhaust gases to improve fuel

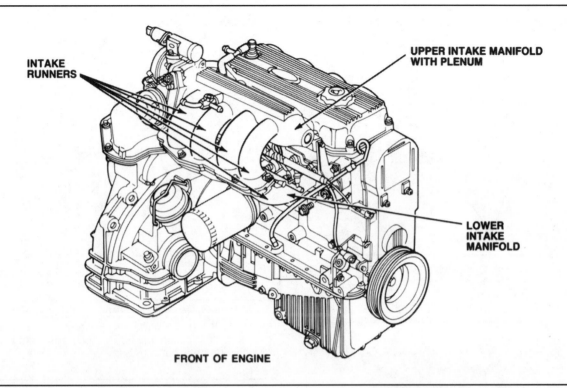

Figure 5-19. The tuned Ford intake manifold on this 4-cylinder engine uses short, separate intake runners.

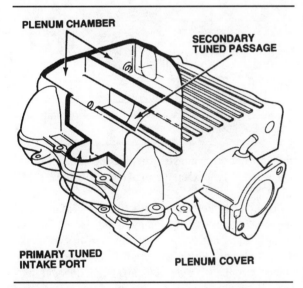

Figure 5-20. This Chrysler V6 intake manifold has a large plenum with a partial separator for intake tuning.

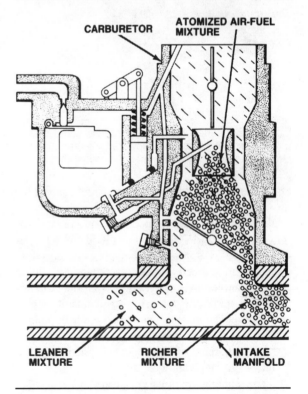

Figure 5-21. The angle of the carburetor throttle valve can affect the flow of the air-fuel mixture.

vaporization in the intake manifold passages. On inline engines with crossflow cylinder heads, a heat jacket on the intake manifold supplies the heat.

The area in the intake manifold heated by exhaust is called the "hot spot." The location, size, and surface area of the hot spot on the manifold floor affect vaporization. The hot spot is primarily a cold start feature.

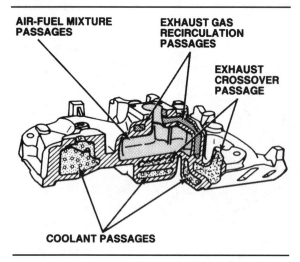

Figure 5-22. On this intake manifold both exhaust gases and engine coolant provide heat for air-fuel mixture vaporization.

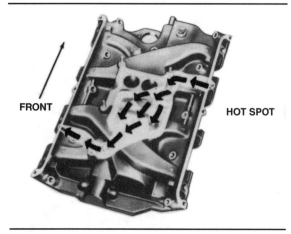

Figure 5-23. Exhaust gases are routed from ports in the cylinder heads through separate passages in the intake manifold to form the manifold hot spot.

Once the manifold warms up, fuel vaporizes through it.

While exhaust gas heat is only used on manifolds for carburetors and TBI, many manifolds for all types of induction systems, including port fuel-injection, route engine coolant through the manifold, figure 5-25. Engine coolant provides little manifold heat during a cold startup. Its job is to provide manifold heat and improve vaporization at all times while the engine is warmed up and running.

Most carbureted and fuel-injected systems also use an intake air temperature control system, figure 5-26. The heat radiating off of the exhaust manifold is fed to the intake to warm the inlet air and help vaporize the fuel just after a cold startup. When the engine warms up, a thermostat closes the passage from the exhaust

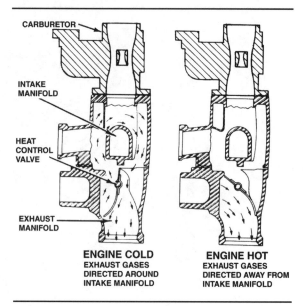

Figure 5-24. In this cross section of an in-line engine manifold, the heat control valve forces the exhaust gases to flow either around the intake manifold, or directly out the exhaust system.

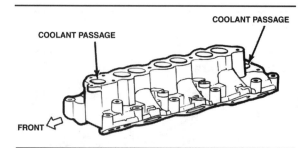

Figure 5-25. An engine coolant passage heats this Lexus 250 fuel-injection manifold.

manifold to the intake. The thermostat constantly cycles slightly to keep the intake air at around 125°F (52°C).

Even with the help of heat from the exhaust and engine coolant, a small amount of the air-fuel mixture usually does not completely vaporize in the intake manifolds of carbureted and throttle-body fuel-injected engines. This results in a slight unequal mixture distribution among the cylinders, with some cylinders receiving more fuel and developing more power than others. This "problem" (although quite small) is greater during engine warmup, when less than normal heat is available to vaporize the fuel.

Basic Intake Manifold Types

The exact design and number of manifold outlets to the engine depend on the engine type, number of cylinders, fuel delivery, and valve port arrangement.

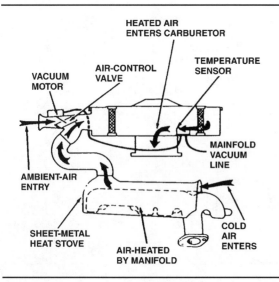

Figure 5-26. Heat radiating off of the exhaust manifolds heats the intake air on fuel-injected and carbureted engines.

Single-plane manifolds

Production V intake manifolds for V engine–equipped vehicles are classified as either single-plane or dual-plane designs. A single-plane manifold, figure 5-27, uses short runners to connect all of the engine's inlet ports to a single, common plenum. The single-plane design permits more equal cylinder-to-cylinder mixture distribution at high rpm, but its large internal volume causes a drop in mixture velocity, reducing airflow in low-to-intermediate speed ranges. Single-plane manifolds can produce more horsepower at high rpm than can dual-plane manifolds because each cylinder can draw mixture from all throttle bores.

In-line engines usually have single-plane manifolds. For the best compromise, designers place the carburetor or throttle body in the middle of the manifold. However, this carburetor or TBI placement can lead to the air-fuel flow condition shown in figure 5-15.

Dual-plane manifolds

A dual-plane intake manifold has two separate plenum chambers connected to the intake ports of the engine, figure 5-28. Each chamber feeds two central and two end cylinders. Mixture velocity is greater at low-to-intermediate rpm than with the single-plane manifold, but mixture distribution is usually even less than that produced by the single-plane manifold at high rpm. When a 4-barrel carburetor is used with a dual-plane manifold, each side of the carburetor (one primary and one secondary barrel) feeds one plenum chamber. When a 2-barrel carburetor is used, each barrel also feeds one chamber. V engine intake manifolds may be single-plane, but dual-plane designs are more common.

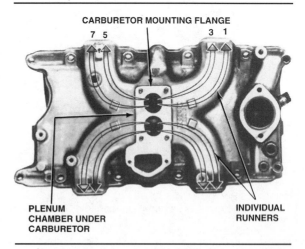

Figure 5-27. This single-plane manifold feeds eight cylinders from a single plenum.

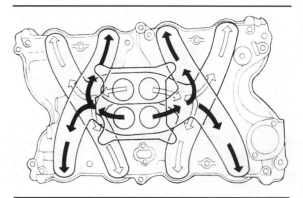

Figure 5-28. The two chambers in a dual-plane manifold are at different heights.

Throttle-body fuel-injection manifolds

Fuel-injected engines with throttle body-injection have manifolding requirements similar to those of a carbureted engine. The TBI unit sits on the intake manifold, and its injector, or injectors, spray fuel into the manifold to mix with incoming air. Theoretically, the injector sprays fuel into the area of maximum air velocity to ensure thorough atomization and ideal distribution. Some two-bore TBI units, however, use individually calibrated injectors to assist in proper distribution. When two TBI units are installed, as in GM's Cross Fire Injection system, figure 5-29, the rear injector must be calibrated differently than the front injector for better distribution.

Port fuel-injection manifolds

Fuel-injected engines using individual port injectors do not deliver fuel through the manifold, so each separate runner can be specifically tailored in size and length. Designers call this sonic tuning, or ram tuning. This improves induction flow to increase cylinder

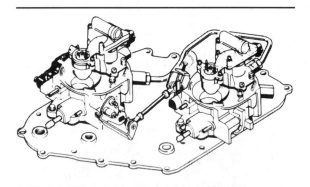

Figure 5-29. Chevrolet's cross-fire injection uses two TBI units mounted on a common manifold.

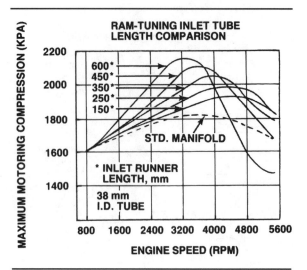

Figure 5-30. This graph shows the effect of sonic-tuning the manifold runners. The longer runners increase the torque peak and move it to a lower rpm. By increasing the peak torque, they also narrow the power band.

charging and improves fuel distribution among the cylinders. The difference in performance between a standard manifold and the sonic-tuned inlet runners on the Chevrolet 2.8-liter V6 is shown in figure 5-30. As an example, the 600-mm inlet runner develops 20 percent more compression, and therefore more power, than the standard manifold.

Unlike carbureted or TBI systems, the port fuel-injection manifolds must provide a place to mount the injectors and fuel rail assembly, figure 5-31. This requires precision casting or drilling to properly locate each injector at its required position. Injectors generally are retained in the manifold by clamps and sealed with O-rings.

Two-piece manifolds
Port fuel-injection manifolds may be one- or two-piece. The two-piece design is used with in-line and some other V engines, figure 5-32, and consists of the upper intake manifold with a plenum and a lower intake manifold with individual runners. Airflow entering the throttle body passes into the upper intake manifold or intake plenum where it is distributed to the individual runners. These runners are designed to a specific length according to cylinder requirements. They route the airflow directly to the individual ports where it mixes with the fuel charge just before entering the combustion chamber.

One-piece manifolds
The one-piece design is used with some V engines and has the plenum, intake passages, and runners within a single unit, figure 5-33. The one-piece manifold works in basically the same way as the two-piece manifold, except that everything is built into the casting. The length of the intake passages and the shape and size of the plenum are designed to produce a denser air mass at each cylinder just before the intake stroke. In the design shown in figure 5-34, a large diameter passage located inside the plenum chamber behind the airflow entry point produces secondary

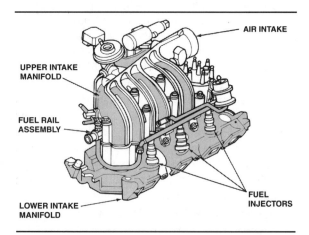

Figure 5-31. A fuel rail connects the fuel injectors installed at individual ports.

tuning. This increases airflow at low speeds for better torque.

Variable induction systems
Intake manifold design has always been a compromise. Traditionally, a manifold designer had to develop a manifold that produced the best power at either high rpm, low rpm, or something in between. A manifold that cannot vary its intake runners could never be best at all engine speeds. The variable induction systems on some port fuel-injection systems change that. These manifolds employ multiple runners and additional throttle valves. With the help of electronic controls, these throttles open at a specified rpm to change the manifold volume and length, providing power over very wide rpm bands. Some variable-induction

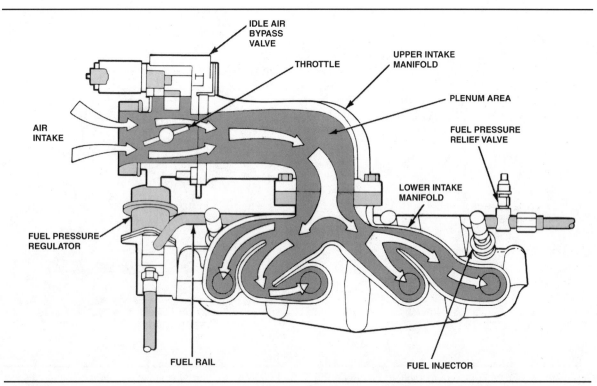

Figure 5-32. Airflow through the upper intake manifold is distributed to individual runners in the lower manifold.

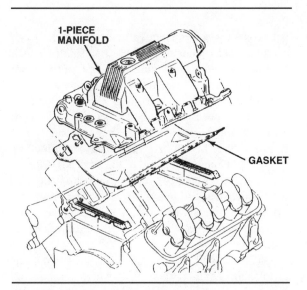

Figure 5-33. One-piece port-injection manifolds, like this Buick example, incorporate the air intake, plenum, and runners.

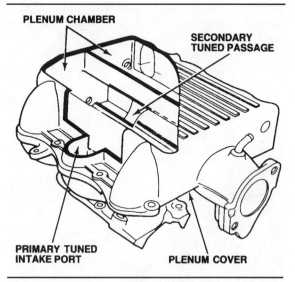

Figure 5-34. Secondary tuning in this Chrysler manifold is provided by a large diameter passage inside the plenum.

systems also work in conjunction with variable camshaft timing systems, which we will detail in Chapter 9.

Ford's SHO engine with port fuel-injection and a variable intake manifold system features two runners for each cylinder (one short and one long runner) that optimize both low- and high-rpm performance, figure

5-35. This system, known as Intake Manifold Runner Control (IMRC), uses the long set of runners to provide a small inlet area to speed the velocity of the airflow, giving good low-end performance. Power output, while running on the long runners, levels off by 3,000 rpm. The short set of runners, or "secondaries," has a valve plate that stays closed until 3,000

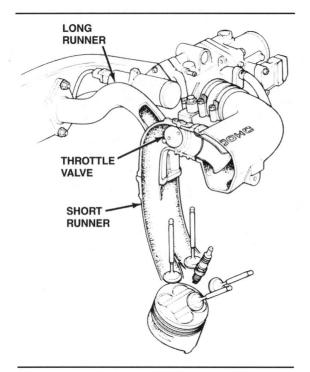

Figure 5-35. Ford's SHO intake manifold has one long and one short intake runner to each intake port.

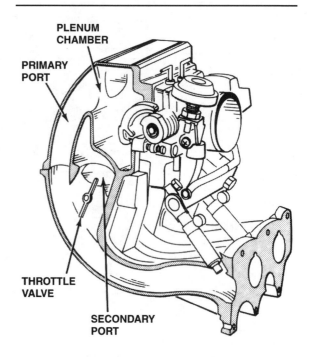

Figure 5-36. Mazda's VRIS system uses an inertia charge effect to improve torque output through interconnected, branched intake paths equipped with throttle valves.

rpm, at which point the engine control computer opens it to provide additional airflow for good high-end performance.

Ford says that the transition point when the short runners kick in is not felt by the driver. At the 3,000 rpm kick-in point, the engine's torque output when running on the long runners or both long and short runners is identical. Therefore, the feeling of power is constant throughout the rpm range. The transition is only apparent when looking at a graph of the engine torque curve.

The Chevrolet ZR-1 Corvette, introduced in 1990, has a high power output as well as docile low-speed operation made possible by an induction system similar to the SHO's. In the ZR-1 system, each cylinder is fueled by two intake runners, a primary and a secondary. Below 3,500 rpm or half-throttle, the engine runs on only the smaller primary runners to provide good fuel efficiency and smooth operation. When the gas pedal is pressed further and rpm rises, a computer-operated throttle valve opens each of the larger secondary runners, allowing the engine to take in more air and fuel and develop more power.

A removable-key-operated "valet-mode" switch prevents unauthorized use of the ZR-1's maximum power. Turning the switch alerts the engine control computer to disable the secondary intake system.

Mazda uses a variable ram-effect induction system (VRIS). This design uses a common plenum chamber

with long curved runners (primary ports) for each cylinder. Each runner also has a second shorter branch (secondary ports), containing a throttle valve, figure 5-36. These branches are interconnected by a passage and the valves remain closed below about 5,000 rpm. Above that speed, the computer activates a vacuum actuator through a solenoid to open the secondary port control valve. This shortens the effective length of the passage and interconnects all of the branches. Closing the intake valves creates a positive pressure in the system, which has a domino effect moving from one cylinder to another to pack in more air. Since the interconnecting passage reduces air intake resistance, flow speed is also reduced.

On Nissan's VG30 V6 engine, air flows into a single plenum chamber, figure 5-37, where it passes into a pair of collector boxes. Each collector box contains a throttle chamber which reduces air resistance at high rpm and load conditions. An interconnecting passage between the collector boxes contains a valve controlled by the computer. At low- and mid-range rpm, this valve remains closed to lengthen the intake path. As rpm increases, the valve opens to connect the two collector boxes and increase the volume of airflow.

Toyota also builds several 4-valve engines with dual-level, variable intake manifolds as part of the computer-controlled fuel-injection system. The

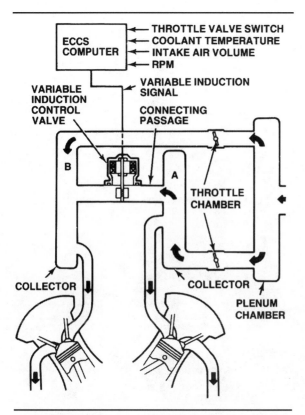

Figure 5-37. Nissan's Induction Control System uses a computer-controlled valve to change the length of the induction path according to the engine speed.

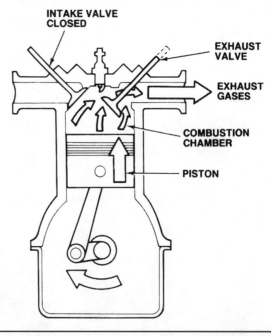

Figure 5-38. Burned gases are pushed out of the combustion chamber by the piston's upward exhaust stroke.

Honda-Acura-Sterling V6 has a dual intake system with 12 manifold runners for 6 cylinders. A solenoid-controlled vacuum diaphragm opens the secondary runners for more airflow at high speed.

While there are many more such induction designs, the intent behind each is the same — to maintain engine torque across a wider power band.

The important principle of variable intake manifolds is that the primary (low-speed) runners are longer and narrower than the secondary (high-speed) runners. Long, small cross section runners increase airflow velocity at low engine speed to deliver more air, faster at engine speeds below approximately 3,000 rpm. Short, large cross section secondary runners allow airflow to travel a shorter distance at engine speeds above 3,000 rpm. This delivers more air at higher speed than long, narrow runners can. Variable manifolds thus avoid the traditional compromises of intake manifold design.

THE EXHAUST SYSTEM

After the air-fuel mixture is burned and escapes through the exhaust valve, it must continue its exit away from the car. The exhaust system routes engine exhaust gases to the rear of the car, quiets the exhaust noise, and, since the mid-1970s, also reduces the pollutants in the exhaust.

Backpressure

The exhaust system's design has a marked effect on engine performance. Proper extraction of exhaust gases through the exhaust manifold design is as important as air-fuel flow through the intake manifold. The flow of exhaust gases from the engine should be as smooth as possible. As the piston moves upward during the exhaust stroke, it forces combustion gas through the open exhaust valve and out of the cylinder, figure 5-38. The exhaust gases are hot and therefore under pressure, which also helps cause them to flow out the valve.

Whenever gas is pushed through a passageway, turbulence and friction along the sides of the passage cause a resistance called **backpressure**. A piston encounters backpressure each time it comes up on the exhaust stroke. Too much backpressure will not allow all the exhaust to leave the combustion chamber

Backpressure: The resistance, caused by turbulence and friction, that is created as a gas or liquid is forced through a restrictive passage.

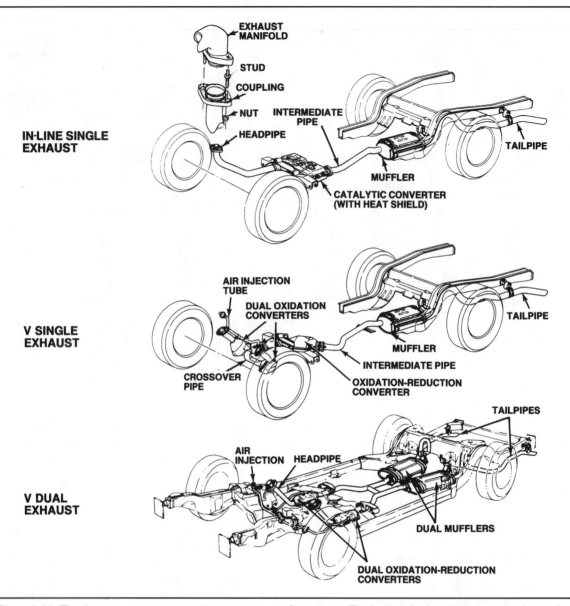

Figure 5-39. The three most common exhaust system configurations. The in-line single exhaust system is used with an in-line engine. The V single exhaust system uses a Y-pipe to connect the two cylinder banks to a single exhaust pipe. The V dual exhaust is essentially two separate exhaust systems for a V engine.

before the next stroke starts. This preheats and leans the incoming air-fuel mixture, reducing efficiency and power. It can also cause engine mechanical failures such as burned valves.

Backpressure in the exhaust manifold and exhaust system can also contaminate the intake manifold's fresh air-fuel mixture. A brief camshaft overlap period, combined with backpressure in an exhaust manifold, can cause one cylinder's intake stroke to draw exhaust gases from a nearby cylinder. A direct and unrestricted flow of exhaust gas causes less backpressure and prevents this unnecessary power loss or engine damage.

Exhaust System Arrangement

The exhaust system is usually arranged to suit the engine design. There are three basic exhaust system configurations, figure 5-39. Since an in-line engine usually has a single exhaust manifold, a single exhaust system is generally used, with the exhaust pipe connecting directly to the exhaust manifold flange. Each cylinder bank on a V engine, however, has its own exhaust manifold. When a single exhaust system is used, the two manifolds will connect with a Y-pipe or a crossover pipe, figure 5-40. With dual-exhaust systems, each manifold is connected to its own exhaust pipe, muffler, and tailpipe.

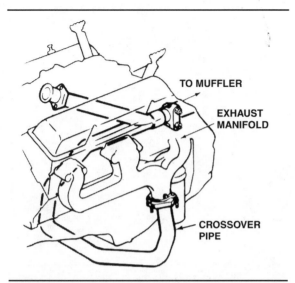

Figure 5-40. This V engine has a crossover pipe connecting the two banks to a single exhaust.

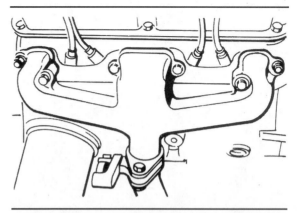

Figure 5-41. Cast-iron exhaust manifolds are quiet and durable, but their design is not the most efficient for producing power.

Dual-exhaust systems reduce exhaust backpressure by splitting the exhaust gas flow into two outlet lines. Since the exhaust manifold is at the beginning of the flow, manifold design is the most important factor in reducing backpressure. However, good muffler design also can help minimize restrictions.

Exhaust System Components

Major parts of the exhaust system include:

- Exhaust manifolds
- Pipes
- Mufflers
- Resonators
- Catalytic converters.

Exhaust manifolds

Exhaust manifolds attach directly to the side of the engine cylinder head, matching their ports to the exhaust ports of the cylinder head, figure 5-41. Exhaust manifolds are the first pieces outside of the engine that absorb the intense heat of the exhaust, so manifolds are usually made of thick and sturdy cast iron. The thick cast iron is long lasting and helps to damp out combustion noise.

Some exhaust manifolds are fabricated from tubular steel, which is lighter than iron and easier to assemble into an efficient manifold with low backpressure. For greater corrosion resistance, other exhaust manifolds are made of stainless steel.

High-performance exhaust manifolds

Although the specialized science of exhaust tuning was once the province of the hot rodder and high-performance vehicles, the problem of exhaust backpressure is an important factor in controlling the emissions of late-model engines. Automotive engineers design backpressure-reducing exhaust systems that can affect the volumetric efficiency of an engine, and yet still leave enough backpressure to regulate the recirculation of exhaust gas in order to reduce NO_x and prevent detonation.

Sharp turns and narrow passages in an exhaust manifold slows the flow of gases from the exhaust ports and increases the amount of backpressure in the system. Cast-iron exhaust manifolds also increase back pressure because they release exhaust gases from each cylinder into a common chamber. The pressure of gases from one cylinder interferes with the flow from other cylinders, figure 5-42. Many original equipment manifold designs have duplicated high-performance manifolding by using the sweptback manifold design, figures 5-43 and 5-44.

Some Ford 4-cylinder engines use a **bifurcated** stainless steel exhaust manifold, figure 5-45. The low-restriction, tuned tubular header design consists of four primary runners. These form into two secondary runners and converge into a single outlet that connects to the exhaust system. By routing pressure pulses from adjacent firing cylinders through different pipes, backpressure is reduced by approximately 30 percent.

Bifurcated: Separated into two parts. A bifurcated exhaust manifold has four primary runners that converge into two secondary runners; these converge into a single outlet in the exhaust system.

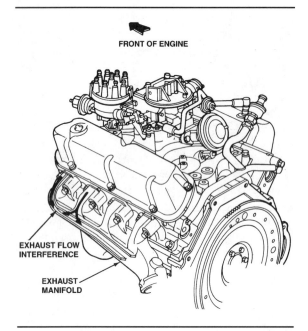

Figure 5-42. Exhaust gas interference from adjacent cylinders increases back pressure.

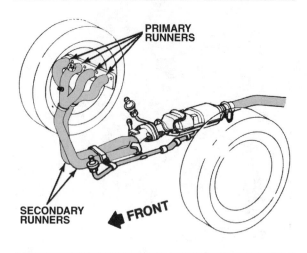

Figure 5-43. Streamlining the exhaust manifold is the first step in producing high-performance exhausts that reduce back pressure.

On in-line 6-cylinder engines, Ford uses two separate cast iron exhaust manifolds, one for the three front cylinders and the other for the three rear cylinders. This design solves two former problems: the single manifold previously used did not seal well against the head and also tended to fail prematurely due to cracking. Like the bifurcated manifold design, the two manifolds connect to a single outlet that leads into the exhaust system.

Pipes

To reduce backpressure, exhaust pipes should be as straight as possible, without sharp turns and restrictions. Pipes must also be able to withstand the constant presence of hot, corrosive exhaust gases, salted roads during winter, and undercar hazards such as rocks.

To improve their strength, the exhaust pipes on many late-model cars are formed with an inner and an outer skin. Occasionally, the inner skin will collapse and form a restriction in the system. From the outside, the exhaust pipe will look normal even when the inside is partially or almost completely blocked. Close inspection is necessary to locate such a defect.

To further improve the exhaust pipe's resistance to corrosion, most exhaust pipes are now made of stainless steel. However, stainless-steel pipes do not last forever. They are more brittle and prone to fatigue breakage than ordinary mild steel.

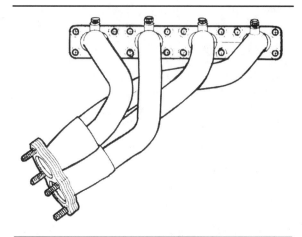

Figure 5-44. The 1991 BMW 318 engine uses an example of the most efficient type of exhaust manifold. The system is streamlined, smoothing exhaust flow. Each cylinder has its own pipe, eliminating exhaust gas interference from other cylinders.

Mufflers and Resonators

The muffler is an enclosed chamber that contains baffles, small chambers, and pipes to direct exhaust gas flow. The gas route through the muffler is full of twists and turns, figure 5-46. This quiets the exhaust flow but also creates backpressure. For this reason mufflers must be carefully matched to the engine and exhaust system.

Some mufflers consist of a straight-through perforated pipe surrounded by sound-deadening material, usually fiberglass. These "glass-pack" mufflers, as they are often called, reduce backpressure but are not nearly as quiet as conventional mufflers.

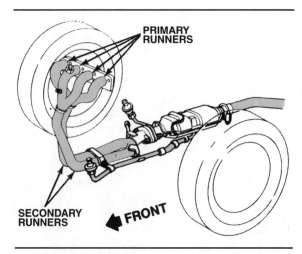

Figure 5-45. Ford's bifurcated exhaust manifold is essentially a set of stainless steel headers.

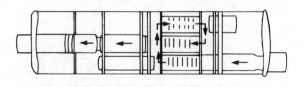

Figure 5-46. Exhaust gases must twist and turn to travel through this muffler.

Some engines also add resonators to the system. These are small mufflers specially designed to "fine-tune" the exhaust and give it a pleasant, quiet, "resonant" tone.

Catalytic Converters

One way of reducing HC and CO is to increase the combustion temperature, which causes more complete burning. As combustion temperatures rise, however, so does the formation of a third pollutant, oxides of nitrogen (NO_x). To meet tightening exhaust emissions standards, vehicle manufacturers incorporated catalytic converters into vehicle exhaust system in 1975. The catalytic converter is installed in the exhaust system between the manifold and the muffler, figure 5-47. By passing the exhaust gas through a catalyst in the presence of oxygen, the HC and CO compounds combine with the oxygen to form the harmless by-products water vapor (H_2O) and carbon dioxide (CO_2).

A catalyst is a substance that promotes a chemical reaction, but is not changed or affected by that reaction. An oxidation (burning) reaction takes place when oxygen is added to an element or compound. Platinum

and palladium are catalytic elements that promote oxidation. If there is not enough oxygen in the exhaust, an air pump or aspirator valve provides extra oxygen.

Since the catalytic oxidation reaction has no effect on NO_x, a separate reaction called reduction is required. The reduction reaction mixes NO_x with CO to form nitrogen gas (N_2) and CO_2. The most common reduction catalysts are platinum and rhodium.

A catalyst does not work when cold; it must be heated to at least 400° to 500°F (204° to 260°C) before it reaches 50 percent effectiveness. When fully effective, the converter reaches a temperature range of 900° to 1,600°F (482° to 871°C). Because of the extreme heat, a converter remains hot long after the engine is shut off. Catalytic converters have no moving parts and never need adjusting.

SUMMARY

The primary parts of the intake system are a manifold and a fuel-delivery system. The pistons pump in air and fuel for combustion and pump out the burned exhaust. They do this by creating a difference in air pressure.

Volatility is a measure of gasoline's ability to change from a liquid to a vapor. The more volatile it is, the more efficiently the gasoline will vaporize. Before gasoline can do its job as a fuel, it must be metered, atomized, vaporized, and distributed to each cylinder in the form of a burnable mixture.

Good manifold design is critical for the smooth performance and low emissions required of today's engines. The intake manifold must evenly distribute the air-fuel mixture to each cylinder, and the exhaust manifold must quickly remove the exhaust from the cylinders so that the next incoming air-fuel charge will not be contaminated. Runner length, smoothness, and the arrangement of headers and exhaust pipes are important for the smooth flow of gases.

A port fuel-injection system does not deliver fuel through the manifold, so each separate runner can be "tuned" to increase efficiency. An intake air temperature control system is used and port fuel-injection manifolds incorporate provisions for mounting the injectors and fuel rail assembly.

Intake manifold designs favor good low-to-middle speed operation, good high-speed operation, or a compromise. Variable air induction systems maintain torque across a wider power band.

The type of fuel-delivery system and its relationship to the manifold affects manifold efficiency. Several methods promote better vaporization in the

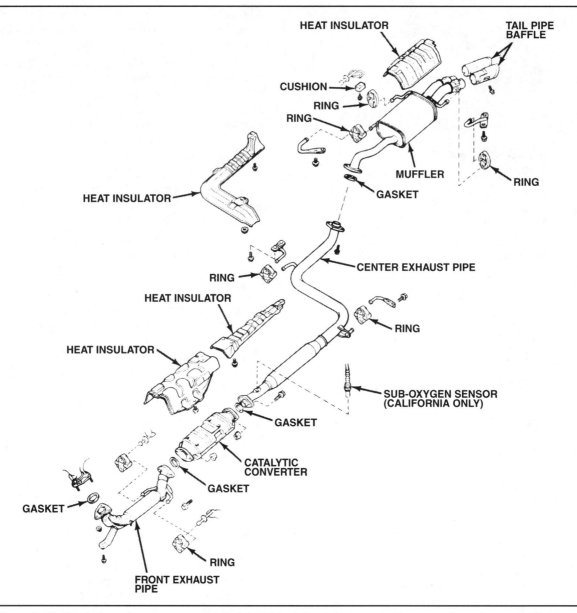

HEAT INSULATOR

TAIL PIPE BAFFLE

CUSHION

RING

RING

MUFFLER

GASKET

RING

HEAT INSULATOR

CENTER EXHAUST PIPE

RING

HEAT INSULATOR

HEAT INSULATOR

RING

SUB-OXYGEN SENSOR (CALIFORNIA ONLY)

GASKET

CATALYTIC CONVERTER

GASKET

GASKET

RING

FRONT EXHAUST PIPE

Figure 5-47. Catalytic converters are typically placed in the exhaust system between the exhaust manifold and muffler.

manifold, including a heat control valve which routes exhaust gases through the intake manifold or routing hot engine coolant through the manifold.

Exhaust systems carry the exhaust away from the combustion chambers through a series of pipes, mufflers, resonators, and catalytic converters. The system must be as straight as possible to reduce back-pressure in the cylinders, and generally must damp engine exhaust noise.

The exhaust system is arranged to suit the engine design. The three major types are: in-line, single V, and dual V. When an engine's cylinders are in-line, all of the exhaust valves are on the same side of the engine. An exhaust system pipe connects to this side of the engine at the exhaust manifold. A V engine has exhaust valves on both cylinder banks, so it requires two separate exhaust systems, one for each bank. Regardless of exhaust pipe arrangement, the pipes connect to one or more units that may include a muffler and a catalytic converter.

Review Questions

Choose the single most correct answer.
Compare your answers with the correct answers on Page 328.

1. Opening the throttle valve to a carburetor or fuel injection system causes:
   a. Air to move from the higher pressure area outside the engine to the lower pressure area of the manifold
   b. The venturi effect
   c. A decrease in pressure inside the engine
   d. None of the above

2. Unequal distribution of the air-fuel mixture may occur because of:
   a. Mixture flow misdirected by the throttle valve
   b. Sharp turns in the manifold
   c. Distance of travel to the cylinders
   d. All of the above

3. A venturi:
   a. Increases air velocity through a carburetor
   b. Lowers air velocity through a carburetor
   c. Improves fuel economy
   d. Is a special fuel injection manifold

4. Technician A says vaporization only occurs when the fuel mixture boils or evaporates rapidly. Technician B says vaporization occurs when the fuel mixture is chilled.
   Who is right?
   a. A only
   b. B only
   c. Both A and B
   d. Neither A nor B

5. Poor vaporization occurs when:
   a. Fuel is too volatile
   b. The intake manifold is warm
   c. Manifold vacuum is high
   d. None of the above

6. The intake manifold may be heated by:
   a. Exhaust manifold heat
   b. Engine coolant
   c. Both a and b
   d. Neither a nor b

7. Technician A says that some intake manifolds are cast as one piece with the cylinder head. Technician B says that all intake manifolds are separate from the cylinder head.
   Who is right?
   a. A only
   b. B only
   c. Both A and B
   d. Neither A nor B

8. Long, small diameter intake manifold runners:
   a. Produce the best power at low to medium rpm
   b. Are necessary for low exhaust emissions
   c. Produce low air velocity at low rpm
   d. Produce the best power at high rpm

9. The plenum chamber:
   a. Lowers exhaust emissions
   b. Is not required on most street-driven automobiles
   c. Stabilizes the incoming air charge
   d. Cools and doubles the air-fuel mixture

10. Technician A says intake manifold heat from exhaust or engine coolant is necessary on port fuel-injected engines. Technician B says that intake manifold heat from one or both of these same sources is not required on port fuel-injected engines.
    Who is right?
    a. A only
    b. B only
    c. Both A and B
    d. Neither A nor B

11. Technician A says that port fuel-injection systems provide better fuel vaporization under most conditions than carburetors. Technician B says that port fuel-injection systems reduce exhaust emissions better than other systems.
    Who is right?
    a. A only
    b. B only
    c. Both A and B
    d. Neither A nor B

12. Technician A says that a TBI system uses one injector at each cylinder port. Technician B says that the TBI unit is installed on the intake manifold where a carburetor would be.
    Who is right?
    a. A only
    b. B only
    c. Both A and B
    d. Neither A nor B

13. Variable-induction manifolds differ in design, but all have one common goal which is:
    a. To reduce production costs
    b. To reduce engine weight
    c. To widen torque across the power band
    d. To help maintain a low hood profile

14. Exhaust backpressure:
    a. Causes power loss
    b. Can contaminate the air-fuel mixture in the intake manifold
    c. Can be minimized by exhaust manifold streamlining
    d. All of the above

15. High-performance exhaust manifolds:
    a. Increase horsepower by 30 percent
    b. Cause engines to operate at higher temperatures
    c. Reduce backpressure
    d. All of the above

16. A common type of exhaust system is the:
    a. In-line
    b. Single V
    c. Dual V
    d. All of the above

17. Catalytic converters help change:
    a. HC and CO to H_2O and CO
    b. HC and CO to CO_2 and H_2O
    c. HC and CO to HC_2O
    d. HC and CO to H_2O and C

6
Engine Lubrication

An automotive machinist must understand the lubrication system and the motor oil that runs through it. The quality of lubrication an engine receives has a direct effect on the life and performance of that engine. Even if you perform the finest machining and assembly, an engine will provide inferior service, or may even fail, if you neglect the lubrication system or use the improper lubricants in the engine.

In this chapter we first cover the purpose of motor oil, how it is rated, and what additives do to help motor oil. Then, we detail how the lubrication system works and the relationship between lubrication and mechanical wear.

PURPOSES OF MOTOR OIL

Motor oil in a car engine performs five major jobs:

- It reduces friction between moving parts, which lessens both wear and heat.
- It acts as a coolant, removing heat from the metal of the engine.
- It carries dirt and wear particles away from moving surfaces, cleaning the engine.
- It helps seal the combustion chamber by forming a film around the valve guides and between the piston rings and the cylinder wall.
- It acts as a shock absorber, cushioning engine parts to protect them from the force of combustion.

These jobs help keep the engine running smoothly and efficiently. If the motor oil fails at any one of them, performance suffers or engine damage results.

MOTOR OIL COMPOSITION AND ADDITIVES

Petroleum-based motor oils consist of hydrogen and carbon molecules. They are complex hydrocarbon compounds, as is gasoline. Petroleum-based motor oil is a product of crude oil, which oil companies extract from underground oil-bearing rock. Oil companies transport the crude oil to refineries where it is separated into fractions, which are the various components of the crude. The lighter fractions become products like gasoline or kerosine, and heavier fractions usually become the base stock for petroleum-based motor oil. Synthetic oils may use only a small amount of petroleum products or none at all in their make-up, so we cover synthetic motor oil in a later section of this chapter.

ACCEPTABLE   BORDERLINE

Figure 6-1. Sludge deposits usually occur because of lubrication system neglect.

Motor Oil Additives

Fifty years ago, motor oil consisted of only the base stock. Change intervals every 1,000 miles were common — as were stuck piston rings and rapid engine wear. To improve the performance of a motor oil, manufacturers blend in chemical additive packages. These packages are secret, propriety formulations, but a common additive is zinc diorganodithiophosphate (ZDDP). Additives make up about 25 percent of motor oil. A typical SJ/CF 10W-30 oil may be 75 percent oil base stock, 11 percent viscosity improver, and 14 percent other additives.

The purpose of a motor oil additive can be to:

* Replace a property of the oil that was lost during refining.
* Strengthen a natural quality already in the oil.
* Add a property that the oil did not naturally have.

A few of the common motor oil additives and their jobs are described in the following paragraphs.

Oxidation occurs when oil reacts chemically with oxygen, just as rust occurs when iron and oxygen react. Heat and contaminants such as copper and glycol, accelerate the oxidation reaction in motor oil. This process of oxidation can leave hard carbon and **varnish** deposits in the engine. **Antioxidants** reduce this problem by preventing the buildup of acids, destroying chemicals that produce undesirable oxidation by-products, and interrupting the oxidation chain reaction.

Varnish deposits that do develop are fought with **detergent** additives. For example, detergents clean piston ring grooves and keep the rings free to seal with maximum effectiveness, which maintains peak engine performance.

In any engine, some of the combustion chamber gases get past the piston rings and enter the crankcase. This is called **blowby**. Blowby gases contain water vapor, carbon dioxide, and unburned fuel that form acids which rust or corrode engine parts. Rust and corrosion preventives are added to motor oil to neutralize acids. Both the water vapor and the fuel in blowby gases tend to mix with cold oil and form **sludge**, figure 6-1. This thick black deposit clogs oil passages and reduces engine lubrication. **Dispersants** reduce sludge formation by keeping sludge particles suspended in the oil, to be removed when the oil and filter are changed.

The oil in an engine is constantly being churned by moving parts. This can mix air and other gases with the oil, causing the oil to foam. Oil foam does not form a good oil barrier between moving parts and may allow metal-to-metal contact. In extreme cases, foam can cause oil pumps to gas lock, or lose their prime, and stop pumping oil. Foam inhibitors reduce the oil's surface tension, the property that allows bubbles to form and stay formed. Oil with low surface tension allows the bubbles to burst, releasing the gas, and reducing the formation of foam.

Viscosity is the tendency of a liquid to resist flowing. Some additives help an oil to flow under wide temperature ranges. These additives are called **viscosity index (VI) improvers** and **pour-point depressants** and they are among the primary ingredients in multigrade oils. VI improvers help the oil resist thinning at high temperatures. Since VI improvers actually increase viscosity at low as well as high temperatures, a 10W-30 multigrade motor oil starts out as slightly less than a 10 weight base stock oil. Oil companies then add sufficient VI improvers so that the oil is able to perform like a 30 weight oil at high temperatures.

Pour-point depressants help the oil flow in colder temperatures. These additives lower the temperature at which the oil solidifies by blocking wax crystal growth in the oil base stock. Surprisingly, some VI improvers also function as pour-point depressants, so the oil company need not add supplemental pour-point depressants.

Additive precautions

At one time it was not considered good practice to mix oil brands in the same engine. Experts thought that since different manufacturers use different additives, the chemicals could oppose each other and decrease the cleaning and lubricating abilities of the oil. Today, it is not as great a concern because most oils are compatible. There are only about half a dozen additive packages available that allow an oil to meet current vehicle manufacturer's recommendations. To be on the safe side, it is still a good practice to always add the same oil, but it is not critical.

Many oil manufacturers advise against the use of other oil additives that are sold separately. A quality motor oil already contains all the additives that it needs. The extra additives are simply not necessary, and in some cases the extra additives can cause problems. Adding a high-detergent additive to an engine with a great deal of sludge may break loose large chunks of that sludge. The oil filter will be quickly filled and as the oil bypasses the filter, large amounts of sludge and dirt will flow through the engine unchecked.

Synthetic Motor Oils

Synthetic-base lubricants have been used in aviation and special-purpose engines for decades. Since the late 1970s, **synthetic motor oils** have attracted a lot of attention and have slowly gained in popularity because they offer several advantages over petroleum-based engine oils.

Synthetic oils are created by causing a chemical reaction in various complex molecules to form a new molecular structure. Petroleum, several types of acids, and alcohols may be used in varying proportions to make synthetic base oils. The process involved is complex, consequently synthetic oils may sell for three to five times the price of their petroleum-based counterparts.

Synthetics contain less wax than petroleum oils so they tend to remain liquid when cold. This permits faster lubrication in very cold temperatures, which is perhaps the greatest advantage of synthetic engine oils.

Some manufacturers of synthetic oils claim that the increased slipperiness of their oil decreases engine friction, thus contributing to small increases in both

■ PCV System Service

When a PCV system becomes restricted or clogged, the cause is usually an engine problem or the lack of proper maintenance. For example, scored cylinder walls or badly worn rings and pistons will allow too much blowby. Start-and-stop driving requires more frequent maintenance and causes PCV problems more quickly than highway driving, as will any condition allowing raw fuel to reach the crankcase. Using the wrong grade of oil or not changing the crankcase oil at periodic intervals will also cause the ventilation system to clog.

When a PCV system begins to clog, the engine tends to stall, idle roughly, or overheat. As ventilation becomes more restricted, burned plugs or valves, bearing failure, or scuffed pistons can result. Also look for an oil leaking out around valve covers or other gaskets. Do not overlook the PCV system while troubleshooting. A partly or completely clogged PCV valve, or one of the incorrect capacity, may well be the cause of poor engine performance.

Varnish: An undesirable deposit, usually on the engine pistons, formed by oxidation of fuel and of motor oil.

Antioxidants: Chemicals or compounds added to motor oil to reduce oil oxidation, which leaves carbon and varnish in the engine.

Detergent: A chemical compound added to motor oil that removes dirt or soot particles from surfaces, especially piston rings and grooves.

Blowby: Combustion gases that get past the piston rings into the crankcase; these include water vapor, acids, and unburned fuel.

Sludge: A thick, black deposit caused by the mixing of blowby gases and oil.

Dispersant: A chemical added to motor oil that keeps sludge and other undesirable particles picked up by the oil from gathering and forming deposits in the engine.

Viscosity: The tendency of a liquid such as oil to resist flowing.

Viscosity Index (VI) Improvers: Chemical compounds added to motor oil to help the oil resist thinning at high temperatures.

Pour-Point Depressants: Chemical compounds added to motor oil to help the oil flow at colder temperatures.

Synthetic Motor Oils: Lubricants formed by artificially combining molecules of petroleum and other materials.

fuel economy and power. Although these claims have been substantiated in some cases, you must weigh the savings in fuel economy and the increase in performance against the higher price of the oil. For example, the higher cost of the synthetic oil may more than equal any money saved due to improved fuel economy.

In addition, many synthetic oils have been shown to deteriorate at a slower rate than comparable petroleum-based oils, and therefore can remain in service longer between oil changes. However, synthetic oils hold contaminants in suspension no better than petroleum-based oils, so you should not assume that the use of a synthetic oil alone justifies postponing scheduled oil and filter changes.

While synthetic oil base stocks and additives may differ chemically from petroleum oils, they undergo the same tests and receive the same SAE viscosities and API Service Classifications. Thus, you select them in the same manner, according to the engine or car manufacturers' specifications.

MOTOR OIL DESIGNATIONS

Because engines and operating conditions vary greatly, oil refiners blend and sell different types of motor oil. The oil used in a heavy-duty diesel truck is different from that used in a high-performance car engine. For example, diesel engine oils must resist the acidity of sulfur in the diesel fuel, which is not a factor in gasoline engines. Both engines need lubrication, and as we shall see, an oil may or may not meet all the requirements of both engines.

Engine oil is commonly identified in two ways: by American Petroleum Institute (API) Service Classification and by Society of Automotive Engineers (SAE) Viscosity Number. You may also see an API specification such as GF-2, which is an international standard that meets the lubrication requirements of certain light and heavy duty vehicle manufacturers.

Another method for identifying oils is by military specification numbers or "MIL-specs." While they are rarely used to specify engine oils for cars and trucks, they often appear on oil containers.

API Service Classification

The **API Service Classification** rates engine oils on their ability to lubricate, resist oxidation, prevent high- and low-temperature engine deposits, and protect the engine from rust and corrosion. API has organized a system of letter classifications with two categories, the S-series and the C-series. The S-series service classification emphasizes oil properties

critical to gasoline engines, while the C-series emphasizes oil properties for diesel engines. In order for an oil formulation to be given a particular classification, the oil is run through a series of tests in specific engines. Some oil companies conduct their own tests; others pay independent laboratories to perform the tests for them. If the oil's performance meets the minimum standard, the oil can be sold bearing that API Service Classification.

The S-Series Oils

The S-series oils come under the following classifications: SA, SB, SC, SD, SE, SF, SG, SH, and SJ. There is no SI classification. No performance tests are required to meet the SA classification.

- The SE classification requires still higher levels of control of high- and low-temperature deposits, wear, rust, and corrosion in gasoline engines over SB, SC, and SD oils, plus some anti-wear performance. This level of protection was required for some cars and some trucks beginning in 1971, and continued through 1979.
- The SF classification requires increased oxidational stability and anti-wear performance relative to SE oils. This classification met new-vehicle warranties from 1980 through 1988.
- The SG classification requires increased control of engine deposits, oil oxidation and engine wear relative to the other oils in this series. SG oils exceed the performance of SF and also CC oils, to be discussed below. This classification met new-vehicle warranty requirements from 1989 through 1993.
- The SH classification requires increased control of engine deposits, oil oxidation, engine wear, rust, and corrosion relative to the SG classification. This classification meets new-vehicle warranty requirements beginning in 1994.
- The SJ classification meets the same performance requirements as GF-2, except for fuel economy. An SJ oil that bears the "Energy Conserving" designation meets the GF-2 requirements, including fuel economy. Most manufacturers recommend oils meeting this classification for 1998 vehicles.

As these designations progress alphabetically they increase in levels of protection. Each classification replaces the one before it, with SJ offering the most protection. Just as an SB oil would not provide enough protection for a 1998 engine, an SJ oil would be better for a 1930s engine than the SB oils available for it in its day.

The C-Series Oils

The C-series oils are CA, CB, CC, CD, CD-II, CE, and CF-4. The CA classification requires that the oil protect against bearing corrosion and ring-belt deposits in naturally aspirated (non-supercharged) diesel engines running on fuel with a minimum sulfur content of 0.35 percent. The requirements of this classification date back to the late 1940s. These oils meet military specification MIL-L-2104A.

- The CB designation requires the same protection as CA, but the test engine is run on fuel with a minimum of 0.95 percent sulfur. These oils meet the obsolete "Supplement 1" specification added to MIL-L-2104A, and were used in gasoline engines as well.

- The CC designation requires that the oil provide protection from high temperature deposits in lightly supercharged diesel engines running on low-sulfur-content fuel, and also protection from rust, corrosion and low-temperature deposits in gasoline engines. This designation was introduced in 1961, and meets the MIL-L-2104B specification.

- The CD classification requires oils to provide protection from bearing corrosion and high-temperature deposits in supercharged diesel engines running on fuels of various qualities. Oils meeting the CD classification meet the Caterpillar Series-3 specification introduced in 1955, and also MIL-L-45199.

- The CD-II classification requires the same performance as CD oils, and is designated for two-stroke diesel engines where deposit and wear control are critical. Two-stroke diesels are most commonly used in intercity and highway buses. This classification came into being in 1988.

- The CE Classification also requires the same performance as CD oils, plus passing the engine oil performance tests for the Mack EO-K/2 specification, and the Cummins NTC-400 test for piston deposits and oil consumption. This Classification was introduced in 1987, although oils meeting its individual requirements were already available for some time. CE oils are intended for heavy-duty diesel engines, such as those used in large on- and off-highway trucks, and construction equipment.

- The CF-4 classification requires performance equal to CE oils in many respects, with significantly improved control of oil consumption and piston deposits. This classification is designed for high-speed, four-stroke diesel engines, particularly those used in on-highway heavy-duty

trucks. Introduced in late 1990, CF-4 oils are part of the overall engine design required to meet the 1991 EPA emission regulations affecting heavy trucks. CF-4 oils may be used where CC, CD, and CE oils are recommended.

- The CG-4 classification requires performance exceeding CF-4 oils, with improved control of high-temperature piston deposits, corrosion, foaming, oxidation, and soot accumulation. This oil has been especially designed to meet 1994 U.S. EPA exhaust emission standards, and may also be used in engines requiring API service classifications CD, E, and CF-4. Introduced in 1995, this classification is designed for high-speed, four-stroke diesel engines, particularly those used in highway and off-road applications.

While each S-series classification exceeds the one before it, this is not necessarily the case in the C-series. For instance, the CB tests are run on higher-sulfur fuel than the CC tests. As a result, a CC oil may not offer equivalent protection from the effects of sulfur on the engine as a CB oil. On the other hand, CD and CE oils must pass tests running on high-sulfur—content fuels, and thus would offer similar protection in this area.

Some oils are identified with dual Service Classifications separated by a slash, for example, SF/CD or SG/CE. These oils meet the requirements of both Service classifications shown, and can be used in any engine that calls for one or the other. Many manufacturers specify oils that meet both "S" and "C" Service Classifications; this is particularly true for heavy-duty and turbocharged engines. Where an oil with a dual Service Classification is specified, do not use oils that meet only a single classification or the engine warranty may well be voided.

SAE Viscosity Grades

The **SAE Viscosity Grade**, expressed as a number, refers to an oil's resistance to flow. Typical oil Viscosity Grade numbers are [0W,] 5W, 10W, 15W, 20W, 20, 30, 40, and 50. Lower Viscosity Grade

---

API Service Classification: A system of letters signifying an oil's performance. It is assigned by the American Petroleum Institute.

SAE Viscosity Grade: A system of numbers signifying an oil's viscosity at a specific temperature. It is assigned by the Society of Automotive Engineers.

numbers indicate thinner oils that flow more easily. Higher numbers indicate thicker oils with a greater resistance to flow.

The viscosity of an oil is greatly affected by its temperature. When an oil is cold, its viscosity increases and the oil does not flow as easily. If an oil's viscosity is too high at low temperatures, there will be a lag between the time the engine is started and when oil actually reaches heavily loaded engine parts. When an oil is hot, it thins out and flows quite easily. If an oil's viscosity is too low at high temperatures, the oil film between critical engine parts may break down and allow them to make contact, causing damage.

To make sure oils are capable of providing adequate lubrication, they are tested for viscosity at both high and low temperatures. Viscosity numbers *without* a "W" suffix indicate that the oil meets certain viscosity requirements at 212°F (100°C) only. Grades *with* a "W" suffix meet the requirements at 212°F (100°C) and they also meet certain minimum flow rates at temperatures ranging from 23°F (-5°C) down to -22°F (-30°C).

Oils with only one SAE viscosity number (**single-grade** oils) were once widely recommended. However, you must change the oil with the seasons to ensure that the oil in the crankcase strikes the proper balance between high- and low-temperature operation. Engines require low-viscosity oils in winter, while high-viscosity oils are needed in summer. Today, some manufacturers still recommend single-grade oils in heavy-duty and diesel engine applications.

Most modern oils have dual Viscosity Grade numbers separated by a hyphen (dash) such as 5W-30, 10W-30, 15W-40, or 20W-50. These **multigrade** oils meet the low- and high-temperature specifications for both grades of oil indicated. They contain VI improver additives that allow them to flow well at low temperatures, yet still resist thinning at high temperatures. These properties allow you to use multigrade oils throughout the year with less concern about ambient temperature fluctuations. Figure 6-2 is an example of one manufacturer's gasoline engine oil viscosity recommendations based on ambient temperature. The multigrade oils allow use in the widest range of temperatures.

Vehicle manufacturers today recommend multigrade oils for most of their engines; however, not all multigrade oils are recommended for all cars. For example, General Motors and Chrysler do not recommend 10W-40 oils. Also, most automakers warn against the use of low-viscosity multigrade (and single-grade) oils, such as 5W-20 and 10W, for sustained high-speed driving, trailer towing, or any other

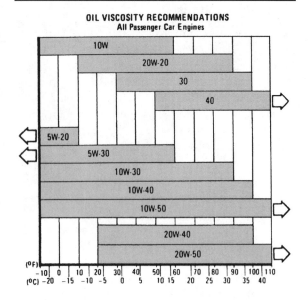

Figure 6-2. Engine oil viscosity recommendations based on ambient temperature. There is much overlapping of the temperatures at which different multigrade oils work.

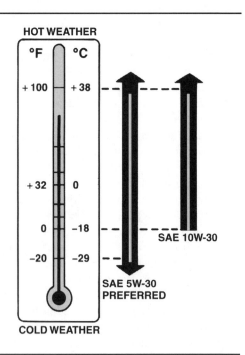

Figure 6-3. The engine oil viscosity recommendations for 1998 Chevrolet car engines shown here are typical of recommendations for many late-model cars.

circumstance in which the engine is placed under constant heavy load. Vehicle manufacturers specify the exact kind of oil to use under certain operating conditions. Figures 6-3 and 6-4 show engine oil charts prepared by two different manufacturers.

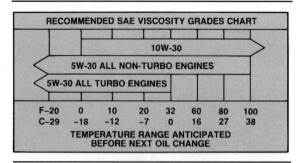

Figure 6-4. Recommended viscosity grades for 1990 Chrysler front-wheel-drive passenger cars. These recommendations reflect the current popularity of lighter oil grades, which promote good fuel economy.

Energy-Conserving Oils

In addition to their Service Classification and Viscosity Grade, the API designates certain oils as "Energy Conserving." Oil makers specially formulate these oils to reduce internal engine friction, and thus improve fuel economy. Laboratories test candidate oils to determine if they really promote fuel economy. Oils achieving a 1.5 percent fuel economy increase may be labelled Energy Conserving. Those achieving a 2.7 percent increase may be labelled Energy Conserving II. These oils may contain friction-reducing additives and usually have a lower viscosity than non-Energy Conserving oils.

Engine Oil Identification

To allow easy identification of engine oils, the API established the Engine Service Classification Symbol, or "donut," figure 6-5. The symbol appears as a label printed on the container of oil. The upper half of the symbol displays the API Service Classifications of the oil, the center of the donut displays the SAE Viscosity Grade of the oil, and the lower half of the symbol contains the words "Energy Conserving" if the oil is formulated to meet those requirements. When selecting an oil, always make sure the oil quality information on the API donut conforms with the vehicle manufacturer's specifications.

Since 1993, oil quality has been further identified by the use of a "starburst" symbol, figure 6-6. This symbol indicates the oil has been certified by the International Lubricant Standardization Approval Committee (ISLAC) as the correct type for gasoline engines in passenger cars and light trucks. To qualify for the starburst symbol, oils must meet the Energy Conserving II requirements. Only multi-viscosity oils with SAE 0W, 5W, and 10W, such as 5W-30 and 10W-30, will qualify.

Figure 6-5. The API Engine Service Classification symbol, or "donut" appears on every container of motor oil. Learn to read this symbol because it gives you essential information about the oil.

Figure 6-6. The Starburst symbol identifies Energy Conserving oils suitable for light duty gasoline engine use.

ENGINE OILING SYSTEM AND PRESSURE REQUIREMENTS

All modern car engines have a pressurized lubrication system. The parts of this system, figure 6-7, are the:

- Oil reservoir and its ventilation
- Oil pump and pickup
- Pressure relief valve
- Filter
- Galleries and lines
- Indicators.

Single-grade: An oil that has been tested at only one temperature, and so has only one SAE viscosity number.

Multigrade: An oil that meets viscosity requirements at more than one test temperature, and so has more than one SAE viscosity number.

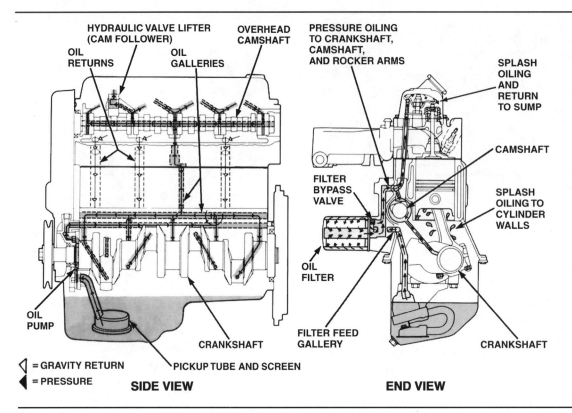

Figure 6-7. Modern engine design uses both pressure and splash methods of oiling. Oil travels under pressure through galleries to reach the top end of the engine. Gravity flow or splash oiling lubricates many parts. A bypass valve is used to prevent oil starvation if the filter clogs.

Oil Reservoir

There must be enough oil in the engine to circulate throughout the system, plus some reserve so that the oil can cool before being recirculated. This oil is kept in the engine oil pan, or sump. Because the oil pan is at the bottom of the engine, the oil drains into the pan after passing through the engine.

The flow of air past the pan when the car is moving helps cool the oil. If this cooling effect is not sufficient, the manufacturer may add an external oil cooler, which is common on diesel, turbocharged, or air-cooled engines, figure 6-8. Heat is removed from oil passing through the cooler by airflow or coolant circulation, depending upon the cooler design.

Engine oil pan capacities vary. Four-cylinder engines usually hold three to five quarts of oil. Most V engines hold four to five quarts. Some high-performance gasoline engines and diesel engines require six or more quarts. Oil is put in the crankcase through a capped oil filler hole at the top of the engine.

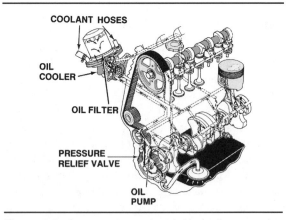

Figure 6-8. A separate coolant jacket allows engine coolant to circulate around the oil cooler of this diesel engine to help reduce oil temperature.

Ventilation

We have seen that blowby gases enter the crankcase from the combustion chambers. The gases can mix with motor oil and cause engine damage. Because the combustion chamber gases are under very high pressure, they increase the pressure within the crankcase. If the crankcase is not vented, the pressure will force

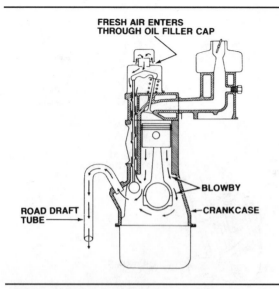

Figure 6-9. Open crankcase ventilation systems used road draft tubes. These systems were efficient at ventilating the crankcase, but added to air pollution.

oil out of the engine at loosely sealed points such as the oil filler cap and the junction of the oil pan and block, among other points. This can be dangerous as a fire hazard or if the oil gets on the brake or clutch materials. To prevent this, the crankcase is ventilated.

On older cars, the oil pan was vented directly to the atmosphere through a **road draft tube**, figure 6-9. Until the 1960s, this was the most common type of crankcase ventilation.

All late-model cars have **positive crankcase ventilation** (PCV) systems, figure 6-10, the first pollution control device. They were required on all California cars in 1964, and were standard nationwide by 1968. These systems are closed to the atmosphere, which keeps blowby from polluting the air. In a PCV system, clean, filtered air is drawn into the crankcase, and crankcase vapors are recycled to the intake manifold. PCV systems also provide better crankcase ventilation than the road draft tube. Since the crankcase is vented to the intake manifold, engine vacuum works to evacuate the crankcase of vapors.

Oil Pump and Pickup

The oil pump is a mechanical device that forces motor oil to circulate through the engine. On most overhead-valve engines and some overhead-camshaft engines, the pump is driven by the camshaft through an extension of the distributor shaft, figure 6-11. On many late-model overhead-camshaft engines, the front of the crankshaft drives the oil pump, figure 6-12. Some overhead-camshaft engines use a separate

■ Dry-Sump Lubrication

The vast majority of engines carry their oil supply in the oil pan or sump at the bottom of the engine. These "wet sump" lubrication systems work very well on most street engines. However, some motorcycles, exotic street cars, and racing cars carry their oil in a separate tank, away from the engine. When the engine sump is freed of oil reservoir duties it is comparatively dry, hence the term "dry-sump."

The key part of dry-sump lubrication systems is an oil pump with two sections. A scavenging section that draws oil from the engine and sends it to a remote tank reservoir and a pressure section that draws oil from the tank and pumps it to the block oil passages. The scavenging section works in direct relation to its shaft speed, the faster it turns, the more oil volume it pumps. The pressure section has a pressure-relief valve to limit the maximum delivery pressure.

With a dry sump system designers can mount the engine much lower in the chassis since there aren't four or more quarts of oil carried beneath it. The oil reservoir is located away from engine heat, reducing oil temperature. Oil control is also improved and this provides the two largest benefits of dry sump lubrication. First, the crankshaft is no longer splashing around in a sea of oil, which reduces oil drag on the crank and can significantly increase horsepower. Second, on wet-sump systems the centrifugal force imposed by high cornering speeds stacks the oil against one side of the oil pan. If the oil is out of reach of the wet-sump oil pickup tube, the system will draw air. The engine will starve for oil while the air bubble works its way through the system. Dry-sump systems avoid this problem and deliver oil to the engine regardless of the cornering loads, improving engine life in racing conditions.

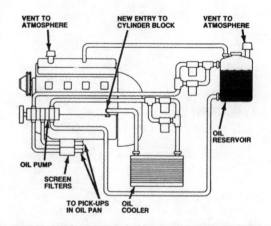

Road Draft Tube: The earliest type of crankcase ventilation; it vented blowby gases to the atmosphere.

Positive Crankcase Ventilation (PCV): Late-model crankcase ventilation systems that return blowby gases to the combustion chambers.

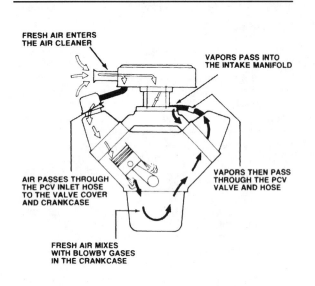

Figure 6-10. Modern PCV systems are closed to the atmosphere.

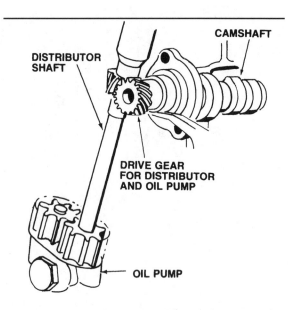

Figure 6-11. This oil pump is driven by the camshaft.

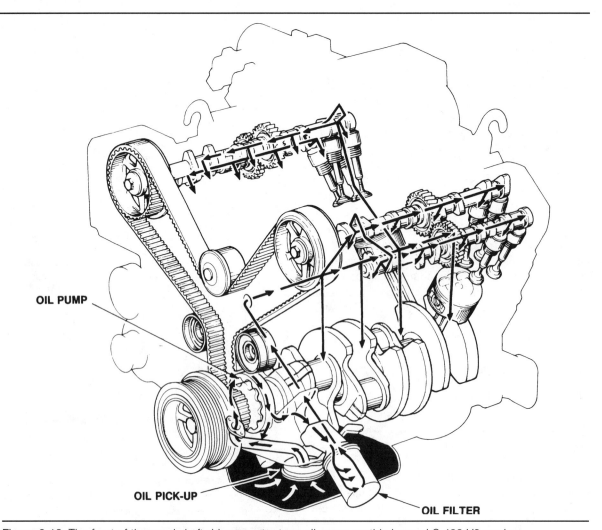

Figure 6-12. The front of the crankshaft drives a rotor-type oil pump on this Lexus LS 400 V8 engine.

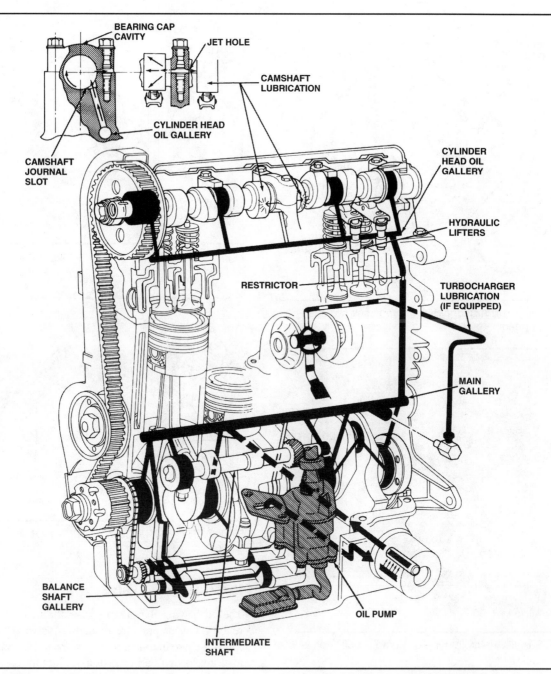

BEARING CAP CAVITY

JET HOLE

CAMSHAFT LUBRICATION

CYLINDER HEAD OIL GALLERY

CAMSHAFT JOURNAL SLOT

CYLINDER HEAD OIL GALLERY

HYDRAULIC LIFTERS

RESTRICTOR

TURBOCHARGER LUBRICATION (IF EQUIPPED)

MAIN GALLERY

BALANCE SHAFT GALLERY

INTERMEDIATE SHAFT

OIL PUMP

Figure 6-13. On this overhead-camshaft Chrysler engine, an intermediate shaft drives the oil pump and distributor.

or intermediate shaft to drive the oil pump and the distributor, 6-13.

Two types of oil pumps are in use today:

- The gear type
- The rotor type.

Traditionally, the most common pump is the gear type, figure 6-14. One oil pump gear is driven by gears and a shaft from the camshaft and is called the drive gear. When the drive gear turns it forces the second oil pump gear, called the idler gear, to turn. As the

two gears turn, the oil between the gear teeth is carried along. At the point where the two gears mesh, there is very little room for oil, so the oil is forced out of the area under pressure.

The rotor type pump works on the same principle of carrying oil from a large area into a smaller area to create pressure. An inner rotor is mounted off-center within an outer rotor, figure 6-15. The inner rotor is driven through gears by the camshaft or off of the nose of the crankshaft as in figure 6-12. The inner rotor drives the outer rotor. As the rotors turn, they

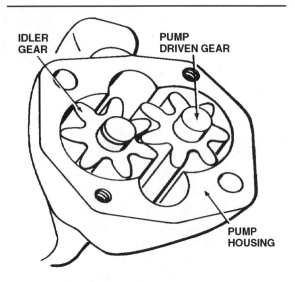

Figure 6-14. A gear-type oil pump uses two spur-type gears.

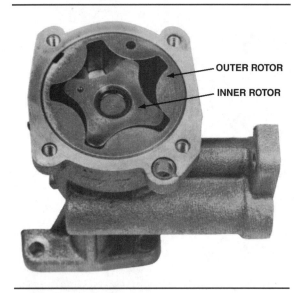

Figure 6-15. A rotor-type oil pump usually contains a single rotor.

carry oil from the areas of large clearance to the areas of small clearance, figure 6-16, forcing the oil to flow from the pump under pressure.

The oil entering the pump comes from the oil pan. An oil pickup tube extends from the pump to the bottom of the pan. A screen at the bottom end of the pickup tube keeps sludge and large particles from entering the oil pump and damaging it, or clogging the oil lines and galleries.

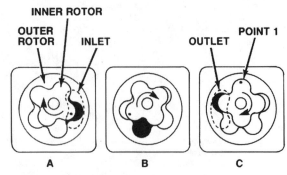

A. Oil is picked up in lobe of outer rotor.
B. Oil is moved in lobe of outer rotor to outlet.
C. Oil is forced out of outlet because the inner and outer rotors mesh too tightly at point 1 and the oil cannot pass through.

Figure 6-16. The operating principle of the rotor-type oil pump.

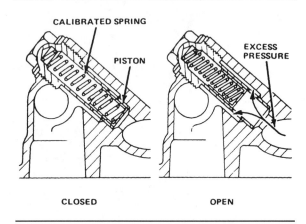

Figure 6-17. Oil pressure relief valves are spring loaded.

Pressure Relief Valve

The oil leaving the pump passes through a pressure relief valve, or pressure regulating valve. The valve limits the maximum engine oil pressure. It consists of a spring-loaded ball or piston set into an opening in the valve body, figure 6-17. Oil pressure forces the ball or piston to move against spring tension and open the hole in the valve body. Oil escapes through this hole to decrease the overall system pressure. The strength of the spring determines the maximum oil pressure the valve permits.

The pressure relief valve is usually built into the oil pump housing. The oil that escapes through the relief hole is sent back to the inlet side of the pump. This arrangement of oil flow from the relief valve reduces oil foaming and agitation so that the pump puts out a steady stream of oil.

Filter

Oil leaving the pressure relief valve flows through the oil filter before reaching the rest of the system. The filter consists of paper or cloth fibers that will pass liquid oil but trap dirt. If dirt plugs the filter, oil flow is restricted.

To prevent a clogged filter from completely stopping oil flow and damaging the engine, the filter or the filter housing contains a bypass valve. It allows oil to flow around the filter element instead of through it when the oil is too thick to go through the filter because of its low temperature, or when the filter outlet pressure drops. You must replace oil filters at specific intervals to prevent clogging. Older cars have replaceable oil filter elements inside permanent housings. Most modern engines use spin-on filters that are completely disposable, figure 6-18.

Galleries, Lines, and Drillways

A modern engine has many areas that must have a constant supply of oil under pressure. To ensure that they do, the lubrication system includes a network of passages that direct oil to these parts of the engine, figures 6-7, 6-12, and 6-13. The main oil passages are called oil galleries, and are cast into the engine block during manufacture. They can also be separate tubes, called oil lines, connected to the engine. Not all engines have separate oil lines, but all have oil galleries.

From the filter, oil flows to a large main gallery. Inline engines and some V engines have one main gallery. Other V engines have two main galleries, one for each bank of cylinders. From the main gallery, smaller passages drilled in the block, called drillways, direct oil to the camshaft bearings and to the crankshaft main bearings. Oil must also reach the crankshaft connecting rod bearings. The manufacturer drills holes through the crankshaft, figure 6-19, so that oil at the main bearings can also flow to the rod bearings.

There are two methods for oiling the cylinder walls. Some engines rely on oil splash created by the crankshaft and connecting rod to throw oil on the cylinder walls, figure 6-7. On other engines the rod bearings caps have small holes that line up with similar holes in the crankshaft and rod bearings. When the holes in the cap and bearing align with the hole in the crankshaft, figure 6-20, a small stream of oil squirts through and hits the lower cylinder wall. This helps lubricate the cylinder wall and piston.

Some turbocharged gasoline engines and many diesel engines have an oil jet that directs a shot of oil directly to the underside of the piston crown, figure 6-21. This spray of oil not only lubricates, but helps

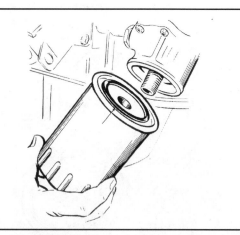

Figure 6-18. Spin-on oil filters are completely disposable and are usually discarded at each oil change.

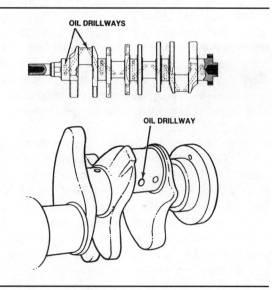

Figure 6-19. Crankshafts have a network of drilled passages so that oil can reach the main and rod bearings.

■ Bypass Oil Systems

Before the full-flow oil filtering system was invented, engines that used oil filters had what is called a bypass system. The filter was mounted on a bracket attached to the engine or anywhere in the engine compartment. Oil lines were connected to a tapped hole in the side of the engine block, and to a drain on the pan or block. The oil was fed under pressure to the filter, and allowed to drain back into the oil pan after filtration.

Because the oil did not have to go through the engine filter before getting to the bearings, a piece of dirt could, theoretically, circulate through the oil system indefinitely until it happened to get into the filter and be trapped. This catch-as-catch-can system was discontinued in favor of the full-flow system.

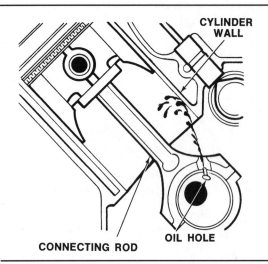

Figure 6-20. Some oil spurts through the connecting rod cap onto the cylinder walls.

reduce piston crown temperatures. Combustion chamber temperatures are always high in turbocharged and diesel engines. The oil spray helps reduce the chance of detonation and preignition in turbocharged gasoline engines and piston damage due to high temperatures in both types of engines.

Different engine designs have different methods of getting oil from the main gallery to the head and valve assemblies. There can be galleries drilled through the block and the head, or the oil can travel through hollow valve train pushrods. Also, oil may flow through an enlarged head-bolt hole. From the valve assembly, the oil drains down through the engine into the oil pan. Engine designers may place the drain holes so that the dripping oil helps lubricate the camshaft.

Indicators

Manufacturers install a dipstick in the oil pan as the primary indicator of engine oil level. Some vehicles use an electronic oil level sensor that indicates when the oil level is low.

To keep the driver informed of engine oil pressure while the engine is running, vehicle manufacturers equip cars with a low oil pressure warning lamp or a gauge that indicates the pressure at all times.

Dipstick

When the engine is at rest, almost all of the oil drains into the oil pan. All engines have a measuring rod, called a dipstick, that extends from the outside of the engine into the pan, figure 6-22. The dipstick has markings on it that indicate the maximum and minimum oil levels for that engine. When you pull the dipstick out of the pan, you can see a film of oil on

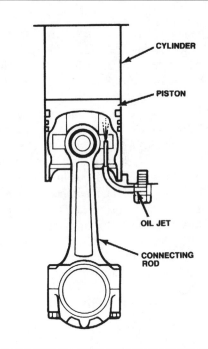

Figure 6-21. Oil jets in diesel and turbocharged engines cool piston crowns.

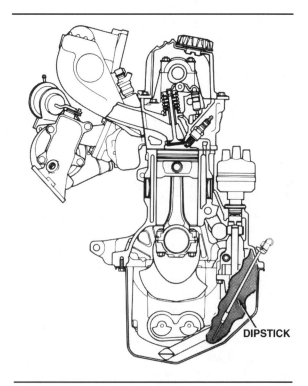

Figure 6-22. The engine dipstick is the main oil level indicator.

the stick. The level of the film relative to the markings on the stick indicates how much oil is in the pan, figure 6-23.

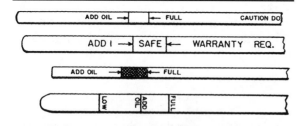

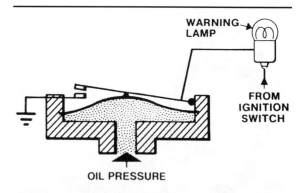

Figure 6-25. The oil pressure switch is connected to a warning lamp that alerts the driver of low oil pressure.

Figure 6-23. Dipstick markings vary between vehicle manufacturers. Here are several different types.

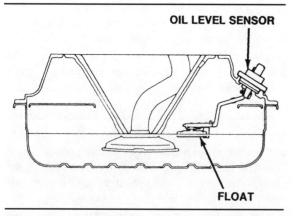

Figure 6-24. A small float in the oil pan indicates the oil level through an electrical sending unit.

Oil level sensor

Some vehicles have an oil level sender that warns the driver when the level is low. A small float and electronic sending unit are installed in the oil pan, figure 6-24. When the level is below a certain level, the float is low enough to complete the electric circuit, illuminating a warning light on the instrument panel.

Oil pressure warning lamp

The oil pressure warning lamp lights when the oil pressure is less than a set level. This happens during cranking and when there is a problem in the lubricating system.

Warning lamps light when a set of electrical contacts close, figure 6-25. The contacts are controlled by a movable diaphragm that is exposed on one side to engine oil pressure. The contacts close when oil pressure is low, and the lamp lights. When pressure increases, the diaphragm moves and the contacts open, turning off the lamp.

The contacts and diaphragm are contained within a sending unit. The sending unit is usually threaded into the side of the engine block so that it reaches an oil passage. The sending unit may be threaded into the oil filter adapter where the filter attaches to the engine.

Oil pressure gauge

Oil pressure gauges operate electrically or mechanically. An electric gauge has a sending unit similar to the warning lamp sending unit. A movable diaphragm varies the current flow through the gauge in proportion to oil pressure. The current flow determines the position of the gauge needle.

Mechanical gauges, figure 6-26, have an oil line running from the engine to a flexible, curved tube inside the gauge. When there is engine oil pressure, oil is forced through the oil line to the tube in the gauge. Oil pressure causes the tube to straighten out and the higher the pressure the more the tube moves. Since the tube is connected to the gauge needle, engine oil pressure determines the position of the gauge needle.

ENGINE LUBRICATION EFFECTS ON MECHANICAL WEAR

When an engine is new, the moving parts fit together very closely. The distance between two moving parts is called clearance. As the engine wears, the clearance increases. After a small amount of wear an engine is considered "broken in." At this point internal friction diminishes, the piston rings seal well against the cylinder walls, and performance improves. When an engine wears far beyond this point, or becomes "loose," it does not operate as efficiently as it once did. When the piston rings no longer seal the combustion chamber well, power is wasted as blowby. Camshaft and valve assembly wear changes valve timing and performance suffers.

The lubrication system helps keep engine wear to a minimum. Using the proper API-rated oil will reduce engine wear because the oil is formulated to match the engine's requirements. Changing the oil and filter at or before recommended intervals will minimize

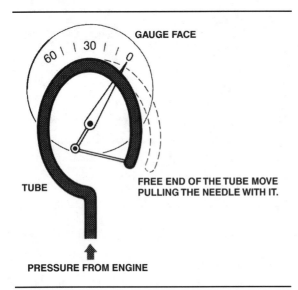

GAUGE FACE

60 | | 30 | | 0

TUBE

FREE END OF THE TUBE MOVE PULLING THE NEEDLE WITH IT.

PRESSURE FROM ENGINE

Figure 6-26. A typical mechanical oil pressure gauge uses a flexible tube that reacts to oil pressure.

engine wear by replacing used additives and by removing harmful dirt and wear particles from circulation.

The SAE oil viscosity rating will affect engine wear and performance as well. If oil with too low a viscosity is used during high-speed or high-temperature operation, the oil film between moving parts may become so thin that it breaks down, allowing metal-to-metal contact and rapid wear.

If oil with too high a viscosity is used during low-temperature operation, the oil may be so thick it is difficult to pump through the oil passages. The result may be inferior lubrication and high wear because not enough oil reaches key critical areas. Also, power will be wasted in forcing engine parts to overcome the

resistance of the thick oil. In extreme cases, a combination of cold temperature and high-viscosity oil will prevent the starting system from cranking the engine at all.

SUMMARY

Motor oil performs five major jobs in an engine: reducing friction, cooling, cleaning, sealing, and absorbing shock. Additives mixed with the oil help it to do these jobs. Two oil rating systems are generally used: the API service classification and the SAE viscosity rating. The API number rates an oil's performance in a laboratory engine, and the SAE viscosity number rates the oil's thickness. An oil may have one or more API ratings. An oil's service classification and viscosity rating must be matched to an engine's requirements for best engine life, performance, and economy.

An engine lubrication system includes the pan, filter, pump, oil galleries or oil lines (or both), dipstick, and pressure warning lamp or pressure gauge. The oil is stored in the oil pan. The pump takes it from the pan, pressurizes it, and sends it through the oil galleries or oil lines to lubricate the moving parts of the engine. The filter traps dirt and other particles that the oil holds in suspension, so the oil and filter must be changed periodically. A dipstick is used to measure the oil level in the pan when the engine is not running, and a pressure warning lamp (for low oil pressure) or a pressure gauge (for continuous pressure reading) is used when the engine is running.

The main purpose of engine lubrication is control of mechanical wear. Excessive wear will hurt performance, economy, and emission control. The use of the proper oil, and regular oil and filter changes, will keep engine wear to a minimum.

Review Questions
Choose the single most correct answer.
Compare your answers with the correct answers on Page 328.

1. Which of the following is *not* a primary job of motor oil?
   a. Cooling the engine
   b. Reducing friction
   c. Reducing exhaust emissions
   d. Cleaning the engine

2. Additives in motor oil can:
   a. Replace qualities lost during refining
   b. Strengthen natural qualities
   c. Add qualities not naturally present
   d. All of the above

3. Blowby gases come from the:
   a. Crankcase
   b. Combustion chamber
   c. Oil filter
   d. Carburetor

4. The oil used in a 1998 passenger-car gasoline engine should have an API service classification of:
   a. SD
   b. CD
   c. CC
   d. SJ

5. The "W" in an SAE viscosity grade indicates that the oil:
   a. Meets viscosity requirements at 0°F
   b. Is for winter use
   c. Has no additives
   d. None of the above

6. Typically, the highest viscosity number given to automotive motor oils is:
   a. SAE 40
   b. SAE 50
   c. SAE 100
   d. SAE 70

7. The term "multigrade" means that an oil:
   a. Has been given two API service classifications
   b. Has many additives
   c. Meets viscosity requirements when both hot and cold
   d. Can be used in gasoline or diesel engines

8. Synthetic motor oils:
   a. May deteriorate at a slower rate than conventional petroleum oils
   b. Cost less than conventional oils
   c. Contain more wax than conventional petroleum oils
   d. Require more frequent drain intervals

9. If the crankcase is not ventilated:
   a. Exhaust emissions will increase.
   b. The oil will foam and not protect the engine.
   c. Too much oil will enter the combustion chambers.
   d. Blowby gases will pressurize the crankcase and force oil out.

10. The oil pump is often driven by the:
    a. Pistons
    b. Camshaft
    c. Fuel pump
    d. Valves

11. The oil pressure relief valve:
    a. Is usually mounted in the oil pump housing
    b. Limits the maximum pressure in the oiling system
    c. Contains a spring-loaded ball or piston
    d. All of the above

12. When an oil filter becomes clogged:
    a. A bypass valve allows dirty oil to lubricate the engine.
    b. The oil pump must slow down.
    c. Engine operating temperature increases.
    d. All of the above.

13. To direct oil to critical points, the oiling system includes:
    a. Internal galleries
    b. Connecting rod squirt holes
    c. Passages drilled in the crankshaft
    d. All of the above

14. Most dipsticks will show you:
    a. What grade of oil to use
    b. If the oil is oxidized
    c. The maximum and minimum oil levels
    d. None of the above

15. Oil pressure warning lamps have a _____ installed in the engine:
    a. Sending unit
    b. Lamp bulb
    c. Oil tube
    d. Pressure gauge

16. Lubrication's greatest effect on an engine is:
    a. Making the engine run cooler
    b. Keeping the compression ratio steady
    c. Controlling engine mechanical wear
    d. Increasing engine horsepower

17. Technician A says that oil pumps use a rotor to force the oil through the oil system. Technician B says that oil pumps use gears for that purpose. Who is right?
    a. A only
    b. B only
    c. Both A and B
    d. Neither A nor B

18. Technician A tells his customers to use the oil recommended the year their engine was manufactured. If SE oil was recommended in 1978 for the customer's Chevrolet Caprice, then the customer may use an SE oil after the engine is rebuilt by a machine shop in 1998. Technician B tells his customers to use the oil with the latest specifications regardless of the year the engine was manufactured. His customer's 1978 Caprice may use SJ oil.

    Who is right?
    a. A only
    b. B only
    c. Both A and B
    d. Neither A nor B

PART TWO

Automotive Engine Construction

7

Cylinder Blocks and Heads

MATERIALS AND CONSTRUCTION

As we explained in Chapter 3, cylinder blocks, heads, and manifolds are usually cast from iron or aluminum. Foundries make the casting by filling a mold with the desired metal. Many aluminum castings are die-cast in permanent metal molds. Most iron castings are sand cast. In a complex iron casting, such as a cylinder block, the foundry fills the interior of the prospective engine with sand cores. After the casting cools, the sand is poured out through the core holes. The interior space becomes the cylinder bores, crankcase, and water jackets, figure 7-1.

The rough engine castings undergo initial machining operations. Next, since new castings tend to warp, they are stored and allowed to "season" for a time. This gives the castings time to fully harden and take final shape. Today, a heat-treating operation speeds the seasoning process. Then, the castings are bored, drilled, tapped, machined to finished size, and assembled, figure 7-2.

Engine covers are cast or made from sheet metal or plastic. Their main purpose is to prevent oil from escaping, but they also aid heat dissipation and may act as oil reservoirs. The front cover of an engine is often cast from aluminum. Oil pans, cam covers, and valve covers may be stamped out of sheet steel or cast in aluminum. Front, valve, and cam covers may also be molded in plastic because they usually do not have to provide any structural strength.

CYLINDER BLOCK DESIGN

The cylinder block is a heavy-duty casting because it is the foundation of the engine, figure 7-3. As we learned in Chapter 1, the automobile cylinder block is usually an in-line, V engine, or horizontally opposed configuration. In-line engines of the past have had two, three, four, five, six, or eight cylinders. In-line threes, fours, and sixes are still made today. In the U.S., the popular in-line eight of the 1930s and 1940s was replaced by the V8, which is shorter, lighter, and stronger. The V8 is still a popular design for luxury and performance automobiles and trucks. However, family cars and economy cars most often use the V6 or in-line 4-cylinder.

In the late 1990s Dodge and Ford introduced V10 engines for light trucks. Dodge also uses a lighter, aluminum version of the V10 engine configuration in its high-performance Viper. The V12 engine, a mainstay of Ferrari, Lamborghini, and Jaguar, has made a resurgence in high-performance and luxury cars built by Rolls-Royce, BMW, and Mercedes-Benz.

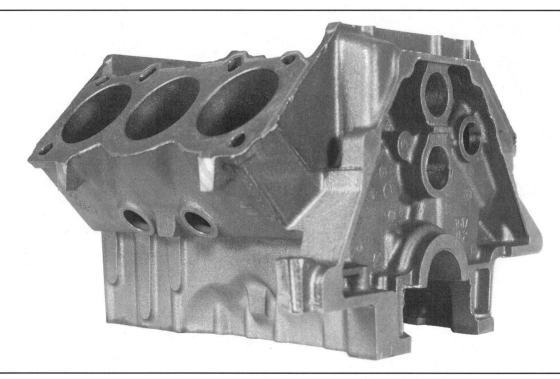

Figure 7-1. A cast-iron cylinder block before the factory has performed any machining operations on it.

Figure 7-2. Quality control technicians inspect a newly machined cylinder block on the Oldsmobile engine line for bore size prior engine assembly.

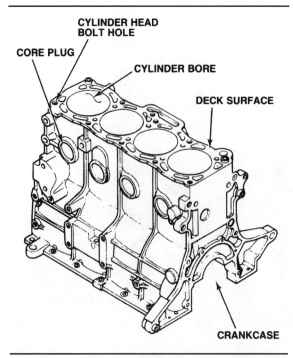

Figure 7-3. The cylinder block usually extends from the oil pan rails at the bottom to the deck surface at the top.

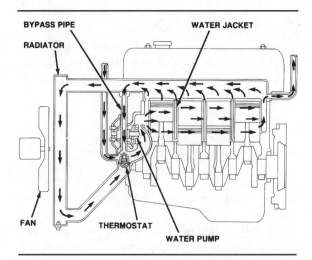

Figure 7-4. Designers try to route coolant flow around the combustion chambers and around the entire length of the cylinders. These are the areas where the engine radiates most of its heat.

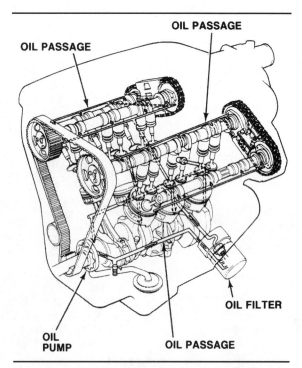

Figure 7-5. The oil is forced under pressure through passages cast or drilled in the cylinder block.

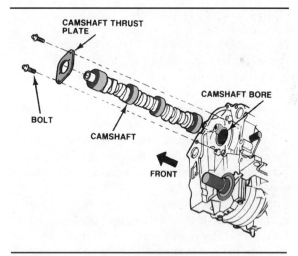

Figure 7-6. This Ford 3.8-liter V6 carries its camshaft in the block, like other overhead-valve engines.

As we have already seen in Chapter 4, water jackets are cast into the blocks and heads of most engines. The water jacket surrounds the cylinders and combustion chambers to provide heat dissipation, figure 7-4. Oil passages, figure 7-5, are cast or drilled into the block, as we showed in Chapter 6.

Overhead-valve and flathead engines carry their camshafts in the cylinder block, figure 7-6. The camshaft usually spins in three or four tubular, one-piece bearings. We detail camshafts in Chapter 9.

Cylinder Block Size

Blocks must be designed as compactly as possible. A large engine is not only heavy, but requires a larger engine compartment which increases the size and weight of the chassis. For this reason, only enough space is allowed inside the crankcase for the parts to

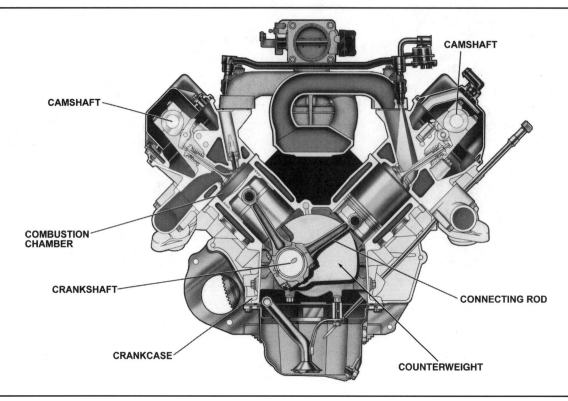

Figure 7-7. The crankshaft counterweights require extra crankcase space. The strongest geometric form is one that is supported on three sides.

move without interference. The room needed by the crankshaft is determined by the length of the piston stroke. A long stroke engine has a longer crankshaft throw, and therefore needs a bigger crankcase.

Crankshafts are usually counterweighted to balance the weight of the crankpin, connecting rods and pistons. The **crankshaft counterweights** are up when the pistons are down, so the designer must allow additional clearance within the crankcase to accommodate them, figure 7-7.

Another critical dimension is the thickness of the cylinder walls. If they are too thick, they will retain heat and the engine will run hot. If too thin, they will be weak and distort or warp when heated and cooled.

The block must be heavy enough to withstand all the clamping and pulling forces that are imposed on it by the engine parts and accessories. Everything that mounts on the engine is bolted to the block or the heads. The heads, of course, are bolted to the block. The block is not only the foundation of the engine, but also its framework, figure 7-8.

Ribs and Webs

Designers add ribs and webs to reinforce both aluminum and iron engine block castings, figures 7-9 and 7-10. Ribs and webs add strength and rigidity, without

adding much extra weight, by tying together opposite ends of a structure. Ribs and webs often work from the principle of triangulation. The strongest geometric form is one that is supported on three sides.

The Crankcase

Most engine blocks are cast in one piece. The cylinders, oil passages, and water jackets usually form the top half of the block and the main bearings and oil pan make up the crankcase area or the bottom half. This method of construction is the lightest way to build an engine and modern metallurgy and foundry technology make these engines strong enough for their intended purpose.

However, many older engines and racing engines have crankcase castings separate from the cylinder portion of the block. The extra crankcase castings require more block material but improve strength. Some designers are returning to this type of engine construction because additional crankcase castings also help form an extremely rigid structure that minimizes vibration and engine noise. The Volvo 3.0-liter in-line six features an additional crankcase casting, figure 7-11. The SR 4-cylinder engine Nissan used in the Sentra and Infiniti G20 has a girdle that ties the main bearings together, stiffening the block, figure

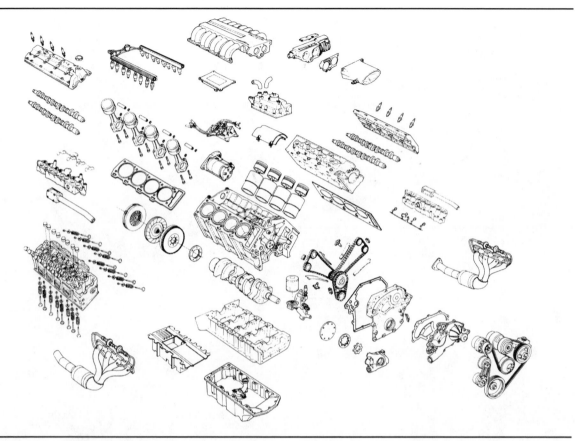

Figure 7-8. The block is the framework of the entire engine. A technician always begins engine assembly with the block. The main bearings, crankshaft, and pistons are installed first. Then the cylinder heads and intake manifold are bolted on.

7-12. The 1998 Ford 4.6-liter V8 has a cast, one-piece oil pan to add rigidity.

Some engines do not have a cylinder block at all, at least in the usual sense that we describe it. The Volkswagen and Porsche horizontally opposed engines use the crankcase as their foundation, figure 7-13. These engines are air-cooled and have individual finned cylinders bolted to a common crankcase. Most air-cooled motorcycle engines are also built this way.

Main Bearings

The main bearings are two-piece inserts clamped in a housing that supports the crankshaft, figure 7-14. They must give support without interfering with the rotation of the crankshaft or the action of the connecting rods. At the bottom of the cylinder block, webs are

Crankshaft Counterweights: Weights cast into, or bolted onto, a crankshaft to balance the weight of the piston, piston pin, connecting rod, bearings, and crankpin.

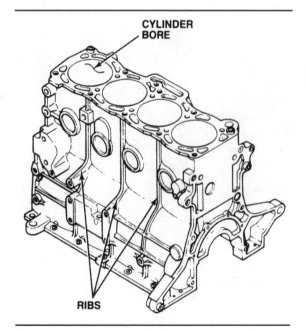

Figure 7-9. The ribs on this Mazda cylinder block run the length of the block, structurally tying the top and bottom of the block together.

Figure 7-10. The webs on this Pontiac in-line 4-cylinder engine block form a triangle. They structurally tie thin portions of the casting to the main cylinder block.

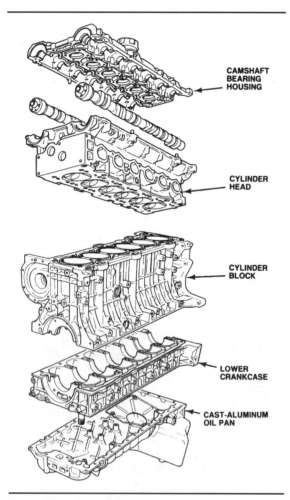

Figure 7-11. The one-piece lower crankcase section of Volvo's 3.0-liter in-line six-cylinder engine and the cast-aluminum oil pan greatly improve block rigidity.

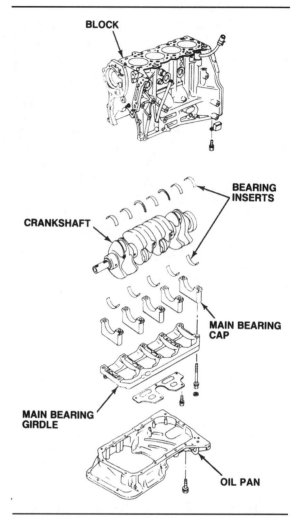

Figure 7-12. The Nissan SR four-cylinder engine has a beam that when bolted in place ties the main bearing caps together, increasing block rigidity.

cast to support the bottom half of the main bearings, called the main bearing saddles, figure 7-15.

The remaining half of the main bearing is a bolt-on cap. Caps are usually made of a stronger material than the material the block is made from. Manufacturers commonly put nodular-iron or steel caps on gray iron or aluminum blocks. Main bearing caps attach to the block with two bolts, figure 7-16, or four bolts for more strength, figure 7-17. In addition to the usual vertical bolts, designers may add cross bolts for extra strength, figure 7-18. Cross bolts pass through the sides of the block and thread into the sides of the bearing caps. With more bolts the caps and block can withstand more crankshaft downward pressure without failure. On some engines, all the main bearing caps are joined into one rigid piece, figure 7-19, to increase bottom end strength. The shell-type bearing

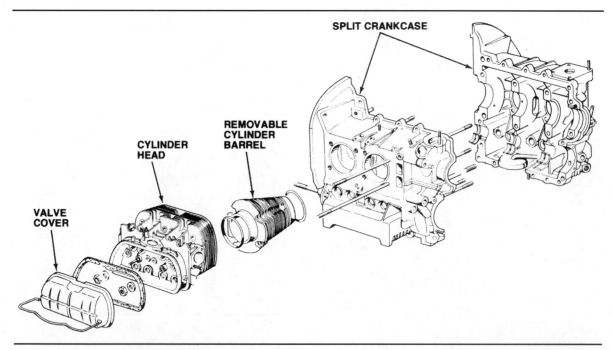

Figure 7-13. Volkswagen flat-four engines are built on a split crankcase.

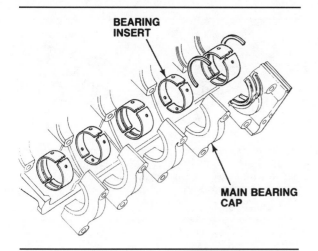

Figure 7-14. The main bearings support the crankshaft.

Figure 7-15. The webs that support the main bearing saddles are cast into the engine block. This V8, like most V8s, has five main bearings.

inserts that fit between the saddles and caps are detailed in Chapter 13.

Main bearing number

Some very early engines had only two main bearings for the crankshaft. However, designers have found that putting a main bearing on both sides of each connecting rod results in longer engine life. A 6-cylinder in-line engine would have seven main bearings in this configuration and an in-line 4-cylinder would have five, figure 7-20. Some in-line 4-cylinder engines have only three main bearings.

Because of the practice of pairing connecting rods on each crankshaft throw in a V engine, main bearings are usually placed on both sides of every rod pair. This results in a crankshaft with five main bearings for a typical V8 engine and four for a typical V6.

Although most engines could get by with the minimum number of main crankshaft bearings, using the maximum number of bearings allowed by the engine configuration gives the structural rigidity necessary for long engine life.

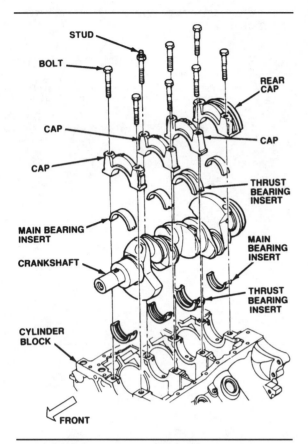

Figure 7-16. Two-bolt main bearing caps provide adequate bottom end strength for most engines. This V6 engine block features two-bolt caps for each of its four main bearings.

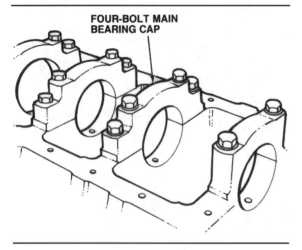

Figure 7-17. High-performance engines and truck engines often have four-bolt main bearing caps for greater durability.

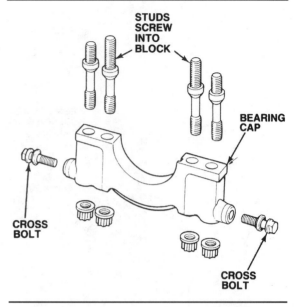

Figure 7-18. This Lexus V8 main bearing cap features four studs and nuts to attach it to the block. Cross bolts pass through the sides of the block for additional bottom-end strength.

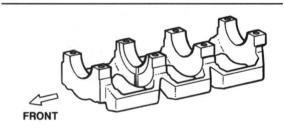

Figure 7-19. The Lexus ES 250 and Toyota Camry V6 use an integrated main bearing cap that improves main bearing strength.

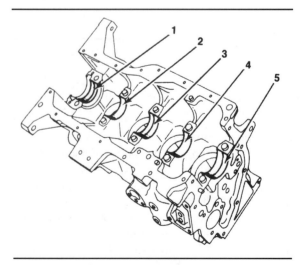

Figure 7-20. This Peugeot 4-cylinder engine has five main bearings.

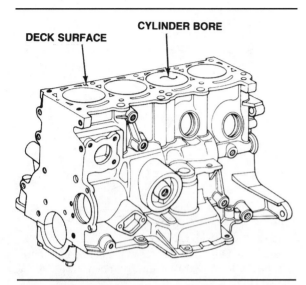

Figure 7-21. Cylinders are part of the cylinder block in most water-cooled engines.

Cylinders

The cylinders are part of the block casting in a water-cooled engine, figure 7-21. The tops of the cylinders are even with the block deck surface. Most air-cooled engines have individual cylinders that bolt to the crankcase, figure 7-22.

In most cases, the rough cast cylinders are bored and honed to a smooth finished size so that the piston will fit with the proper clearance. The rings wear away the cylinder bore. When the cylinder wall becomes tapered from the ring wear, a machinist bores it out to fit the next larger size piston.

Siamesed cylinder bores

Some engines are built with **siamesed cylinder bores**, figure 7-23. The cylinder walls of adjacent cylinders are cast attached to each other. Therefore, there is no water jacket between the cylinders, a drawback to efficient cooling. Engines with siamesed cylinders require the manufacturer to drill extra steam holes in the deck surface, figure 7-24, and matching holes in the cylinder head. These holes reach through to the water jackets, creating an extra flow path. The holes aid water circulation at low rpm and vent steam and air pockets. The steam holes would not be necessary if the engine ran only at higher rpm because increased water circulation purges the air and steam pockets. Engines with siamesed cylinders have relatively thin cylinder walls between adjacent bores, so they do not usually allow more than one overbore before the block is sleeved or replaced.

Despite the drawbacks, there are two reasons to build an engine this way. First, it allows the engine designer to enlarge the bore of an existing engine, yet

■ Cylinders in a Circle

Although we commonly think of engines with their cylinders arranged side by side, there isn't any reason for limiting an internal combustion engine to a longitudinal design. Aircraft engine designers commonly specified radial engines — those in which the cylinders protrude around a central axis like the spokes of a wheel. Radial engines powered many civilian and military aircraft throughout the propeller-powered airplane era, and no longer appear today only because the modern turboprops are so much more economical for their power.

Radial engines can be designed with any number of cylinders, although the fewest you will usually ever see is five. During the Second World War, U.S. radial engines used as many as nine cylinders, sometimes in double rows for a total of 18. The BMW 803 engines had 28 cylinders; two 14-cylinder units mounted in-line, each containing two banks of seven cylinders. Its two counter-rotating propellers pumped out 960 horsepower.

Radial engines had an extremely unusual connecting rod arrangement. Because the cylinders were concentric, all the connecting rods had to attach at the same throw on the crankshaft. Rather than lengthen the journal to accommodate all the rods, the designers mounted a single extra-large master rod on the crankshaft. Then, bolted smaller link or articulator rods to the big end of the master rod. As many as nine connecting rods therefore shared the same big end bearing support on a single journal. The crankshafts for an eighteen-cylinder radial aircraft engines might be no longer than those for a two-cylinder motorcycle.

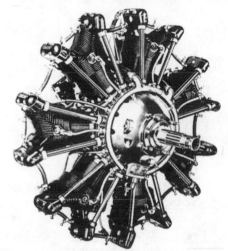

This Jacobs radial air-cooled engine featured seven cylinders, a one-piece forged-steel master rod, and forged-aluminum link rods.

Siamesed Cylinder Bores: A bore design where the cylinder walls of adjacent cylinders are attached to each other. This design prevents the water jacket from going all the way around the cylinders.

Figure 7-22. The Volkswagen engine uses four air-cooled, individual cylinders like these.

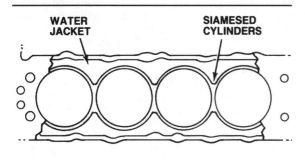

WATER JACKET   SIAMESED CYLINDERS

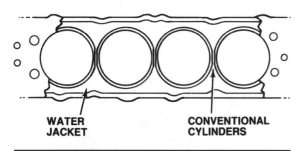

WATER JACKET   CONVENTIONAL CYLINDERS

Figure 7-23. You must cut the decks off the blocks to compare siamesed cylinders with conventional cylinders. Siamesed cylinders have no water jackets between them. On conventional cylinders coolant flows around the entire cylinder.

keep the bore centers the same as the smaller, original engine. It can then be machined by the manufacturer on existing tooling. The cylinder block does not have to be any longer, so it fits into existing cars. The block is also no heavier, but can provide considerably more displacement. Chevrolet took this approach when it built the 400 cubic inch (6.6-liter) small-block engine, used from 1970 to 1976. It allowed the engineers to increase the cylinder bore size from 4.0 inches (102 mm) to 4.125 inches (105 mm) and retain the same block. Second, the siamesed bores add strength and rigidity to the block.

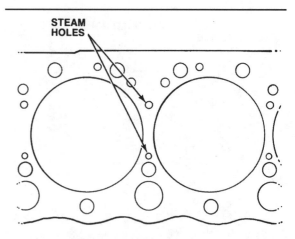

STEAM HOLES

Figure 7-24. Extra steam holes in the block and heads provide better cooling for the siamesed cylinders at low engine speeds.

Cylinder Sleeves

Cylinder sleeves, also called cylinder liners, are either a standard engine part or used to make a repair. In some engines, the cylinders are originally designed to accept a sleeve. When the sleeve wears out, a machinist replaces it. In theory, the block should last forever because installing new sleeves makes the cylinder bore like new. This feature is valuable in high mileage industrial or truck engines which would, otherwise, have to be replaced every year.

In engines that are not originally equipped with sleeves, a machinist may bore the block to accept a sleeve that will work with a new, standard size piston. Machinists make this type of repair if the original cylinder is too badly worn to successfully rebore (assuming that the cylinder block is worth saving otherwise), or a new block is unavailable.

Types of sleeves

There are three kinds of automotive cylinder sleeves. The two most common types of sleeves are cast iron, while the third is a carbon fiber composite. A cast-iron **dry sleeve** does not contact the engine coolant, figure 7-25. Some aluminum cylinder blocks have iron, dry sleeves cast in as a sliding surface for the pistons. A dry sleeve may also be used as a repair in cast-iron blocks.

The **wet sleeve** is so named because it does contact the coolant. Many aluminum block car engines and most large diesel trucks have pressed-in, iron wet sleeves. A rubber or copper seal at the bottom of the wet sleeve keeps coolant from leaking into the cylinder, figure 7-26. Wet sleeves are either screwed into the block or use a retaining device at the top of the block to hold the sleeve in place, figure 7-27. The head gasket keeps the coolant from leaking into the cylinder at the top.

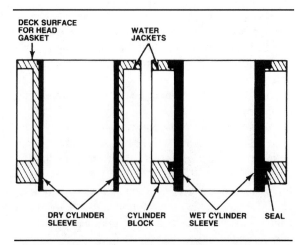

Figure 7-25. A dry sleeve is supported by the surrounding cylinder block. A wet sleeve must be thicker to withstand combustion without total support from the block.

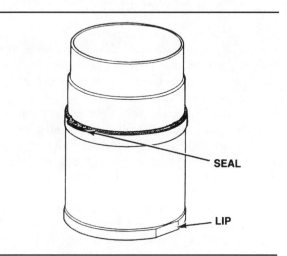

Figure 7-26. This Peugeot cast-iron wet cylinder sleeve uses a rubber seal.

The third type of sleeve is a carbon fiber composite, invented by Honda and first used in the 1991 Prelude. It is made of carbon and alumina fibers and cast into the Prelude's aluminum block. Honda allows one overbore and sells oversize pistons in 0.010 inch (0.25 mm) size.

Sleeveless cylinders

Reynolds Aluminum developed a method of making aluminum cylinder blocks without cast-iron sleeves in the late 1960s. Iron-plated aluminum pistons slide directly in the bores of these high-silicon-content aluminum cylinder blocks. The 1970 Chevrolet Vega was the first passenger car to use the aluminum-silicon bores. Although the Vega went out of production after 1977, Mercedes-Benz and Porsche still use this method of cylinder construction because it results in a

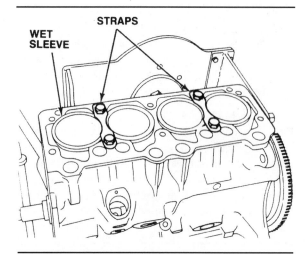

Figure 7-27. Metal straps held down by bolts secure the wet sleeves in this Peugeot block.

very lightweight block. The silicon goes very deep in the aluminum and both M-B and Porsche allow overbores and provide two sizes of oversize replacement pistons. These blocks may also be renewed by installing dry, iron cylinder sleeves. We detail aluminum-silicon cylinder construction in Chapter 3.

Block dimensions

The major axis, or datum, line for the engine is the centerline of the crankshaft. Cylinder bores must be perpendicular to the crankshaft centerline. The centerline of the camshaft and the gasket surface at the top of the cylinders must be parallel to the crankshaft. Ideally, a machinist should bore the cylinders and deck the surface of the block with the equipment indexed in relation to the crankshaft. And many machines do locate the cylinder block this way. This ensures that the cylinders are bored and block deck resurfaced parallel to the crankshaft.

The second best, but acceptable method indexes the boring or decking equipment on the oil pan rail at the bottom of the block. In theory, these surfaces were originally made parallel to the crankshaft at the factory. Therefore, cylinders bored and blocks decked in

Cylinder Sleeve: The liner for a cylinder which provides a good surface for the piston rings. Some sleeves can be replaced when worn to provide a new cylinder surface.

Dry Sleeve: A sleeve that does not come in direct contact with the engine coolant.

Wet Sleeve: A sleeve that comes in contact with the engine coolant.

relation to the pan rail will be perpendicular to the crankshaft.

A third type of machinery locates the block in relation to the deck or gasket surface. This type of equipment was used for many years but is not generally accepted as being accurate enough for engines today.

Crankshaft offset

Designers plan most engines to rotate clockwise as you look at them from the end opposite the flywheel. The design blueprints of some in-line engines show that the centerline of the crankshaft main bearings does not line up with the centerline of the cylinders. When looking at the front of the engine, the crankshaft may be offset slightly to the left (the passenger side of the car). Not all engines use the offset, but its purpose is to reduce the angle of the connecting rod during the power stroke.

To understand this, look at the illustration of the crank throw that is halfway through the power stroke, figure 7-28. The piston is being pushed down by the burning mixture, but the rod is at such an extreme angle that the piston is also being forced into the side of the cylinder. If the crankshaft is offset, the angle of the rod is reduced. This lets the piston push without as much side thrust, and there will be less pressure on the connecting rod bearing and cylinder wall.

When designers first built engines with crankshaft offsets, they were as great as three-quarters of an inch (17 mm), but modern engines use an offset of only about $\frac{1}{16}$th of an inch (1.5 mm), or none at all.

Basically the same effect, reducing side thrust, can be achieved by piston pin offset, the practice of offsetting the connecting rod's link to the piston. We detail this design feature in Chapter 12, the chapter that focuses on pistons.

Cylinder offset

V engines require a cylinder offset because of crankshaft design. Since two connecting rods share each crankshaft throw, designers shift one bank of cylinders in front of the other, figure 7-29. The distance one bank of cylinders moves in relation to the other is not great — only about the width of a connecting rod. Shifting the cylinders this way centers the cylinders over each rod and piston.

CYLINDER HEADS

Cylinder heads for overhead valve and overhead camshaft engines do much more than provide a lid for the combustion chamber. They also provide a place for the spark plugs, support the rocker arms of the valvetrain, provide guides and seats for the valves, and in the case of overhead-camshaft heads, furnish a place for the camshaft, figure 7-30.

Passages for the intake and exhaust gases go through each head to the intake and exhaust valves. The flanges around the exhaust and intake ports must support the weight and clamping action of the intake and exhaust manifolds. In effect, overhead-valve and overhead-camshaft cylinder heads are the upper half of the engine.

An entire book could easily be written on cylinder head design (and many have!). Not all is known about cylinder head design, especially of the combustion chamber and port. This is where the greatest advances in engine efficiency have come and will continue to come for some time. In the following sections of this chapter we will touch on the important aspects of combustion chamber and port design, but this is by no means an exhaustive study. Later chapters of this book will cover the valves, camshafts, and other parts of the valvetrain and cylinder head.

Combustion Chamber Design

The combustion chamber consists of the hollowed out portion of the cylinder head, the upper edge of the cylinder wall, and the top of the piston, figure 7-31. Combustion chamber design affects engine breathing and combustion efficiency, which help determine power, fuel economy, and exhaust emissions. Engineers have worked with combustion chamber design since the automobile was first invented. It is probably the most important part of the engine for the designer and the way in which engines differ the most from one another.

In the years prior to 1967, before exhaust emission controls, much of the experimentation and combustion chamber design work was done with racing engines in an effort to make them go faster. The need to reduce exhaust emissions, however, refocused attention on the combustion chamber. Efforts were made to promote rapid, uniform burning of the air-fuel charge to control emissions and improve fuel economy.

Combustion of the air-fuel charge in a cylinder is not an instantaneous explosion, but rather, a controlled burning of the charge by a flame started by the spark from the spark plug. When the spark ignites the air-fuel mixture, a flame front spreads out across the combustion chamber to consume the mixture. The time it takes the flame front to consume the charge is called burn time and requires only a few milliseconds.

However, combustion chamber design, high engine temperature, pressure, or poor gasoline quality can cause an unwanted, violent explosion of the air-fuel charge. A single one of these factors or a combination of them may cause the abnormal combustion, detonation or pre-ignition, which we discussed in Chapter 2.

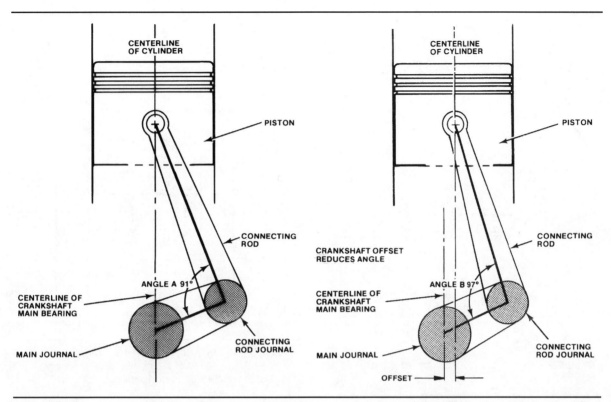

Figure 7-28. The crank throw is halfway through the power stroke in both illustrations, and the pistons are being pushed down by the burning mixture. A lack of crankshaft offset produces rod angle A. This sharp angle forces the piston against the cylinder with side thrust. Crankshaft offset produces rod angle B. This angle causes less piston side thrust.

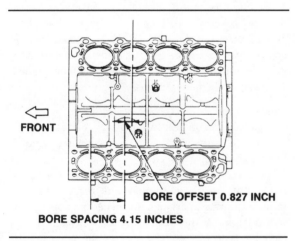

Figure 7-29. The Lexus LS 400 V8 has its cylinder banks offset by less than an inch.

In the ideal combustion chamber design, the entire air-fuel charge would burn completely, leaving no unburned areas to be exhausted and eliminating the possibility of detonation. In actual practice, however, there is always some part of the mixture that does not completely burn.

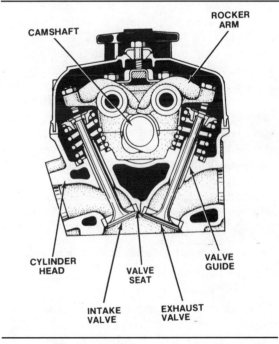

Figure 7-30. The seats and the guides for the valves are in the cylinder head.

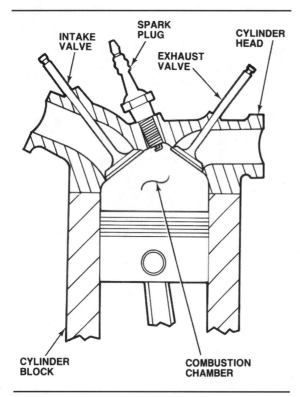

Figure 7-31. The hollowed out portion of the cylinder head makes up most of the combustion chamber. The piston top seals the bottom of the chamber.

Combustion efficiency and breathing ability

A combustion chamber designer must balance combustion efficiency with breathing ability to produce the best head for its intended purpose. Sometimes an element of combustion chamber design that improves combustion efficiency harms breathing ability and vice versa.

Most design factors affect both combustion and breathing, at least to a small degree. However, we can point to certain factors as being more important in one area than the other. The combustion chamber design factors that most affect the efficiency of combustion are:

- Squish area
- Quench area
- Spark plug placement
- Surface-to-volume ratio.

The factors that most affect breathing ability are:

- Valve placement
- Valve number.

Other factors such as combustion chamber shape and valve shrouding affect *both* combustion efficiency and breathing ability *equally*. Port design has probably the greatest affect on an engine's breathing ability, but it is not part of the combustion chamber, so we detail it later in this chapter.

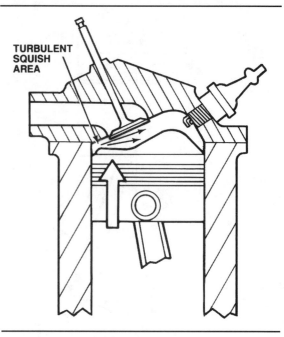

Figure 7-32. The squish area is usually created by a very shallow portion of the combustion chamber. A flat part of the piston goes up to meet it.

Squish area

The **squish area** is a region in the combustion chamber where the piston very nearly contacts the cylinder head, figure 7-32. As the piston nears TDC, the air-fuel mixture is rapidly pushed out of the squish area, causing turbulence. Turbulence eliminates dead pockets and mixes the air and fuel, ensuring more uniform and complete combustion. The squish area also helps push the mixture back toward the spark plug in some combustion chamber designs.

Quench area

The squish area may also double as a **quench area**. As you'll remember from our discussion in Chapter 2, knocking or detonation can occur if the end gases reach a very high temperature and ignite before the flame front reaches it. To prevent detonation, the area farthest from the spark plug quenches or cools the air-fuel mixture. As the piston comes very close to the cylinder head, the excess heat of the end gases is quenched or drawn into the relative coolness of the metal.

The engineer may design the squish-quench area to provide less squish or quench in engines where low emissions are important. Emission-controlled engines need squish for efficient combustion, but too much quench can cool the end gases and leave some of the gases unburned, raising hydrocarbon emissions. To prevent this, a designer may increase the distance between the top of the piston and the cylinder head in

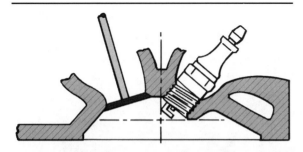

Figure 7-33. Central location of the spark plug in the combustion chamber reduces the distance the flame front must travel to reach the edges of the chamber.

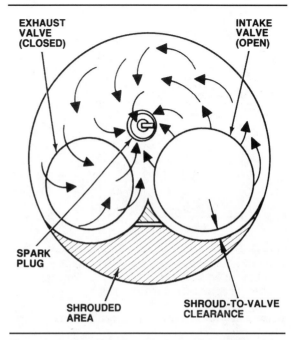

Figure 7-34. Masking the area around the intake valve with more metal causes the intake mixture to swirl as it enters the combustion chamber.

the squish area at TDC. This design would still provide some squish and turbulence without risking over quenching the end gases.

Spark plug placement

The best place for the spark plug is the center of the combustion chamber, figure 7-33. The closer it is to the center, the shorter the flame travel to all edges of the chamber. With a center-mounted plug, resistance to detonation and preignition is greatest. Some combustion chamber designs allow center-mounted spark plugs, but others do not because of chamber shape, valve placement, and valve size.

Surface-to-volume ratio

The surface-to-volume ratio is important in combustion chamber design. A typical surface-to-volume ratio might be 7.5:1. This means that the surface area, divided by the volume, is 7.5. If the surface-to-volume ratio is too high, there will be a lot of surface to which fuel can cling. The fuel next to the walls of the chamber may not burn completely because the chamber walls cool the mixture below ignition temperature. When the exhaust valve opens, the unburned fuel goes out the exhaust. Unburned fuel includes hydrocarbons, a contributor to smog. So, there is a direct relationship between the surface area of the combustion chamber and the number of unburned hydrocarbons. To minimize the amount of unburned hydrocarbons, the designer must keep the surface area as small as possible. This is done by eliminating ripples and angles, but at the same time keeping a shape that is easy to manufacture.

Valve shrouding

Designers say valves are shrouded or masked when the walls of the combustion chamber are very close to the valve heads. The shrouded valve area in the combustion chamber, figure 7-34, causes the intake mixture to swirl and increases mixture turbulence as the piston moves down in the cylinder. The result is more complete combustion. However, shrouding around the intake valve reduces breathing ability by restricting

intake flow. The incoming mixture has a harder time getting around the valve and into the chamber because the shrouding restricts its path. The shrouded areas also reduce breathing by limiting the combustion chamber room available for valves, limiting valve size.

Valve placement

Valve placement in the cylinder head is an important factor in breathing efficiency. Probably the best example of a poor breathing valve placement is in the flathead engine, figure 7-35. The incoming mixture must make a 90-degree turn to get to the piston. After combustion, the exhaust has to make another 90-degree turn to get out.

Placing the valves overhead allows the port to be a much straighter shot to the combustion chamber. However, if the valves are placed side by side, as they

Squish Area: A narrow space between the piston and the cylinder head. As the piston approaches TDC, mixture squishes or shoots out of the squish area across the combustion chamber. Mixture turbulence results and improves combustion.

Quench Area: An area in the combustion chamber that has only a few thousandths of an inch clearance from the piston at TDC. The close proximity of the cool cylinder head and piston crown prevent the end gases between them from becoming hot enough to autoignite.

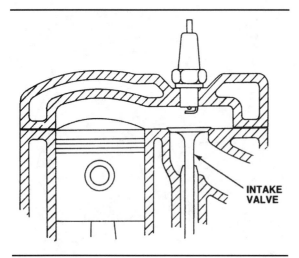

Figure 7-35. In a flathead, the valves are placed next to each other in the cylinder block. The camshaft pushes them up to open them.

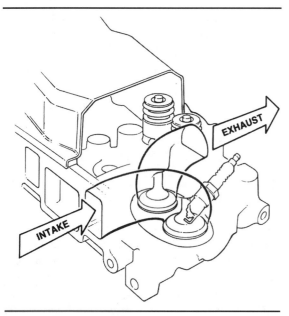

Figure 7-36. Overhead-valve cylinder heads allow a better route for the intake mixture into the combustion chamber. However, if the valves are placed side-by-side, mixture scavenging during overlap is not as effective.

are in many OHV engines, the mixture enters in a straight line, but the exhaust must make a turn before it can exit, figure 7-36.

The best valve placement for breathing efficiency is with the intake and exhaust valves overhead and on opposite sides of the combustion chamber. This design furnishes an easy path from the intake port through the combustion chamber to the exhaust port, figure 7-37. This is of primary importance during the camshaft's overlap period, which we discuss in

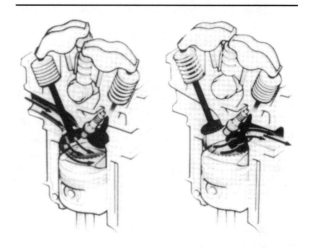

Figure 7-37. Valves opposite each other in the combustion chamber allow good breathing, especially during camshaft timing overlap.

Chapter 9. Briefly stated, during overlap, the intake valve opens before the exhaust valve has closed. The exhaust gases exiting the exhaust valve create a pressure drop in the cylinder that helps pull in fresh mixture through the intake valve.

When the intake and exhaust ports are on opposite sides of the head, the design is known as a **crossflow cylinder head**, figure 7-38. The incoming mixture also tends to cool the exhaust valve, increasing the exhaust valve's life. However, an in-line engine with a crossflow head needs more engine compartment room, and its greater complexity makes it more expensive to produce.

On the crossflow design, both sides of the cylinder head must be machined, whereas on the design with intake and exhaust valves on the same side of the head, machining is simpler. In spite of this, new engines are appearing with the crossflow design because it is must more efficient and it lends itself to transverse positioning of a 4-cylinder engine. With the cylinders running in-line in a transverse position, the intake fits easily at the front and the exhaust at the rear, or vice versa.

Valve number
An engine must have at least two valves per cylinder, one for intake and one for exhaust. However, additional valves per cylinder provide definite breathing

Crossflow Cylinder Head: A head with the intake port and exhaust port exiting from opposite sides of the head. The design provides an almost straight path for the mixture to flow across the top of the piston at overlap, for scavenging.

■ Plus-One Performance (If Four Valves Are Good, More Must Be Better)

Heads with four valves per cylinder have been around a long time in motorcycles and automobile racing engines. Although the 4-valve design was refined in racing applications, it has appeared in production vehicle engines only in the past decade.

The ideal combustion chamber configuration would be a 2-valve head with a high compression ratio and valves large enough to allow enough air to flow through to produce power at high rpm — if the combustion chamber could be a perfect sphere, with the spark plug in the center. Unfortunately, this combustion chamber, while it sounds simple, would leave no room for a cylinder or a piston to compress the mixture. Designing the perfect combustion chamber isn't simple.

The pentroof combustion chamber with two intake valves on one side and two exhaust valves on the other is a fine design but may not be the ultimate. Yamaha has built a 7-valve head with three intake and three exhaust valves around the outside of the

chamber, with a fourth intake valve in the center between two spark plugs. While it performed well, it was too complex and expensive to manufacture and sell, so the design was downgraded to six valves and one spark plug. Power output and fuel consumption were still good, but valve train complexity remained, and cooling the center exhaust valve proved difficult.

Removing the center exhaust valve solved the cooling problem, and opening each valve with its own cam lobe and bucket-type tappet let the cam make greater surface contact with the lifter. While the total valve area of the 5-valve head is less than that of four valves, the effective valve perimeter around the three intake valves is about 14 percent greater. This increased air intake allows higher power output.

Maserati has applied multiple valve technology to its 2.0-liter Biturbo V6 for an amazing 43-percent gain in power — increased from 180 horsepower (134 kW) at 6,000 rpm to 243 horsepower (181 kW) at 7,200 rpm. This high-performance powerplant was a bit unusual to begin with, because it used one large exhaust and two different-sized intake valves. The redesigned version, designated the 6.36 (for 6 cylinders and 36 valves), is equally unusual because it contains three intake and three exhaust valves arranged concentrically around the bore and inclined at different angles for greater mixture swirl. Each valve trio is operated by the same cam lobe through a wide follower.

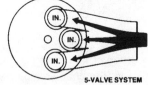

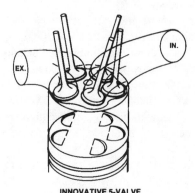

INNOVATIVE 5-VALVE CONFIGURATION

Yamaha's five-valve motorcycle head

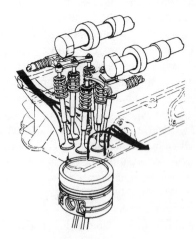

Maserati's six-valve head

The 6.36 is also a complex and expensive engine to manufacture, but Maserati says the performance makes it ideal for limited production use in a 2-seat sports car.

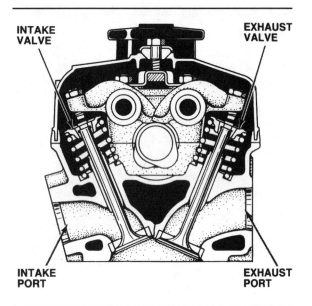

Figure 7-38. A crossflow head has valves and ports on opposite sides of the combustion chamber.

and valvetrain improvements. The following points summarize the benefits of three- and four-valve-per-cylinder engines:

- Two small valves in place of one large one can provide a larger total inlet or exhaust area. This improves volumetric efficiency and increases top-end power.
- Two small valves can provide more total perimeter than a single valve of the *same* area. Even at low valve lifts, multiple valves can flow more air. Therefore, camshaft timing need not be as aggressive, allowing the same total power but improving idle quality and fuel economy.
- Smaller individual valves, springs, and retainers reduce the weight of each valve assembly. This reduces valve train inertia, reducing valve float to allow higher maximum engine speeds.
- Smaller valves allow the engineer more flexibility in combustion chamber design. Valve angles and ports can be designed for improved air-fuel turbulence and combustion.
- Separated intake and exhaust ports can be designed, or "tuned," for optimum intake and exhaust velocity at different rpm. Individual, smaller diameter ports of larger total area can provide increased intake and exhaust flow at higher speed.
- Electronically controlled fuel injection combined with two intake valves allows engineers to design induction systems with separate air-fuel flow for different engine speeds and loads. On some engines, the two intake valves open and

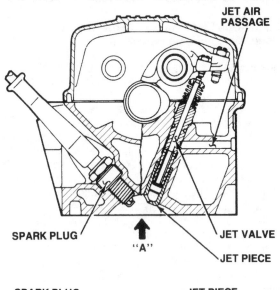

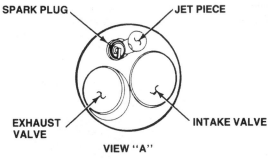

Figure 7-39. The Mitsubishi MCA engine uses a small auxiliary valve to admit a high-speed jet of air on the intake stroke.

close at different times to fine-tune the air-fuel metering for different operating conditions.

Auxiliary valves

To understand the principles of 3-valve, 4-valve, and other similar designs, it is best to distinguish between auxiliary valves and main intake and exhaust valves. Mitsubishi engines with the Mitsubishi Clean Air (MCA) combustion chamber design have a third small valve in each chamber, figure 7-39. Mitsubishi calls this a "jet" valve because it admits a high-speed jet of air to create a swirl effect that promotes complete combustion. On the intake stroke, jet air also removes residual gases from around the spark plugs to improve ignition. This MCA jet valve design has been used in the engines of many Mitsubishi and Chrysler imports since the mid-1970s. In the mid-1980s Mitsubishi redesigned the head and renamed it the "cyclone" combustion chamber.

Toyota builds a 3-valve engine with a smaller, secondary intake valve that opens later than the primary intake valve. The two intake valves, combined with a unique port design, promote mixture turbulence and complete combustion of lean mixtures. The Honda, Mitsubishi, and Toyota 3-valve designs are examples of auxiliary-valve engines.

Combustion chamber shape

Engines come in different configurations, such as V6s or in-line fours. However, the configuration of a V6 or in-line four from one manufacturer is very much the same as the V6 or four from another. Combustion chamber design and shape are how engines really differ. The engineer must match the chamber design to the intended purpose of the engine. Some chamber designs are better for fuel economy, while others are better for high performance. Designers have tried many different shapes. The following designs are the most common from the last 40 years, but all are still in production.

Wedge-shaped combustion chambers

The most popular design for domestic overhead-valve engines of the 1950s through the 1970s was the wedge, figures 7-40 and 7-41. The wedge provides good combustion efficiency because it has squish and quench areas and the air-fuel mixture is swirled about providing turbulence.

In the wedge-shaped chamber, the squish area is farthest from the spark plug, and it doubles as a quench area. This is known as a squish-quench design. Designers may also route the water jacket to bring coolant to the outside of the cylinder, reducing the temperature of the quench area.

Another advantage of the wedge shape is that the valves are in-line, making the combustion chamber compact, decreasing burntime and lessening the chance of detonation. The cylinder head is also small and lightweight, making it inexpensive to manufacture. Lack of an efficient breathing ability is the wedge's main drawback, due to its in-line valves.

Hemispherical combustion chamber

The hemispherical combustion chamber is so named because it roughly resembles half a sphere, figure 7-42. It is successfully used on both overhead-camshaft and overhead-valve engines. The hemi provides the greatest breathing efficiency because the valves are on opposite sides of the chamber. This placement provides space for very large valves, and allows the intake and exhaust to work together effectively during camshaft overlap, 7-43. With the valves on opposite sides of the chamber, the spark plug can also be centrally located.

The hemi breathes well, but its drawback is combustion efficiency. The *classic* hemi chamber has no squish or quench areas so there is little turbulence and no area to cool the mixture. Some versions of the hemi have a squish band designed into the perimeter edge of the piston to improve combustion.

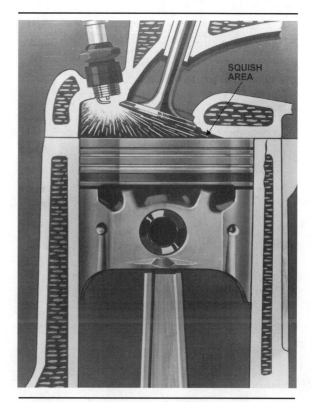

Figure 7-40. The wedge combustion chamber provides good combustion efficiency. It features a very effective squish area.

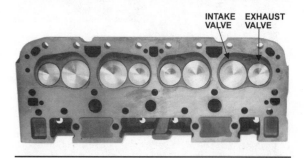

Figure 7-41. Wedge chambers place intake and exhaust valves side by side.

Also, while the spark plug is centrally located, hemi chambers are so large, the spark has a relatively long way to go to reach the air-fuel mixture.

The large hemi chambers also require a tall piston top or dome that extends deeply into the combustion chamber to provide a moderate to high compression ratio. The tall dome acts like a dam, dividing the combustion chamber and impeding flame travel after ignition. With current fuel octane levels, detonation is a problem in hemi combustion chambers. One way to deal with this problem is to add a second spark plug to the combustion chamber.

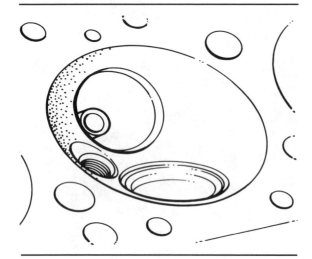

Figure 7-42. The hemi (short for hemispherical) combustion chamber allows large valves and good breathing.

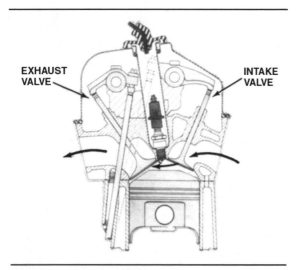

Figure 7-43. The hemi chamber allows optimum valve placement.

Modified wedge chamber designs

The wedge combustion chamber provides efficient combustion, but does not breath as well as the hemi. Some manufacturers modified the wedge design to give better breathing and produced a chamber with some of the advantages of the hemi.

The valves remained in-line, but manufacturers tilted, or canted them away from each other to provide better flow. This was an attempt to approximate the hemi's favorable valve placement. Next, they opened up the combustion chamber, making it closer to the hemi in shape. This unshrouds the valves, providing better flow at higher valve lifts.

Big-block Chevrolet V8 engines used what was called the porcupine head because of its canted valves

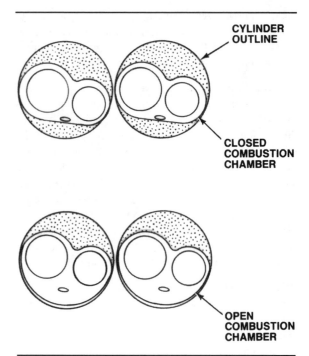

Figure 7-44. Chevrolet opened the combustion chambers and unshrouded the valves in its big-block engine to produce more horsepower.

or the semi-hemi for its open combustion chamber. Chevrolet tried both open and closed combustion chambers. The engine produced more horsepower with the open chambers. The open chamber walls slanted so that much of the top of the piston was uncovered, figure 7-44.

Ford also used canted valves to good effect in the Boss and Cleveland 302 and 351 V8 engines. Ford canted the valves about 25 degrees from vertical, figure 7-45. Ford also tried open and closed chambers.

In the modified wedge chambers, the canted valves made the heads more expensive to produce but caused no functional problems. Open chambers increased combustion chamber volume, lowering the compression ratio. The engines ran fine with the lowered compression, and sometimes the improved breathing raised power enough to offset the compression loss. If high-performance is the goal, high-domed pistons are needed to recover the lost compression. The domes interfere with the flame front, increasing the likelihood of detonation. But more important to everyday operation, the open chambers also reduced squish and quench areas.

Pentroof combustion chambers

The best combustion chamber design currently in use is the pentroof, which resembles a wide, inverted "V", figure 7-46. It is fairly flat for a small volume, allowing high compression ratios with flat-topped pistons.

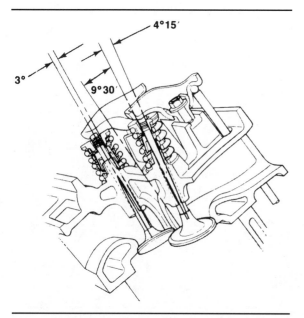

Figure 7-45. Canting the valves in a wedge combustion chamber improves mixture flow.

The small volume also decreases the chance of detonation, permitting high compression ratios with today's relatively low octane fuel.

The shape provides a favorable surface to volume ratio for low emissions of unburned hydrocarbons. But it grants plenty of room for large valves to help breathing. The intake and exhaust valves are opposite one another like the hemi, to further enhance breathing. Also like the hemi, the spark plug is centrally located, figure 7-47. However, like the wedge, the pentroof provides squish areas for turbulence and combustion efficiency.

Pentroof combustion chambers are always equipped with two intake valves in one half of the chamber and one or two exhaust valves in the other half. Since the mid-1980s, engines with pentroof chambers and four valves per cylinder have become common on Japanese and European cars. By the 1990s, almost all imported engines had pentroof chambers and three or four valves per cylinder.

The first domestic production engine with pentroof chambers and four valves per cylinder was the General Motors Quad 4, introduced in 1987. In 1989, Ford brought out the V6 SHO engine with four valves per cylinder and GM introduced the ZR1 Corvette with a V8 engine featuring 32 valves, four per cylinder. In 1991, GM introduced the Twin Dual Cam V6, a four-cam, 24-valve engine. By 1998, Chrysler, Ford, and GM produced more pentroof and four-valve-per-cylinder engines than any other design.

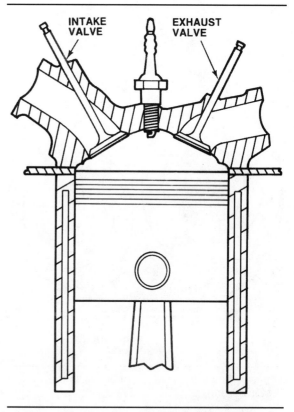

Figure 7-46. The pentroof chamber is fairly shallow and allows high compression ratios with flat-top pistons.

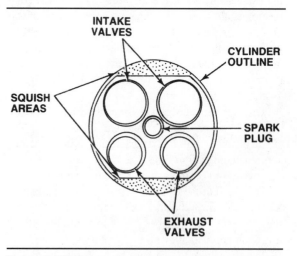

Figure 7-47. The pentroof chamber allows a centrally mounted spark plug and provides squish areas.

The pentroof combustion chamber has been the first choice for racing engines for many years. The triple goals of obtaining best economy, emission control, and performance, plus the competitive nature of the auto market in the 1980s, finally led manufacturers to adopt these designs for production vehicles.

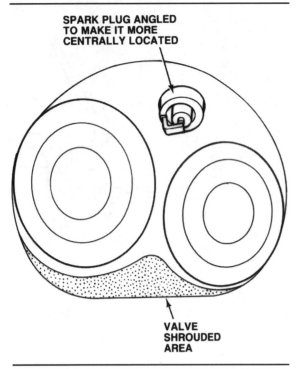

Figure 7-48. Ford's fast burn chamber in the 1.9-liter 4-cylinder engine.

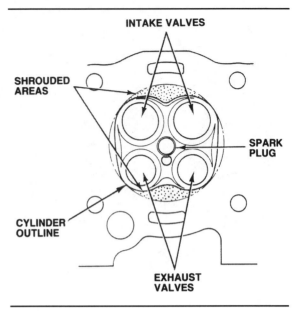

Figure 7-49. Lexus added metal to shroud the valves on the 1990 ES 250 engine.

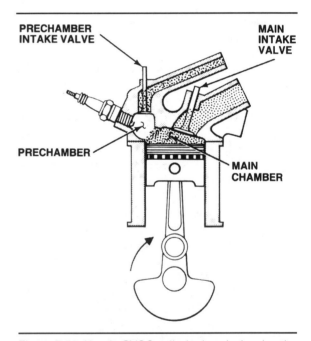

Figure 7-50. Honda CVCC cylinder head, showing the main combustion chamber and precombustion chamber (prechamber) with extra intake valve.

Fast-burn combustion chamber

In the interest of improving fuel economy and reducing exhaust emissions the fast-burn or high-swirl combustion chamber was developed. The main goal of this design is speeding up and improving combustion. Designers apply fast-burn combustion chamber technology to either two-valve, modified wedge-type chambers or four-valve pentroof chambers, figures 7-48 and 7-49.

The fast-burn design consists primarily of a compact combustion chamber and shrouded valves. The small chamber volume reduces the flame front and shortens combustion time. The shrouded valves swirl the mixture to promote complete combustion.

The fast-burn chamber uses a central spark plug location to reduce the distance the flame must travel to the edges of the chamber. The rapid combustion of this chamber also allows a higher compression ratio.

The drawback to fast-burn technology also lies in one of its virtues. It values combustion efficiency to the detriment of breathing ability.

Stratified charge combustion chambers

The concept of a stratified charge engine has been around in many forms for many years. However, Honda's Compound Vortex Controlled Combustion (CVCC) design was the first stratified charge gasoline engine used in a mass-produced car. Introduced in 1975, it continued through the 1987 model year.

The CVCC engine has a separate small precombustion chamber located above the main combustion chamber and contains a tiny additional valve, figure 7-50. Except for this feature, the CVCC is a conventional 4-stroke piston engine. However, it uses a 2-stage combustion process.

CVCC engines use three-barrel carburetors. The third barrel provides a rich air-fuel mixture to the

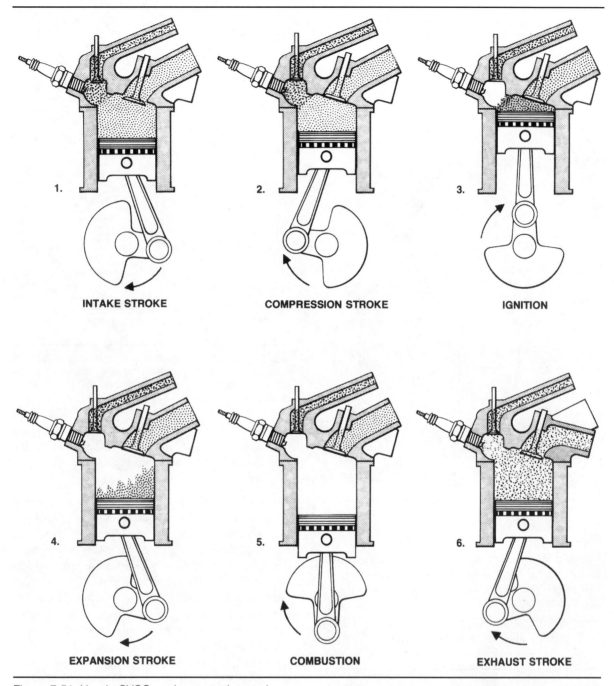

Figure 7-51. Honda CVCC engine operating cycle.

small precombustion chamber. This rich mixture is very easy to ignite. The other two barrels in the carburetor supply a very lean mixture to the main combustion chamber. This lean mixture is difficult to ignite with a spark but easily lit by the flame began in the small chamber. The lean mixture in the main chamber allows good fuel economy and low exhaust emissions.

Figure 7-51 shows the stages in the operating cycle. The first stage is one of precombustion, in which the air-fuel mixture is ignited in the precombustion chamber. In the second stage, the flame front created moves down into the main combustion chamber to ignite a mixture with less fuel in it. The stratified charge engine takes its name from this layering, or stratification, of the air-fuel mixture just before combustion. At that time, there is a rich mixture (with lots of fuel) near the spark plug, a moderate mixture in the auxiliary combustion chamber, and a lean mixture (with little fuel) in the main chamber. The result is a more

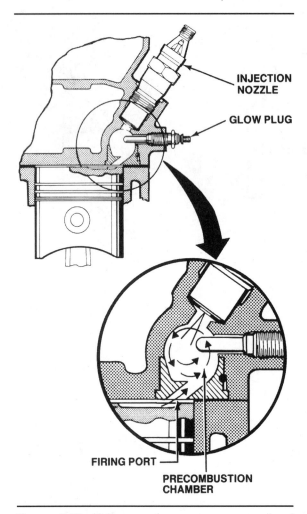

Figure 7-52. Volkswagen's passenger car diesel engine uses a precombustion chamber swirl chamber.

complete combustion of the air-fuel mixture, which keeps unburned fuel and emissions to a minimum.

Charge stratification is applied both to reciprocating diesel engines and to rotary gasoline engines. Most diesel engines used in cars have a precombustion chamber into which the fuel is injected, figure 7-52. This allows the combustion to occur in two stages: in the precombustion chamber and in the main chamber. This improves cold starting and combustion efficiency and reduces engine noise and vibration.

Intake and Exhaust Ports

Ports are the passages that bring the mixture to the intake valve and take the exhaust out past the exhaust valve, figure 7-53. Port design has a big effect on how much mixture will pass through the intake and exhaust valve openings and helps determine the volumetric efficiency of the engine. If the intake port restricts the flow, the chamber will not develop maximum power. If the exhaust does not flow freely out of

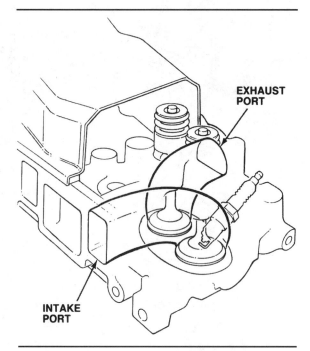

Figure 7-53. Intake ports supply the combustion chambers with the air-fuel mixture. Exhaust ports allow the burned mixture to leave the combustion chamber.

the chamber, some of it will remain and take space that should be filled with fresh mixture. The engine then produces less power than it could.

Port design
Selecting a port size is not as simple as it first appears. It may seem logical for the designer to use the biggest port possible because it can flow the most air. However, the biggest ports do not necessarily perform the best. Much of what we said about intake manifolds in Chapter 5 also applies to ports. If the ports are large, the mixture will travel slowly at low rpm, allowing fuel particles to drop out of suspension. Smaller ports keep the velocity of intake and exhaust gases high. When the throttle is opened, the engine response is quicker with smaller ports. However, smaller ports do not pass as much mixture as the engine needs at wide open throttle.

The angles at which ports turn are also critical. When they are too sharp, fuel may separate out of the mixture.

The efficiency of a port is determined by its shape, interior surface, and size. Port design is usually a compromise. Short, large diameter ports produce best power at high rpm, but long, smaller diameter runners produce the best power at low to medium rpm. Passages that are too large or too small reduce efficiency. Ports should be without sharp corners, bends or turns to interfere with mixture flow. Intake ports

Figure 7-54. Port and valve design is often done on a flow bench.

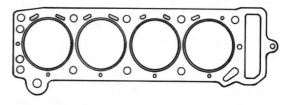

Figure 7-55. Holes in the cylinder head gasket match the holes in the head itself.

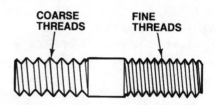

Figure 7-56. When a stud is screwed into a casting it should have coarse threads to make sure it will hold.

CYLINDER HEAD INSTALLATION

Cylinder heads are designed to fit against the top of the cylinder block with a head gasket in between to prevent leaks. Holes in the gasket match the cylinder bores and coolant passages, figure 7-55.

The cylinder heads are usually attached to the block with bolts, but sometimes designers use nuts and studs. The nuts and tops of the studs usually have fine threads, which help to distribute tightening pressure evenly, and stay tight. This prevents warping. The bolts or the ends of the studs that screw into the block casting usually have coarse threads. Coarse threads are bigger and stronger, so there is less chance of stripping them, figure 7-56.

The gasket surfaces of the head and the block must be flat so that the gasket will be compressed equally everywhere. If the head or block is warped so that the gasket is compressed more at the edges than in the middle, then the middle may leak combustion gases or coolant. When designing an engine, the headbolts are positioned as close together as required to equalize the clamping forces over the entire gasket surface. In some places the clamping force may be weak if the design of the engine did not allow the bolts to be close enough together. In that case, the gasket may leak or blow out.

Headbolts must always be torqued starting in the center of the gasket and working out to the edges. As a gasket is compressed, it becomes longer. If the tightening starts at the edges, the center will wrinkle because the edges prevent it from stretching out.

may be slightly rough to aid vaporization by reducing puddling and breaking up mixture flow. Exhaust ports should be smooth to reduce the buildup of carbon.

The designer must consider the operating rpm range of the engine when designing ports. An engine running at higher rpm needs the greater air flow that larger ports can supply.

Designers must also consider engine size. Port size is relative. A port size considered large on a small engine may be small to medium on a large engine. Large engines need larger ports for adequate intake and exhaust flow.

Much of port and valve testing is done on a flow bench, figure 7-54. The port shape is made out of plastic or clay and mounted on a bench with a blower. Turning on the blower and measuring how much air comes out the other end of the port will tell whether or not the port is efficient. Port design is not an exact science.

MISCELLANEOUS ENGINE COVERS

The cylinder block and heads are sealed against outside elements by several covers, such as the valve cover, oil pan, and timing chain cover. These covers keep oil in the engine and dirt out.

One of the most important of these covers is the oil pan, figures 7-57 and 7-58. It covers the bottom of the crankcase and is also the oil reservoir for the engine. The part of the pan that contains the oil is known as the sump. Within, or near, the sump there may be baffles to keep oil from splashing around during hard braking, cornering, or acceleration. At the bottom of the sump is the oil drain plug.

The oil pan bolts to the engine block and seals with a gasket. You must remove it to inspect bearings or repair the oil pump.

The timing cover protects the timing sprockets and chain or pulleys and belt at the front of the engine. The cover may be a simple piece of sheet metal or plastic with nothing attached to it, figure 7-59. On some applications, the timing cover can be a large aluminum casting that provides water passages and a mounting for the water pump, figure 7-60. On some engines with front-mounted distributors, the front cover also provides a place to mount the distributor and oil pump.

On most overhead-valve engines, the front cover holds the front crankshaft oil seal. The oil pan then seals against the bottom of the front cover.

On a V8 engine, there is a large area at the top of the engine between the cylinder banks that must be covered. The intake manifold acts as a cover for this valley in the engine. On other engines a sheet metal valley cover is used. The intake manifold then sits above the valley cover.

On most engines, the valve lifters can be lifted out after removing either the valley cover or the side cover and the pushrod or valve. On overhead valve, V engines you must remove the intake manifold before you can remove the valley cover.

A bellhousing covers the clutch and flywheel, figure 7-61. On an engine with an automatic transmission the bellhousing covers the torque converter, or fluid coupler. On most automatic transmissions, the bellhousing is made in one piece with the transmission, but on manual transmissions the bellhousing is usually separate. The bottom side of many bellhousings is left open for inspection and repair access. Technicians sometimes remove clutches through this bottom opening. To prevent stones or other debris from hitting the moving parts inside the bellhousing, the bottom opening is covered with sheet metal or plastic.

Valve covers and overhead camshaft covers are stamped out of sheet steel, cast out of aluminum, or

Figure 7-57. This stamped-steel oil pan is one of the most common styles.

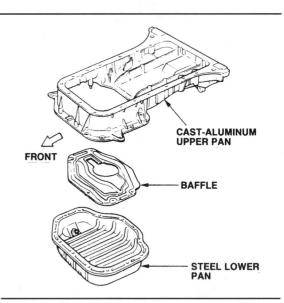

Figure 7-58. Some oil pans are cast in aluminum and may be two-piece.

molded from plastic. They are mounted with a gasket to prevent oil leakage. They may be lightweight and fragile, and require care when installing. Overtightening the bolts that hold the covers in place could cause warping, leaks, or cracks.

SUMMARY

Cylinder blocks, heads, and manifolds are usually cast from iron or aluminum. The rough engine castings undergo initial machining operations, are allowed to "season" for a time, then the castings are bored, drilled, tapped, machined to finished size, and assembled.

■ Blueprinting

If you read the Sunday want ads for used cars, you'll occasionally come across a high-performance car, one in which the owner exclaims, "Balanced and blueprinted!" You can also hear the same terms used by everyone from street racers to engine wholesalers, as they describe the faster — and pricier — powerplants in their shops or in their catalogs. Engine balancing is pretty straightforward, and we describe it in detail elsewhere in this book and in the *Shop Manual*, too. But what exactly is "blueprinting"?

In short, blueprinting is a technique for building or rebuilding an engine to stricter and more precise tolerances than the factory uses on their own production line. A properly blueprinted engine will run smoother, last longer, and will have considerably more horsepower than the best factory-built version. It will also cost considerably more to construct, and will take much more time to assemble.

Consider a typical cylinder bore. On the factory assembly line, multiple-head boring bars and honing machines cut and finish all the bores simultaneously. The bores are quickly measured, and one of several select-fit pistons with standard rings is rapidly installed. A few minutes later, the job is completed and the engine is sent down for the next assembly operation.

A skilled blueprinter may spend all day performing the same operations. He will first choose pistons that are carefully matched for weight, and spend several hours smoothing and contouring their crowns to aid combustion and prevent carbon build-up. Then he will measure each one, and bore and hone each cylinder for that particular piston. He will likely use expensive deck plates during the honing to duplicate the distortions an installed cylinder head will create in the assembled block. He will hand-file the ring end gaps so that they will close exactly, to seal combustion pressure when the engine warms up. The result is cylinders that will hold more pressure for longer than a factory-finished bore.

All the other engine dimensions receive similar care during blueprinting. For example, the block may be Magnafluxed® and sonic tested to ensure that it is structurally sound. Blueprinters use a number of special fixtures that mount to machining equipment, which allow the machinist to correct deck or cylinder bore surfaces that were not made perpendicular to the crankshaft by the factory. Crankshafts have their oil holes chamfered and may be nitrided or hard chromed to improve strength. The connecting rods are polished to remove stress risers and improve reliability. Machinists performing a blueprinting job give the heads very careful valve jobs and may grind five or more valve seat angles, rather than the more common three. These are just a few examples of the many exacting operations a skilled machinist performs to blueprint and engine.

The "BLOK-TRU" from B-H-J Products Inc. is a precision indexing plate for all 90-degree V engines. It allows machinists performing blueprinting jobs to true twisted blocks, uneven deck clearances, and other dimensional errors caused by factory machine tolerances and production line inaccuracies.

Blocks must be designed as compactly as possible, but be strong enough to support the rest of the engine, and allow room for the counterweights of the crank and the selected bore size and stroke length. Ribs, webs, and sometimes separate crankcase castings improve block strength.

The main bearings give support without interfering with the rotation of the crankshaft or the action of the connecting rods. Bolt-on main bearing caps are usually made of a stronger material than the block material. Main bearing caps attach to the block with two bolts or four bolts. Some designers add cross bolts for extra strength.

Putting a main bearing on both sides of each connecting rod results in longer engine life. A 6-cylinder in-line engine would have seven main bearings in this configuration.

The cylinders are part of the block casting in a water-cooled engine, while air-cooled engines usually have individual cylinders that bolt to the crankcase. Some engines are built with siamesed cylinder bores: the cylinder walls of adjacent cylinders are cast together.

Cylinder sleeves, also called cylinder liners, are either a standard engine part or used to make a repair. Most sleeves are cast iron. They may be dry sleeves or wet sleeves. A new type sleeve is made of a carbon fiber composite, invented by Honda and used in the 1991 Prelude. Some aluminum engines have no sleeves, but rely on a high amount of silicon in the

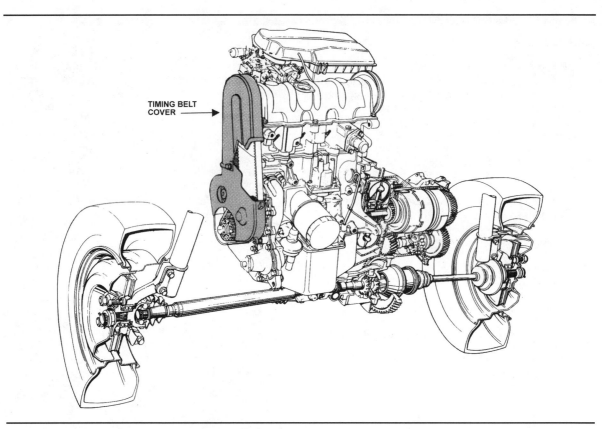

Figure 7-59. This stamped steel cover protects the timing belt and sprockets.

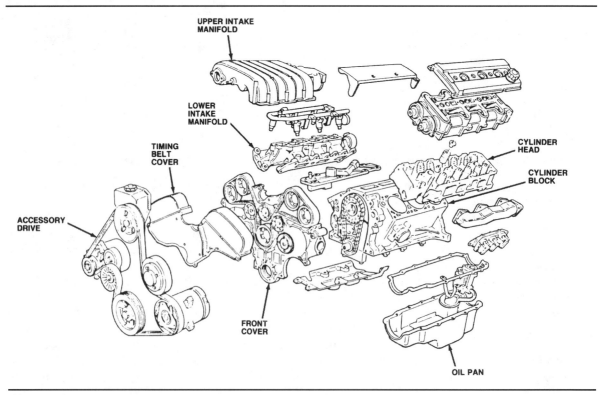

Figure 7-60. This cast-aluminum front engine cover acts as a mount for the camshaft timing belt and sprockets, water pump, and accessory drive belt. A plastic cover protects the timing belt.

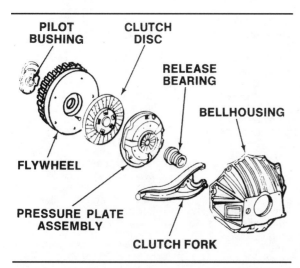

PILOT BUSHING
CLUTCH DISC
RELEASE BEARING
BELLHOUSING
FLYWHEEL
PRESSURE PLATE ASSEMBLY
CLUTCH FORK

Figure 7-61. A bellhousing covers the flywheel and clutch.

block to give an acceptable sliding surface for the pistons.

Crankshaft offset reduces the angle of the connecting rod during the power stroke. Cylinder offset helps center the cylinder over the piston and connecting rod.

The combustion chamber consists of the hollowed out portion of the cylinder head, the upper edge of the cylinder wall, and the top of the piston. Chamber design affects engine breathing and combustion efficiency, which help determine power, fuel economy, and exhaust emissions.

The combustion chamber design factors that determine the efficiency of combustion are squish area, quench area, spark plug placement, and surface-to-volume ratio. Breathing ability is determined by valve placement and valve number. Combustion chamber shape and valve shrouding affect both combustion efficiency and breathing ability equally.

Some Honda, Mitsubishi, and Toyota combustion chamber designs use auxiliary valves to improve combustion efficiency.

Combustion chamber design and shape are how engines really differ. Although other designs were used, the most common from the last 40 years include the wedge, the hemi, the modified wedge, the pentroof, and the fast-burn chambers.

The stratified charge combustion chamber has a separate small precombustion chamber located above the main combustion chamber and contains a tiny additional valve.

Ports are the passages that bring the mixture to the intake valve and take the exhaust out past the exhaust valve. Port design has probably the greatest affect on an engine's breathing ability. Generally, large ports flow large amounts of air to provide high horsepower, but air velocity is low, hurting throttle response and low-rpm torque. Small ports increase flow velocity to improve throttle response and low-rpm torque, but restrict total air flow and reduce maximum horsepower. Port size is relative. Large engines need larger ports for adequate intake and exhaust flow.

Cylinder heads are designed to fit against the top of the cylinder block with a head gasket in between to prevent leaks. They are attached to the block with bolts or with nuts and studs.

The cylinder block and heads are sealed against outside elements by several covers, such as the valve cover, oil pan, and timing chain cover. These covers keep oil in the engine and dirt out.

Review Questions

Choose the single most correct answer.
Compare your answers with the correct answers on Page 328.

1. The main bearings in an engine support the:
   a. Crankcase
   b. Crankshaft
   c. Camshaft
   d. Connecting rods

2. Technician A says that 2-bolt main bearing caps are the strongest type of main bearing cap.
   Technician B says that iron main bearing caps are often used on aluminum cylinder blocks.
   Who is right?
   a. A only
   b. B only
   c. Both A and B
   d. Neither A nor B

3. Technician A says that crankshaft offset reduces the angle of the connecting rod during the power stroke.
   Technician B says that some in-line engines do not have crankshaft offset.
   Who is right?
   a. A only
   b. B only
   c. Both A and B
   d. Neither A nor B

4. Technician A says siamesed cylinders make a cylinder block more rigid.
   Technician B says that these types of cylinders offer more efficient engine cooling at low rpm.
   Who is right?
   a. A only
   b. B only
   c. Both A and B
   d. Neither A nor B

5. A V8 engine with paired rods on each throw usually has _____ main bearings.
   a. 3
   b. 5
   c. 7
   d. 9

6. In a wedge-shaped combustion chamber, squish and quench areas are:
   a. On the same side of the combustion chamber
   b. Necessary for good breathing
   c. Not important in the hemi combustion chamber
   d. Not necessary for good combustion

7. Which of the following most directly affects combustion efficiency?
   a. Port size
   b. Squish area
   c. Valve number
   d. Port length

8. Technician A says 4-valve-per-cylinder engines breath better than 2-valve-per-cylinder engines because they provide more valve area.
   Technician B says they breathe better because they provide more valve perimeter.
   Who is right?
   a. A only
   b. B only
   c. Both A and B
   d. Neither A nor B

9. Technician A says the best location for the spark plug is opposite the exhaust port.
   Technician B says the best location is opposite the intake port.
   Who is right?
   a. A only
   b. B only
   c. Both A and B
   d. Neither A nor B

10. Which of the following is NOT a type of cylinder sleeve?
    a. Dry sleeve
    b. Reusable sleeve
    c. Wet sleeve
    d. Carbon-fiber composite sleeve

11. Technician A says that aluminum-silicon cylinder blocks cannot be overbored.
    Technician B says that aluminum-silicon blocks cannot be fixed by installing iron sleeves.
    Who is right?
    a. A only
    b. B only
    c. Both A and B
    d. Neither A nor B

12. Surface-to-volume ratio should ideally be:
    a. As low as possible
    b. As high as possible
    c. Above 7.5:1
    d. As close to 7.5:1 as possible

13. Valves placed opposite one another in the combustion chamber, such as in a hemi:
    a. Help cool the exhaust valve
    b. Improve breathing
    c. Usually make the combustion chamber larger
    d. All of the above

14. Technician A says that designers place a main bearing on either side of a crankshaft throw or connecting rod for best engine strength.
    Technician B says that main bearings consist of the saddle, cap, and bearing inserts.

Who is right?
    a. A only
    b. B only
    c. Both A and B
    d. Neither A nor B

15. Head bolts should be tightened:
    a. Starting at the edges, and working in toward the center
    b. Starting at the center, and working out toward the edges
    c. Evenly, to approximately 75 foot-pounds of torque
    d. None of the above

16. Small intake and exhaust ports:
    a. Reduce the velocity of intake and exhaust gases
    b. Reduce throttle response
    c. Increase the flow of gases at wide open throttle
    d. Keep the velocity of intake and exhaust gases high

17. A crossflow head:
    a. Requires a wedge-shaped combustion chamber
    b. Requires that the mixture make a U-turn before it can exit
    c. Has the intake and exhaust ports on opposite sides of the combustion chamber
    d. Needs less room in the engine compartment

18. Pentroof combustion chambers:
    a. Can provide good squish areas
    b. Have very limited valve room
    c. Have side-mounted spark plugs
    d. Are used in non-crossflow heads

19. Fast-burn combustion chambers:
    a. Feature large volume
    b. Have centrally mounted spark plugs
    c. Is another name for hemispherical combustion chambers
    d. Unshroud the intake valve for better breathing

20. The stratified charge engine:
    a. Is another name for the 2-cycle engine
    b. Has a hemi combustion chamber
    c. Is not used in modern vehicles because it gives poor gas mileage
    d. Has a small precombustion chamber

8

Valves, Springs, Guides, and Seats

VALVES

All 4-stroke piston engines use intake and exhaust valves, figure 8-1. The intake valve opens to allow the air-fuel mixture to enter the cylinder. The exhaust valve opens to let the burned mixture out of the cylinder. In the closed position, the valve rests on a seat, which shuts off the port. To open the port, the stem of the valve is pushed by the **valve train**, figure 8-2. This raises the head of the valve off the seat and allows the air-fuel mixture to flow in or the exhaust to leave the combustion chamber. We will detail the valve train in Chapters 9 and 10.

As we pointed out in Chapter 7, engines use at least one intake and one exhaust valve per cylinder. The size of the valve is a factor in how much airflow there will be when the valve head is raised off its seat. To get more airflow, designers often use two small intake and two small exhaust valves in each cylinder instead of one large intake and exhaust valve, figure 8-3. These smaller valves have the advantage of low mass over the larger, heavier valves. The greater mass of the large valve may cause it to bounce when it hits the seat, and when it is pushed open it does not stop easily when it reaches its full opening point. This causes slack in the valve train and increased wear. In extreme cases, the valve may stay open long enough for the piston to hit it, resulting in major engine damage.

Valve Design

The poppet valve is used in all 4-stroke engines. It consists of a large head and a long stem, figure 8-4. The head seals the port to the combustion chamber. The head of the intake valve is always larger than the head of the exhaust valve because it must seal a larger port. Unless the engine is supercharged, there is only atmospheric pressure pushing the mixture into the engine. It is harder to get the mixture into the engine than to get it out, so the intake port must be larger. On the exhaust stroke, great heat and pressure, as well as the piston itself, are pushing the exhaust out. Therefore, the exhaust port and valve need not be as large.

The stem slides in and out a small distance inside the valve guide; the distance is called the valve lift. The end of the stem, or valve tip, is the part pushed by the camshaft or rocker arm to open the valve. Just down from the tip, keeper grooves machined in the stem allow the keepers or valve locks to hold the valve and spring in place.

The curved portion of the valve where the stem meets the head is called the fillet. The shape of the **fillet** affects the flow around the valve. A valve with a steeply raked fillet, figure 8-5, is sometimes called a

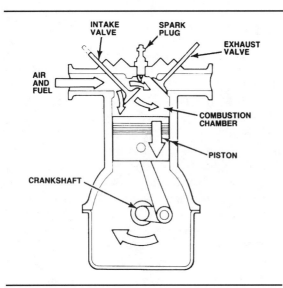

Figure 8-1. The intake valve opens to allow the air-fuel mixture to enter the cylinder. The exhaust valve opens to let the burned exhaust gases out of the cylinder.

tulip valve. The tulip shape is thought to flow better than a flatbacked valve at low valve lifts. Once the camshaft raises the valve to full lift, the port flows air as if the valve were not there, making valve shape irrelevant at that point.

On the stem side of the valve head is a precision ground **valve face**. The face is circular and, on most valves, it is at a 45-degree angle to the top of the head. The valve face seals against a seat in the cylinder head on overhead-valve and overhead-camshaft engines. The seat is in the block on flathead or L-head engines.

The side of the valve head opposite the stem is called the combustion surface. When the valve is closed, this part of the valve forms part of the combustion chamber.

A small **margin** runs around the edge of the valve head between the end of the head and the valve face. The margin is essential to valve strength. Never grind the face down to the head of the valve to a sharp edge, figure 8-6. A thin edge heats up quickly and melts—automotive machinists and technicians call this "burning a valve." An edge also would not have the strength to stand up to the pounding of opening and closing the valve, leading to warping and breakage.

Valve materials

Manufacturers sometimes make intake valves from plain carbon steels, but for higher strength and better resistance to heat, they specify special alloy steels. The greater the heat and the higher the cylinder pressures the valves must withstand, the more alloying elements designers add. Designers raise the amount of carbon to improve strength, add chromium

and silicon to gain corrosion and oxidation resistance, and use molybdenum, tungsten, and vanadium to increase valve life at high temperatures. Cobalt and nickel improve wear resistance.

Exhaust valves have to withstand even more heat than intake valves. Therefore, standard exhaust valve material is equal to high-quality intake valve alloy. Manufacturers may make heavy-duty exhaust valves from nickel-based superalloys, similar to that used for gas turbine blades. Often, exhaust valve alloys have such large amounts of alloying elements they are non-magnetic.

Manufacturers sometimes weld valves together because they need metals with different characteristics for the various parts of the valve, figure 8-7. The entire valve stem and valve head may be welded together from two dissimilar metals. Wear is the primary problem with stems and heat is the primary problem with valve heads. A simple cap may be welded onto the tip for improved wear resistance.

Any small imperfection, or **stress riser**, in the valve could develop into a crack. Valves are ground to remove the stress risers, casting irregularities and machining marks that could cause valve failure.

Valve coatings

Many vehicle manufacturers cover valve stems and valve faces with special materials to extend valve life. If the valve stem or guide wears, the valve can come down on its seat in a cocked position, figure 8-8. The valve actually has to slide to reach its final position, accelerating wear and leakage. For this reason, some manufacturers apply a hard chromium plating or nitriding process to the valve stems. Both of these finishes improve wear characteristics.

Some manufacturers deposit a thin layer of aluminum oxide on the valve fillet, face, and combustion surface to help the valve resist corrosion, figure 8-9. Aluminum oxide covered valves look as if the aluminum oxide has been sprayed on, but the process is done so that the aluminum actually fuses with the valve metal. Corrosion occurs when the metal in the valve combines with the oxygen or other gases or chemicals that pass by. Once the metal has corroded or rusted, it has very little strength and rapid wear occurs, figure 8-10. Many vehicle manufacturers use aluminized valves because of their longer life.

Manufacturers often weld a hard overlay of cobalt, chromium, and tungsten to the valve face. This combination is commonly known by its most familiar brand name, **Stellite**. Designers specify Stellite for valve faces because its hardness slows valve recession, the loss of metal from the sealing surfaces of the valve and seat which causes the valve to slowly sink in the cylinder head, figure 8-11. When the

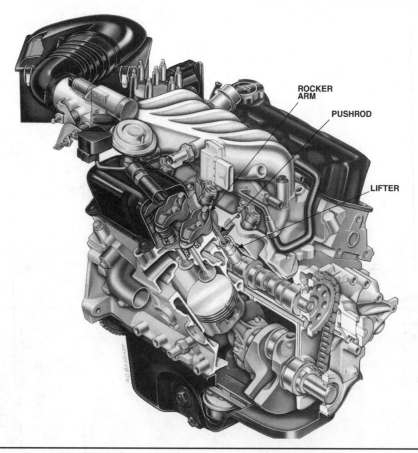

Figure 8-2. A linkage, or valve train transmits the motion of the camshaft to open the valves. In this Ford overhead-valve V6 engine, the camshaft opens the valve through a lifter, pushrod, and rocker arm.

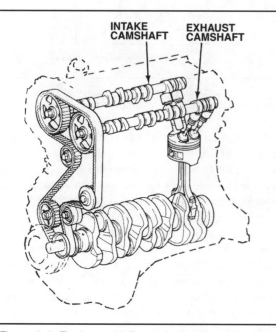

Figure 8-3. Engines with four valves per cylinder usually have a camshaft to operate the intake valves and a camshaft to operate the exhaust valves.

Valve Train: The assembly of parts that transmits force from the camshaft lobe to the valve, including the camshafts and valves themselves.

Fillet: The curved portion of the valve where the stem meets the head.

Valve Face: A precision ground surface of the valve that seals against the valve seat.

Valve Margin: A small margin around the edge of the valve head between the end of the head and the valve face. The margin adds strength and heat resistance to the valve.

Stress Riser: A groove, scratch, or other imperfection in the metal that could develop into a crack.

Stellite: The brand name for a very hard alloy of chromium, cobalt, and tungsten applied to valve faces for longer wear.

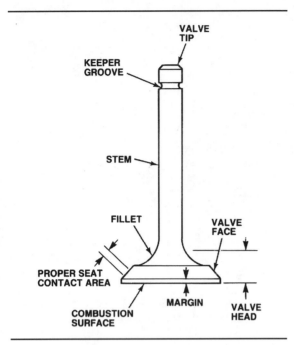

Figure 8-4. All poppet valves share these basic parts.

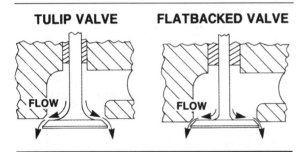

Figure 8-5. A tulip valve allows better flow than a flat-backed valve.

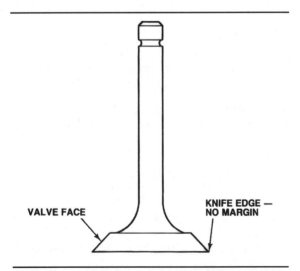

Figure 8-6. Never grind the valve face to a sharp edge. This weakens the valve and increases the chance for burning the valve.

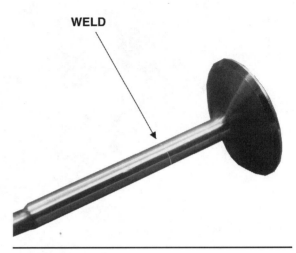

Figure 8-7. Some valves are welded together from two different pieces of metal.

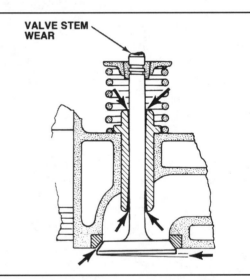

Figure 8-8. Excessive wear on the valve stem or guide can cause the valve to seat in a cocked position.

valve and seat contact in a running engine, heat and pressure are high enough to cause a local welding between the two surfaces. When the valve opens, the weld breaks and pulls metal from both surfaces. The obvious symptom of valve recession is the steady loss of valve operating clearance or valve lash.

Valve recession used to be controlled with lead in gasoline. Gasoline provided a steady stream of soft metallic lead that coated the valve faces and seats. When the valves opened and closed, the lead welded and broke instead of the valves and seats themselves. The gasoline continually renewed the coating and valve recession was slow.

Unleaded gasolines were introduced in 1974 for 1975-model vehicles with catalytic converters, which

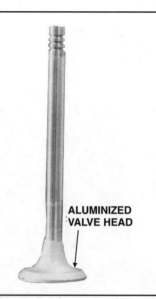

Figure 8-9. Applying aluminum oxide to the valve head helps the valve resist corrosion.

Figure 8-10. Corrosion occurs when the metal combines with oxygen.

were damaged by lead additives. Valve recession can be a major problem for older vehicles running on leaded gasoline, so valves and seats with hard, wear-resistance sealing surfaces are indispensable on cars that run on unleaded fuel. Automotive machinists can retrofit older vehicles fairly easily with new hard valves and seats to make them compatible with unleaded fuel.

Valve face angles
All valve face angles are measured from the horizontal on a line drawn level with the top of the valve head, figure 8-12. This is known as the included angle. The angle itself includes or contains the head of the valve.

Many engines used to have a 30-degree face and seat angle. This angle was thought to allow a better flow of the intake mixture over the edge of the seat. 30-degree valve faces have been almost universally abandoned in favor of 45-degree valve faces. The 45-degree valve face is better at self-centering the valve as it closes and wedges it tighter in the seat for a better seal.

Actually, the 45-degree figure is a general dimension. Some machinists grind the valve face at 44 degrees and the seat at 45 degrees, figure 8-13. The difference in these two angles is called the **interference angle**, and it creates a smaller contact area between the valve and seat. The purpose of the interference angle is to improve valve sealing when the engine is first started after a valve job. Also, the interference angle helps cut up and wedge dirt or carbon from the seat.

Other machinists believe that the interference fit is a temporary way to cover up sloppy work. The small

contact area seals quickly and tightly at first, but it creates a higher loading area between the valve and seat, increasing wear. These machinists believe carefully grinding both seat and valve face to 45 degrees on quality equipment provides the best seal and longest life.

Valve cooling
Because the valve heads form part of the combustion chamber, they get very hot. The intake valve runs

---

■ Why the Poppet Valve?

Have you ever wondered why the poppet valve is so widely used in automotive engines? In the early days of the industry, there were other types of valve arrangements. The Stearns-Knight car used the Knight sleeve valve engine. The valves consisted of two concentric sleeves surrounding the piston located between the piston and the cylinder walls. A separate shaft moved the sleeves up or down to open and close the intake and exhaust ports.

There were also rotary valve engines. A long shaft with holes in it allowed air and exhaust to flow when the holes aligned with the intake and exhaust ports. There were even slide valve engines, which had a sliding bar with holes in it that aligned with ports in the cylinders.

But these valve designs proved to be inferior to the simple poppet valve, mainly because they had a tendency to leak. The poppet valve wedges into its seat when it closes, which produces a good, tight seal. And the higher the combustion pressure, the tighter it seals. But it still opens easily with a push. It's no wonder that the poppet valve won out over the other designs.

Interference Angle: The angle or difference between the valve face angle and the valve seat angle.

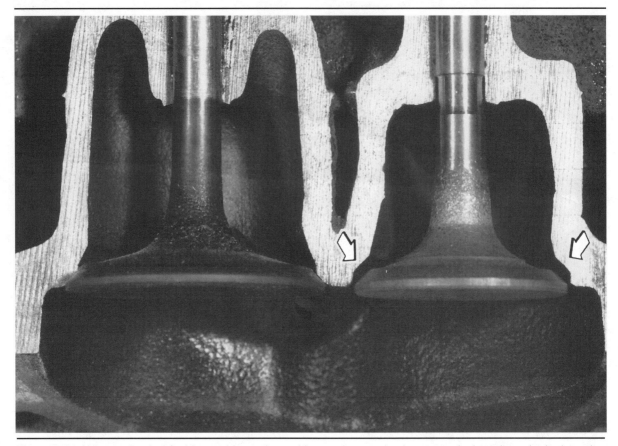

Figure 8-11. When unleaded fuel is used in engines without valves and valve seats designed to take it, the valve seats wear so badly that they appear to recede into the head.

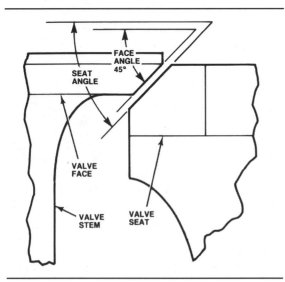

Figure 8-12. Valve angles are measured at right angles to the valve stem.

cooler than the exhaust valve. The incoming air-fuel mixture holds the intake valve head temperature down to about 800°F (425°C). Exhaust valves run hotter because the hot exhaust gases flow past them to escape from the combustion chamber. The exhaust valve temperature can average 1,200°F (650°C).

This heat must be removed or the valve will warp, leak, or burn. Some heat is transferred to the valve stem and then to the guide, but the maximum heat transfer is from the valve to the valve seat. Of course, this can only happen when they are in contact with each other. It is essential that the valve be held tightly against its seat. If the valve springs are weak and not able to do this, the valve will overheat and burn.

A valve will burn if a leak develops at any point around the valve seat. The area around the leak will not cool because it is not in contact with the seat and so cannot transfer heat. As the valve gets hotter, it will start to warp, which will, in turn, result in more heat buildup. This chain reaction quickly melts the edge of the valve and the engine loses efficiency and power. A burned valve has a notched and channeled edge, as if it has been melted by a cutting torch.

Filling valve stems with sodium is a method of cooling the valve to reduce wear. The stem, and sometimes part of the head, is hollow. About half of this cavity is filled with metallic sodium, figure 8-14. Sodium melts at just below the boiling point of water, 208°F (98°C).

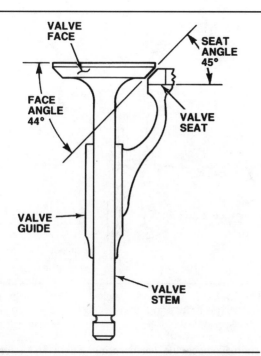

Figure 8-13. Some machinists grind the valve face at a 44-degree angle and the seat at a 45-degree angle. This one-degree difference is known as the interference angle.

During normal engine operation, the valves are much hotter than this, so the sodium stays liquid.

Sodium conducts heat well. During engine operation, the sodium bounces back and forth in its cavity and transfers the heat from the valve head to the valve stem. From the stem, the heat goes into the guide. On a normal valve, very little heat goes into the guide, so a sodium-filled valve is an excellent way of providing additional valve cooling.

Few passenger cars use sodium-filled valves because their engines are not highly tuned and do not produce excessive amounts of heat. Notable exceptions are the air-cooled Porsche engines and some turbocharged engines, which have increased combustion pressure. These engines can reach extremely high temperatures. You will also find sodium-cooled valves on turbocharged truck engines.

Sodium will explode if it comes into contact with water. If sodium comes in contact with your skin, it will cause a serious burn. If the hollow stem of the valve is cracked or broken, you have a potentially dangerous situation. Be very careful when handling sodium valves. Not all sodium valves are marked as such, but they generally have oversized stems.

SPRINGS

The valve spring must hold the valve closed on its seat whenever the lifter is on the heel of the camshaft lobe,

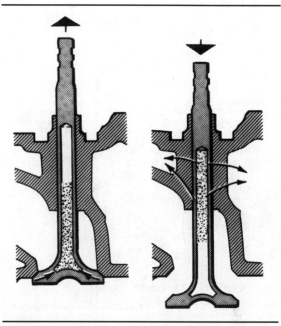

Figure 8-14. The stem of some valves is hollow and partially filled with metallic sodium to conduct heat away from the valve.

figure 8-15. It must also maintain tension in the valve train when the valve is open to prevent valve float, but must not exert so much tension that the cam lobes and lifters begin to wear.

Valve Float

Valve float occurs when the spring cannot keep the valve or lifters in contact with the cam lobe. As the cam lobe starts to open the valve, it sets the components of the valve train in motion. As the valve reaches the fully open position, the inertia of the valve train tends to continue to open the valve. If the inertia is great enough, it will continue opening the valve until the spring is completely compressed. At this point, the spring will attempt to close the valve with great force. Because the cam lobe has moved out from under the lifter, there is no resistance against the spring, and the parts of the valve train will come crashing together. This phenomenon of valves opening through inertia is called **valve float**, and it will quickly damage the valves, seats, and springs.

Valve Float: The condition in which the valve continues to open or stays open after the cam lobe has moved from under the lifter. This happens when the inertia of the valve train at high speeds overcomes the valve spring tension.

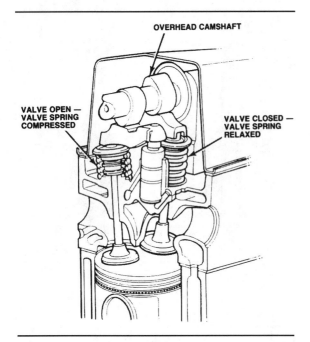

Figure 8-15. When a valve is open, its spring is compressed. When a valve is closed, its spring is only partially relaxed, maintaining tension in the valve train.

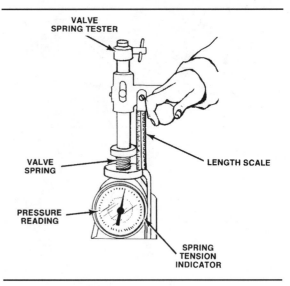

Figure 8-16. Check the valve spring length and strength with a spring compressor.

Valve float may also cause **lifter pump-up** if the engine has hydraulic lifters. Valve float creates a sudden slack in the valve train. The hydraulic lifter quickly takes up this slack because the hydraulic pressure in the lifter causes it to expand. This holds the valve open longer than necessary and the engine loses power. These conditions occur when an engine is run much faster than the speed for which it was designed. In Chapter 10 we detail hydraulic lifters.

Spring Material

Valve springs are made from wire. Valve spring manufacturers select metal that can withstand heat and resist loss of tension. The metal must be able to go through millions of compression and expansion cycles without losing tension. The better springs are made from steel alloys containing chromium and vanadium. Springs are usually shot peened to toughen the outer surface and ground to blend in any stress risers, which lessens the possibility of cracking.

In spite of the precautions designers and manufacturers take, valve springs do break. The pieces of the spring may still exert some pressure on the valve so the cylinder will continue to operate, but at reduced power. It is possible for a broken valve spring to allow the valve to drop down and damage the piston.

You should replace valve springs not only when they are broken, but also when they are not up to specifications. When you are repairing or rebuilding an

engine, check the valve spring length and strength with a valve spring tool, figure 8-16. Compare your measurements to those given by the car manufacturer.

Spring Installations

Valve springs are held in place by a spring seat, retainer, and two split keepers, figure 8-17. The spring seat can be a circular area cut out of the head to keep the spring from shifting sideways, or a washer with a ledge to locate the spring, figure 8-18. The valve stem end of the spring is held by the retainer and split keepers. During assembly, the retainer is put onto the valve spring and the valve spring is compressed; then the keepers are placed in the grooves near the valve stem tip. When the spring pressure is released, the keepers are wedged into the retainer. At the same time, the spring pressure pushes the valve against its seat.

Spring retainers

Although all retainers and split keepers work on the same principle, they are not interchangeable. There are many different arrangements of grooves on the valve stem, figure 8-19. The keepers are designed to fit these grooves. Also, the retainers must be the correct diameter for the spring. The retainer is usually stepped to locate the end of the spring and keep the spring from sliding on the retainer. When dual springs are used, the retainer might have two ledges, one for each spring.

Spring height

The distance between the retainer and the spring seat on the cylinder head determines the tension of the spring when it is installed. Refacing the valve or its seat

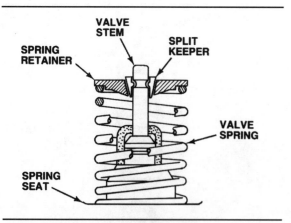

Figure 8-17. A retainer, two split keepers, and a spring seat hold the valve spring in place.

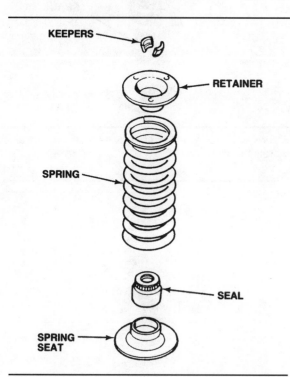

Figure 8-18. This Honda spring assembly uses a steel spring seat.

will sink the valve farther into the seat. This moves the tip of the valve and the retainer away from the spring seat, and reduces valve spring tension. The dimension from the spring seat to the bottom of the retainer, called **spring height**, must be measured whenever springs are installed, figure 8-20. If the measurement is too great, spring spacer washers must be used between the spring and its seat to restore the tension.

Spring Design

Springs appear to be very simple parts but in reality they must operate under very severe conditions. They

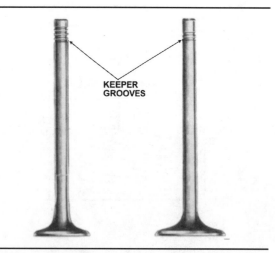

Figure 8-19. The keepers fit the number of grooves on the valve stem.

are exposed to great heat. They are also subject to vibrations or shock waves that travel through the spring like waves in the ocean. On a test fixture, using a strobe light, these waves can actually be seen traveling through the spring coils. At certain engine speeds, these shock waves coincide with the opening push of the cam lobe and cause harmonic vibrations in the springs.

When the vibrations become severe, the spring fails to exert a constant pressure on the valve train; this causes the valves to remain open when they are supposed to be closed. It also creates play or excessive tension in the valve train.

Through experimentation, engine designers have discovered that a simple valve spring does a marginal job on a modern engine.

Dual springs

Several design improvements increase spring performance. One system is the dual spring, figure 8-21. It

Valve Float: The condition in which the valve continues to open or stays open after the cam lobe has moved from under the lifter. This happens when the inertia of the valve train at high speeds overcomes the valve spring tension

Lifter Pump-up: The condition in which a hydraulic lifter adjusts or compensates for the additional play in the valve train created by valve float.

Spring Height: The dimension from the spring seat to the bottom of the retainer. Automotive machinists measure it when springs are installed.

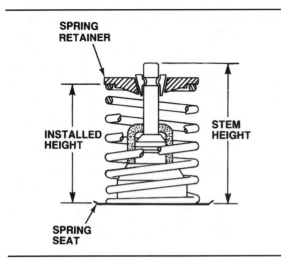

Figure 8-20. Measure the distance from the spring seat to the bottom of the retainer to find the installed height.

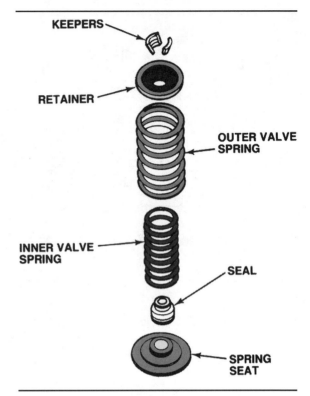

Figure 8-21. Dual valve springs reduce valve train vibrations.

is a pair of springs, an inner and an outer. The two springs not only have different diameters, but slightly different lengths. Because of their differences, they vibrate at different speeds and their harmonic vibrations cancel each other.

Dampers

Another way to reduce harmonic vibrations is to use a damper inside the spring, figure 8-22. This is a flat wound coil that rubs against the inside of the spring. The rubbing tends to subdue the harmonics. Dual round-wire springs may also rub together for the same effect.

Variable rate spring

A spring with unequally spaced coils at one end is called a **variable rate spring**, figure 8-23. The more it is compressed, the harder it pushes for the distance it is compressed. Whereas an evenly spaced coil spring might increase its pressure 50 pounds for each tenth of an inch of compression, the variable rate spring will increase its pressure 50 pounds the first tenth of an inch; 60 pounds the second tenth of an inch; 70 pounds the third tenth of an inch and so forth. When the most closely spaced coils touch each other, the rate of increase becomes less. This prevents excessive pressure on the tip of the cam lobe when the valve is fully open. Because the variable rate spring increases its pressure very quickly when it is opened, the pressure when the valve is closed can be reduced. This helps to prevent the valve from slamming onto its seat and reduces cam lobe wear.

When installing variable rate springs, always install them with the closely spaced coils against the cylinder head spring seat. This puts the lighter end of the spring next to the valve stem tip, and reduces some of the weight that must be removed in the valve train.

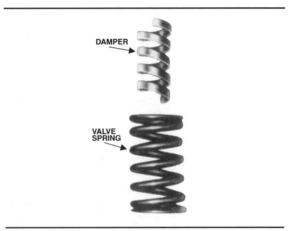

Figure 8-22. A damper inside the spring is another way of reducing vibrations.

Coil bind

The amount of space between the spring coils limits how much the spring can compress, and is determined by the wire diameter and how steeply the spring is wound. If the space is inadequate, the coils will jam against each other. This is called **coil bind**, figure 8-24.

With a properly designed spring, when the cam lobe lifts the valve fully open, some space will remain between the spring coils. However, if the engine is over-revved and the valves float, the coils can bind.

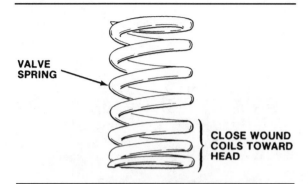

VALVE SPRING

CLOSE WOUND COILS TOWARD HEAD

Figure 8-23. Variable rate springs help reduce valve float by preventing the valve from slamming into its seat and bouncing up again.

Coil bind puts a tremendous strain on the parts because it is a case of an "irresistible force meeting an immovable object." Rocker arms and their supports can fracture and break, pushrods can bend, and lifters and cam lobes may be damaged.

Valve Rotators

The exhaust valve is especially prone to valve burning and wear because of the extremely high temperatures of the exhaust gases. If you could rotate the exhaust valve a little each time it opens, that little bit of friction caused by the rotation could help solve the problems of valve burning and sticking by scraping off the accumulation of carbon and emission deposits.

Many heavy-duty trucks, vehicles equipped for towing, and some cars are equipped with valve rotators. The rotators come in two different types: the release-type valve rotator and the positive valve rotator.

Release-type rotators allow the valve to move freely as the camshaft begins to lift the valve, figure 8-25. In this case, the valve turns slightly because of the vibration involved.

The positive valve rotator actually turns the valve several degrees every time the valve is opened. Small ramps inside the rotator cause the valve and spring to rotate. A design that uses small balls inside the rotator to turn the valve retainer, which turns the valve itself, is called Rotocap, figure 8-26. Another design that uses a coil spring instead of balls is called Rotocoil, figure 8-27.

Valve rotators are no longer as common or as necessary as they once were since hard alloy metals were introduced to cover valve faces and seats. Another reason for the phasing out of automobile valve rotators was the advent of unleaded gasoline. Without the lead to protect the valve seat, the friction of the slight rotation increased seat wear and recession.

■ **Distribution Pneumatique**

Leave it to the French. The British have been saying that they are full of hot air for years. Now, they prove it by using it in their valve trains. *Distribution Pneumatique* is a clever pressurized gas system that eliminates the valve springs, developed by French carmaker Renault for its Grand Prix race cars. The origins of this radical system can be traced to the early 1980s when the Renault racers ran into valve spring problems. Renault designers looked into desmodromic systems, but were put off by its complexity. By 1984, they debuted a system that used compressed nitrogen to close the valves. It immediately allowed reliable valve operation at 12,500 rpm and greater flexibility in choosing camshaft profiles. Further development raised that figure to more than 15,000 rpm in the late 1990s. All Formula 1 engines now employ a similar system.

Distribution Pneumatique used a flange on the valve stem to form a piston that ran in a nitrogen filled chamber. Each chamber was part of a network pressurized to 17 to 26 psi (117 to 179 kPa). The flange on the valve compressed the nitrogen as the cam lifted the valve. The compressed nitrogen then closed the valve very securely when the camshaft allowed it. Increasing the gas pressure is equivalent to fitting stiffer valve springs, but without the consequence of the heavier weight of the bigger springs. A further benefit of the gas is a self-damping effect, reducing vibration in the valve train.

A danger with pressurized valves is a loss of gas pressure, which would have the same result as a weakening or broken valve spring. The Renault system monitored the nitrogen network and an automatic valve let in more gas from a reserve bottle stored in the central valley of the engine.

The system was as popular with the drivers as it was with the engineers because it granted an extra margin of valve train safety. Engines with *Distribution Pneumatique* survived missed down shifts that sent the engine revs soaring yet allowed the car to continue and the driver to win the race.

Variable Rate Spring: A spring that changes its rate of pressure increase as it is compressed. This is achieved by unequal spacing of the spring coils.

Coil Bind: The condition in which all the coils of a spring are touching each other so that no further compression is possible.

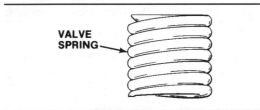

Figure 8-24. If the space between coils is not adequate, coil bind occurs.

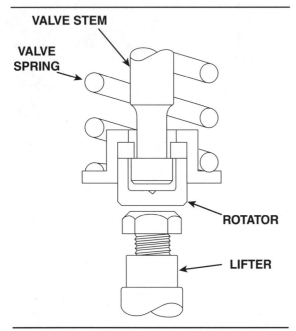

Figure 8-25. The release-type rotator on this flathead engine allows the valve to move freely when the camshaft begins to lift the valve.

GUIDES

Valve stems are supported by valve guides. There are two types of guides. A removable tube which is pressed or driven into the head is an insert guide, figure 8-28. The second type is an integral guide, figure 8-28, and it is simply a hole bored through a projection in the casting.

Insert Guides

Insert guides are made of cast iron or silicon bronze, figure 8-29. Although the silicon bronze is more expensive, it also has better wear characteristics. Designers always specify insert valve guides for aluminum heads because the aluminum could not function as its own guide material. It would wear too quickly.

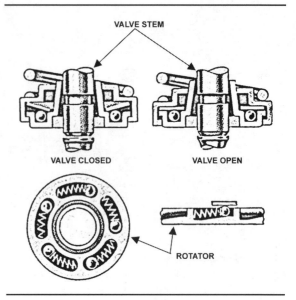

Figure 8-26. A positive valve rotor actually turns the valve.

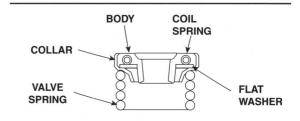

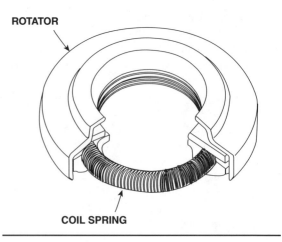

Figure 8-27. This valve rotator has a coil spring to rotate the valve.

Integral Guides

Because integral guides are part of the engine casting they cannot be removed if worn, although they can be reconditioned. Machinists repair integral guides by knurling, by machining the guide to accept an insert

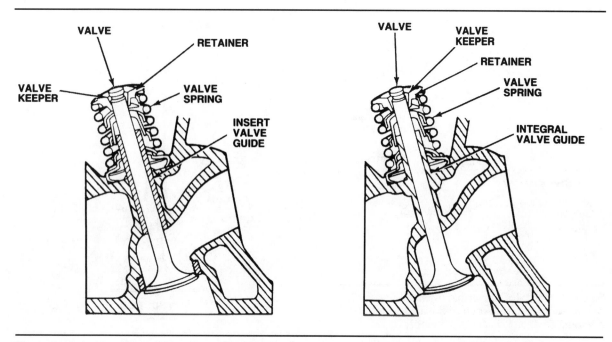

Figure 8-28. Insert valve guides are removable tubes driven into the head. Integral valve guides are part of the head casting.

Figure 8-29. Insert guides are usually made of cast iron or silicon bronze.

type guide, or by tapping the guide bore and screwing in a special coil. These procedures are covered in your *Shop Manual*.

Valve Guide Clearances

Guide to valve-stem clearances must be kept small to prevent oil from entering the combustion chamber or exhaust system through the space between the valve and its guide. Intake valve guide clearances are normally smaller than exhaust valve guide clearances, to prevent the intake port vacuum from pulling in too much oil. Also, because exhaust valves operate at higher temperatures they expand more and require larger initial clearance when cold.

Intake valve stem to guide clearances normally range from 0.001 to 0.003 inch (0.02 to 0.08 mm) while exhaust valves vary from 0.0015 to 0.0035 inch (0.04 to 0.09 mm). These figures represent the average ranges of tolerance; in all cases, the minimum possible clearance within the manufacturers' specifications is most desirable for maximum valve and guide life.

VALVE GUIDE SEALS

In modern engines, valve stems and guides are lubricated by oil dripping from above. In an L-head engine, the valve stems are lubricated by oil vapor and splash. This is possible because the valve chamber is close to the spinning crankshaft, which throws off large amounts of oil.

In an overhead-valve or overhead-cam engine, the valves are a long way from the crankshaft. Oil is pumped through passages or hollow pushrods to the rocker arms or the camshaft. From there the oil runs onto the valve stems and into the guides. In the process, the wear points on the pushrod and rocker arm or cam bearings, and valve stem and guide are lubricated.

A seal is installed on the stems and guides to prevent too much oil from traveling down the guide and onto the fillet and back of the valve head. When the intake valve opens, engine vacuum draws this oil into the combustion chamber where it oxidizes into hard carbon that clings to the chamber walls, figure 8-30. Oil that runs down an exhaust valve is pulled through the guide by a light vacuum in the exhaust. The

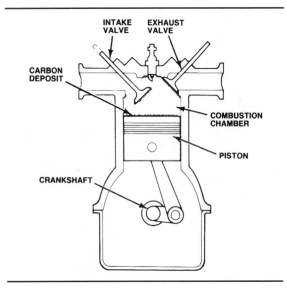

Figure 8-30. If oil is drawn into the combustion chamber, it oxidizes into carbon and clings to the valves, combustion chamber, and piston top.

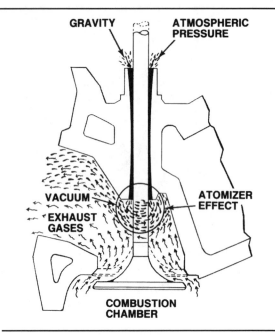

Figure 8-31. The oil that runs down the exhaust valve is pulled in by vacuum.

vacuum is caused by the rush of exhaust gases past the end of the guide, figure 8-31. Because of the high valve temperature, oil can oxidize on the fillets and the backs of the valve heads. The deposits can become so thick that they restrict flow, figure 8-32. Deposits in the chamber raise the compression level and may cause detonation, or become red hot and cause pre-ignition.

However, premature wear occurs if too little oil gets to the stems and guides. Unlike other seals or gaskets, the valve stem seal must regulate the flow of oil, not stop it altogether. It must provide a controlled leak — a much more difficult job. Valve guide seals are small synthetic rubber or plastic components that seal the valve guides from excess oil. Auto manufacturers use three types:

- O-ring
- Umbrella
- Positive.

O-Ring Seals

Because the valve retainer is cupshaped, it will collect oil and funnel it down the valve stem. To prevent this, a small squarecut rubber O-ring is positioned in a groove on the valve stem, figure 8-33. If you examine a keeper and valve stem you will notice that there is one extra groove on those valves that use retainer seals. The upper groove or grooves are for the split keeper. The groove below is for the ring seal. When properly installed, the ring seal will prevent oil from funneling down through the retainer.

Figure 8-32. Oil can oxidize on the valve fillet and the back of the head, obstructing the flow of the air-fuel mixture.

In conjunction with the O-ring seal, vehicle manufacturers use deflectors or shields to keep the oil away from the upper end of the valve guide. On some engines, the shields are installed directly under the spring retainer, figure 8-34. The shield covers the top few coils of the spring. Oil that splashes on the shield runs down the outside of the spring and misses the end of the guide.

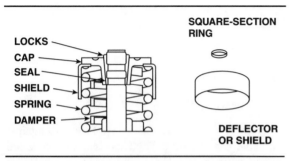

Figure 8-33. A small, square-section rubber ring is often used on the valve stem to prevent oil seeping into the combustion chamber.

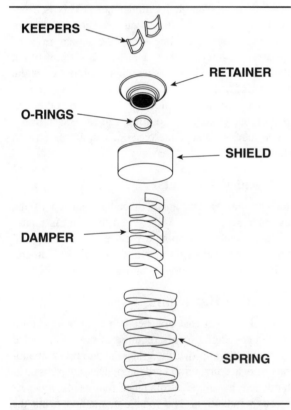

Figure 8-34. Shields may be installed under the spring retainer.

Umbrella Seals

Umbrella seals are simple rubber or plastic seals that fit loosely over the end of the valve guide, figure 8-35, covering it like an umbrella. The valve stem fits through the hole in the umbrella seal, centering it over the guide. The seal is a loose fit over the guide, but a tight fit on the valve stem, so it goes up and down with the valve. Any oil that splashes on the valve stem above the guide will run off the seal instead of going between the valve stem and guide.

This type of seal can be responsible for too much oil going down the guides. If you install an incorrect

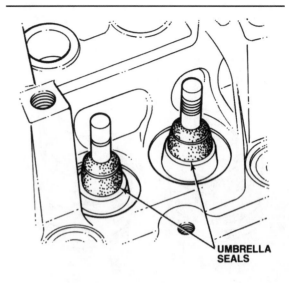

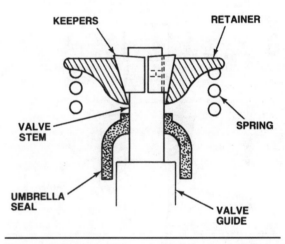

Figure 8-35. Umbrella seals install over the valve guides.

seal with a length that is too short, it will allow oil underneath the seal. As the seal goes up and down with the valve, a pumping action forces the oil between the stem and guide and into the combustion chamber.

Positive Seals

Other types of seals are free to move up and down with the valve stem, but positive seals fit snugly onto the top of the valve guide. They may be rubber with a Teflon® insert or completely Teflon®, figure 8-36. A Teflon® insert rubs on the valve stem and allows a measured amount of oil to reach the stem and guide, but prevents too much oil from passing through the guide into the combustion chamber. Some of the guide seals are held in place with a light spring. Others go over the end of the valve guide and are held in place

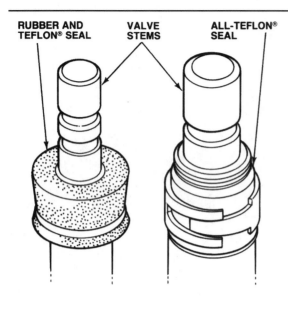

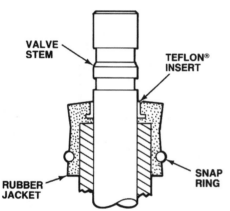

Figure 8-36. Positive valve seals are the most effective and most expensive type.

by a spring clip or by being pressed onto the guide. Because of the large amount of oil necessary to lubricate overhead-camshafts, manufacturers of these types of engines generally use positive seals on both intake and exhaust valves.

The aftermarket sells positive guide seals for engines originally equipped with O-ring or umbrella seals. These positive guide seals require machining of the end of the valve guide. You must remove the valves from the cylinder head to do this.

Some vehicle manufacturers use O-ring seals and oil shields on the exhaust valve and positive oil seals on the intake valves. Positive seals are more important on the intake valve guides because the intake vacuum tends to draw oil down the intake valve.

Unexplained high oil consumption can be caused by lack of oil seals on the exhaust valve guides. The rush of exhaust creates enough suction to draw oil down the guide. For maximum oil control, positive guide seals can be installed on both valves. The low viscosity (5W-30 in most cases) of the oils recommended by vehicle manufacturers these days allows them to slip through more easily.

VALVE SEATS

When the valve is closed, the valve face rests against the valve seat, figure 8-37. If the seat and valve are ground to fit each other, the resulting seal prevents any leakage. A good seal is also important for valve heat dissipation. When the valve seats, it transfers some of the heat it absorbed during the combustion process to the seat and cylinder head. This is especially important for exhaust valves. The intake valve gets most of its cooling from the intake flow of the fresh air-fuel mixture. Unless it is sodium-filled, the exhaust valve head has no other cooling except the transfer of heat to the valve seat.

Valve seats are called integral seats when they are part of the cylinder head casting, and valve seat inserts when they are a separate piece of metal pressed into the head.

Integral Valve Seats

Integral valve seats are a machined portion of the cylinder head, figure 8-38. When the head is manufactured, cutters machine a precise seat in the combustion chamber at the opening to the port. Integral valve seats are common on cast-iron heads.

Induction Hardening

Integral seats in cast-iron heads had sufficient durability until engines were required to run on unleaded gasoline. To improve integral seat durability, manufacturers use induction hardening. The integral seat is hardened by an electromagnet that heats the seat through induction. The induction method heats the valve seat to a temperature of approximately 1,700°F (930°C). This hardens the valve seat to a depth of about 0.060 inch (1.5 mm), and gives it a more durable finish. Sometimes only the exhaust seats are hardened because they have the most difficult service life. Chrysler, Ford, and General Motors all use induction hardening in their cast-iron car engines.

Valve Seat Inserts

If a separate seat is inserted in a groove in the head, it is called a valve seat insert, figure 8-39. Manufacturers always use valve seat inserts in aluminum heads for the same reason insert guides are always used in aluminum heads — the aluminum is not strong enough for the job and would wear too quickly.

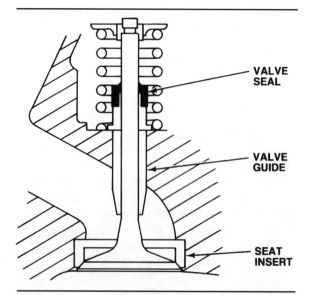

Figure 8-37. The seal between the valve face and valve seat is very important to good engine operation.

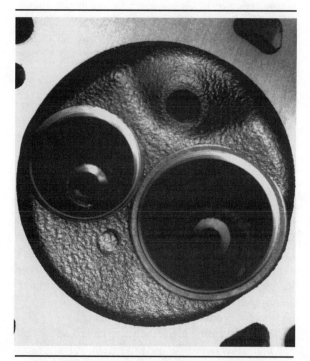

Figure 8-38. A machined portion of the cylinder head serves as the valve seat on most cast-iron heads.

A standard cylinder head repair procedure for machinists is to install valve seat inserts. Valve seat recession may progress to the point where it is impossible to grind a good seat in the cylinder head that you are performing a valve job on. The remainder of the old insert is pulled or pried out or the old integral seat is machined out; then, a new insert is pressed in, making the head like new.

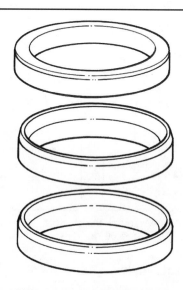

Figure 8-39. Insert valve seats are rings of metal driven into the head. They are made of hard steel alloys.

Sometimes seat inserts are installed into a cast-iron cylinder head to provide longer seat life. In many light truck and most diesel applications, manufacturers install stronger insert seats in cast-iron heads at the factory. Also, machinists are called upon to upgrade an older cast-iron cylinder head with seat inserts so that it can run on unleaded gasoline, or so that it can be converted to run on alcohol or propane. None of these fuels provides the necessary seat and valve lubricating properties.

Hard seat inserts are the best, but also the most expensive way of prolonging seat life. Installation of the seat inserts requires several machining operations, which raises production costs.

Valve Seat Insert Material

Valve seat inserts are made of cast iron, bronze, or steel alloys. The cast-iron seats are only for the repair of cast-iron heads not originally equipped with inserts.

A few factory-installed inserts in aluminum heads are bronze. Designers claim better heat transfer from the valve to the head with this material, prolonging valve life.

Alloy-steel valve seat inserts come in several grades of stainless steel with several degrees of hardness. These seats are often factory installed, may be used to make a repair, or used to upgrade a cylinder head. The steel valve seat alloys are able to withstand high temperature, the lack of lead lubrication, and more pounding with less wear. One of the hardest seats is made of Stellite or a combination of cobalt, chromium, and tungsten.

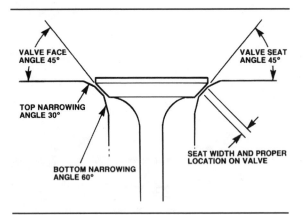

Figure 8-40. A typical three-angle valve job uses angles of 30, 45, and 60 degrees.

Valve Seat Angles

Ideally, machinists grind valve seats to three separate angles, figure 8-40. Machinists grind the middle part of the seat first. This is where the valve actually seals, and it is ground to the same angle as the valve, a 45-degree seat for a 45-degree valve.

The other two angles are 30 degrees and 60 degrees for a typical 45-degree valve. These two angles serve three functions. First, they help narrow the valve seat to the proper width of about $3/32$ inch (2.4 mm) for exhaust valves and $1/16$ inch (1.6 mm) for intakes. A narrow seat seals well but does not allow proper heat transfer; nor is a narrow seat long lasting. A wide seat is just the opposite, so the width figures listed above are a good compromise.

Second, one of a machinist's objectives in a valve job is to make the seat contact the valve in the outer third of the valve face, improving heat dissipation. This gives the heat a shorter flow path from the hottest part of the valve (the combustion face) to the valve seat. The 30- and 60-degree angles help a machinist position the seat contact area at the right place on the valve face. The second angle a machinist grinds is the 30-degree angle on top of the seat. Grinding away material from the top of the seat helps lower the surface where the seat contacts the valve face. The third and last angle a machinist grinds is the 60-degree angle. Removing material from below helps raise the contact area.

Third, the seat is part of the port and the entrance to the combustion chamber. Having three angles smoothes the flow in through the intake port and out the exhaust port. Improved flow raises horsepower.

Some modern valve grinding equipment cuts all three angles at one time. The *Shop Manual* will cover valve jobs in more detail.

VALVE, SPRING, GUIDE, AND SEAT RELATIONSHIPS

The valve, spring, guide, and seat work as a team. If something is wrong with any one of them, it will affect the others. For example, it does little good to put new valves in worn guides. Clearance between the stem and guide may still be excessive, allowing too much oil consumption. The valve will wobble in its guide and not seat properly, figure 8-8. This will cause rapid wear and valve burning.

If a seat or valve is ground or cut to renew it, the valve stem will protrude through the head by a distance equal to the amount of metal that was removed. It will change the angle between the tip of the valve stem and the rocker arm. The original angle can be restored by grinding the end of the valve.

The height of the spring when the valve is installed should be within the manufacturer's specifications or the spring pressure will not be correct. With less spring pressure, the valve will burn more easily. If spring pressure is not correct, you can install spacer washers under the spring to bring the spring to the correct, assembled height. You must also check the springs for the correct tension on a spring testing machine, figure 8-23.

SUMMARY

All 4-stroke piston engines use intake and exhaust poppet valves. The main parts of poppet valves are the head, combustion surface, stem, keeper grooves, tip, fillet, face, and margin. Some engines use valves with 30-degree sealing faces, but most use 45-degree faces.

Intake valves may be made from plain carbon steels, but, for higher strength, can be made of special alloy steels. Exhaust valves are always made of high-quality alloys. Manufacturers may build heavy-duty exhaust valves from nickel-based superalloys. For longer life, some valves are hard chromed, nitrided, aluminized, or faced with Stellite. Exhaust valves may be filled with sodium to improve cooling.

The valve spring holds the valve closed on its seat and maintains tension in the valve train. Valve springs are made from steel wire alloyed with chromium and vanadium. Springs are usually shot-peened to toughen the outer surface and ground to blend in any stress risers. Valve springs are held in place by a spring seat, retainer, and two split keepers. The distance from the spring seat to the bottom of the retainer is the spring height, which you must measure when springs are installed. To control spring vibrations, manufacturers use dual springs, spring dampers, or variable rate springs.

Valve stems are supported by valve guides. There are two types of guides: insert guides and the integral guides. Insert guides are made of cast iron or silicon bronze. A seal is installed on the stems and guides to prevent too much oil from traveling down the guide and onto the fillet and back of the valve head. Vehicle manufacturers use three types of seals: O-ring, umbrella, and positive.

When the valve is closed, the valve face rests against its seat. The seat helps seal the port and allows the valve to dissipate heat. There are two types of seats: integral and insert. Integral seats may be induction hardened. Insert seats are made of cast iron, bronze, or steel alloy.

■ Desmodromic Valves

A desmodromic valve system employs the camshaft to both open and close the valves, rather than using springs for closing valves as most engines do today. If valve springs are used on desmos at all, they are small, lightweight springs for initial sealing only. Desmodromic valve trains add a second cam lobe and cam follower that closes the valve via a projection on the valve stem. Some of its main drawbacks are the added complexity and the expense of producing such a complicated system. Another drawback is the difficulty in setting the valve lash clearances—a painstaking process that can require several hours! But racers and builders of high-performance machinery once thought the drawbacks were worth the benefits.

When desmodromic valves were first used, valve train design and valve spring metallurgy were poor. Valve float greatly limited maximum rpm and engine power, and valve spring breakage from high-rpm running was a constant quandary. A desmodromic valve train eliminated the valve springs by positively closing as well as opening the valve. That immediately allowed more aggressive cam profiles without valve float, raised rpm limits, and ended the problem of broken valve springs. When valve train technology finally caught up with lightweight valves and durable valve springs, desmodromic valve trains faded from the scene.

The idea of a desmodromic valve train was first patented in 1910. The system was used in the French Delage GP cars during the 1914 season and in a few other racing cars in the 1920s and 1930s. The all-conquering Mercedes-Benz Wl96 Grand Prix cars and 300 SLR sports-racers used desmodromic valve gear in the 1950s, the last time desmodromic valves were used on a car. The only production vehicles using desmodromic valves today are certain Italian Ducati road and racing motorcycles.

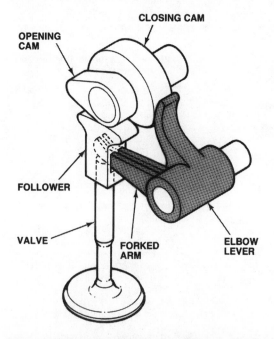

OPENING CAM
CLOSING CAM
FOLLOWER
VALVE
FORKED ARM
ELBOW LEVER

Review Questions
Choose the single most correct answer.
Compare your answers with the correct answers on Page 328.

1. Most engines now use:
   a. 30-degree valve faces
   b. 35-degree valve faces
   c. 40-degree valve faces
   d. 45-degree valve faces

2. If the heat is not removed from the exhaust valves, they will:
   a. Leak
   b. Warp
   c. Burn
   d. All of the above

3. In modern engines, valves are lubricated in their guides by oil:
   a. Vapor and splash
   b. Dripping from above
   c. Injection
   d. Under pressure

4. The curved portion of the valve where the stem meets the head is called the:
   a. Fillet
   b. Flow ramp
   c. Combustion surface
   d. Tip

5. In lubricating the valve train, which of the following wear points may also lubricated?
   a. Pushrod
   b. Rocker arm
   c. Guide
   d. All of the above

6. The combination of chromium, cobalt, and tungsten used for valve faces is called:
   a. Chrome alloy
   b. Stellite
   c. Coballite
   d. Chrome flashing

7. If the valve spring height is greater than specification, you should:
   a. Grind the valve stem
   b. Grind the valve seat
   c. Grind the valve face
   d. Install spacer washers on the spring seat

8. Integral guides cannot be repaired by:
   a. Knurling
   b. Tapping the guide bore and screwing in a special insert
   c. Machining the guide to accept an insert-type guide
   d. Welding the guide to reduce clearance

9. Valve guide seals may be installed on:
   a. Only intake valves
   b. Only exhaust valves
   c. Both intake and exhaust valves
   d. Neither intake nor exhaust valves

10. When rebuilding an engine it is important to replace the valve seat inserts in:
    a. A different location
    b. The same location but at a slightly different angle
    c. Exactly the same location and at the same angle
    d. A different location at exactly the same angle

11. 4-stroke engines use:
    a. Slip-joint valves
    b. Poppet valves
    c. Rotary valves
    d. Reed valves

12. The greater mass of a large valve:
    a. Causes it to seal better when it hits the seat
    b. Reduces the possibility of valve float
    c. May cause it to bounce when it hits the seat
    d. Reduces wear in the valve train

13. Technician A says that maintaining a margin around the valve head is important because it makes the valve flow air better. Technician B says the margin makes the valve stronger and more resistant to burning. Who is right?
    a. A only
    b. B only
    c. Both A and B
    d. Neither A nor B

14. To perform a typical three-angle valve job, a Technician grinds which three angles on the valve seat?
    a. 90, 45, 110
    b. 30, 45, 60
    c. 7, 45, 130
    d. 35, 45, 65

15. Technician A says that manufacturers may weld two different metals together to make a valve. Technician B says exhaust valves are often made of alloys superior to the metals intake valves are made from. Who is right?
    a. A only
    b. B only
    c. Both A and B
    d. Neither A nor B

16. To fight corrosion, manufacturers coat the head of the valve with:
    a. Chromium
    b. Molybdenum disulfide
    c. Aluminum oxide
    d. Stellite

17. Lifter pump-up occurs:
    a. During valve float
    b. If the oil pressure is too high
    c. When the lifter relief port is blocked
    d. During extreme engine load conditions

18. Designers may reduce valve seat wear by using:
    a. Soft-faced valves
    b. Unleaded gasoline
    c. Aluminum valve seat inserts
    d. Induction hardened valve seats

19. Technician A says valve recession can be a major problem when lead is removed from gasoline. Technician B says that older cars can be updated with valves and valve seats that make them compatible with unleaded gasoline. Who is right?
    a. A only
    b. B only
    c. Both A and B
    d. Neither A nor B

20. Technician A says you must be careful with sodium-cooled valves because if the sodium leaks out of a damaged valve and comes in contact with your skin, it can cause a serious burn. Technician B says you must be careful with sodium-cooled valves because sodium bursts into flame when it contacts motor oil. Who is right?
    a. A only
    b. B only
    c. Both A and B
    d. Neither A nor B

9

Camshafts, Camshaft Drives, and Valve Timing

The camshaft controls the opening and closing of the valves. It operates in time with the engine's needs, which the engine designer determines. It is the heart of an engine's "state of tune" and determines to a great degree engine performance. Understanding the camshaft is essential to knowing why an engine runs the way it does. Changing it can have a big effect on how an engine runs or performs.

Automotive machinists do not perform much work on camshafts or camshaft drives, but they must inspect them for condition and proper function, and they must know the correct way to assemble this part of the valve train.

CAMSHAFTS

The camshaft is a series of egg-shaped lobes, or **cam lobes**, ground onto a long shaft, figure 9-1. Camshafts commonly spin in plain bearings or directly within the aluminum casting of the engine. Some camshafts spin in needle bearings or ball bearings. The camshaft bearing journals are precision ground surfaces that must be as smooth as a crankshaft journal.

There is usually one cam lobe for each valve in the engine. However, some 4-valve-per-cylinder engines use one lobe to operate both intake valves and another to operate both exhaust valves. Another exception is the flat, 4-cylinder engine used by Volkswagen. The engine had a total of eight overhead valves but only four cam lobes. Its layout allowed a single cam lobe to operate valves in opposing cylinders. Regardless of the number of cam lobes, the camshaft rotates and the cams push on the valve train to open the valves.

The cam lobes must open and close the valve at the proper time in relation to the position of the pistons. This is called valve timing. In a 4-cycle engine, the camshaft must operate at one-half crankshaft speed. The camshaft must be correctly timed to the crankshaft or the engine may not run at all. We detail valve timing later in this chapter.

Camshaft location and number are the primary differences between engines. They greatly affect the engine's size, weight, design complexity, cost, and power output. When designing an engine, engineers consider a vehicle's proposed retail price, its intended use, and how much power the engine will need before deciding where to place the camshaft and the number of camshafts to use. From a design standpoint, engineers change camshaft location and number to improve breathing and simplify and lighten the valve train. There are only two locations for camshafts: in the cylinder block or mounted in the head.

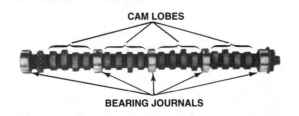

Figure 9-1. The camshaft consists of a series of cam lobes on a long shaft.

Camshaft in the Cylinder Block

Engines with camshafts in the block include L-head and overhead-valve (OHV) engines. The L-head or flathead engine design, figure 9-2, is small, simple, and lightweight. Since the cylinder head carries no valves or camshaft, the head is inexpensive to produce and relatively "flat," giving it its name. The valve train is also very simple and efficient; it is short and requires no pushrods or rocker arms. Camshafts with radical valve timing profiles have been used in flatheads without valve float. The great drawback of the flathead is its inefficient breathing and low power output for its size. The breathing problems are caused by the poor placement of its intake and exhaust ports, as explained in Chapter 7.

OHV engines keep their camshafts in the block, but locate all the valves in the cylinder head, figure 9-3. As we said in Chapter 7, the overhead valves allow a better placement for the intake and exhaust ports. The improved breathing gives the OHV higher power output for its weight than the flathead.

The OHV head is heavier, more expensive, and physically larger. The valve train is also more complex than the flathead. The OHV requires long pushrods to transmit the movement of the cam lobe to the valves. The OHV design is still common in some domestic V8, V6, and in-line 4-cylinder engines. These engines are also called "pushrod engines." The pushrods add weight and flex to the valve train, which increases the chances of valve float.

Domestic vehicle manufacturers have refined the modern OHV design since the late 1940s. Engineers have dealt with the theoretical drawbacks of OHV valve train so that they are not a problem on a street engine. Two popular, domestic V8 engines, which were originally designed in the 1950s and 1960s, the Chevrolet 5.7L and Ford 5.0L, are OHV powerplants.

Camshaft in the Cylinder Head

Engines with camshafts mounted in the head include the single-overhead-camshaft (SOHC) and

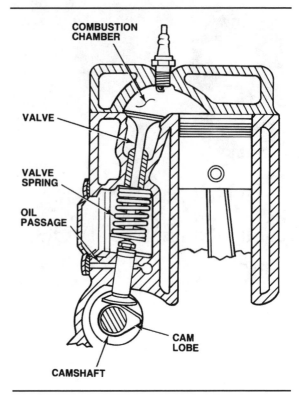

Figure 9-2. The L-head, or flathead, engine has a simple, lightweight valve train.

dual-overhead-camshaft (DOHC) engines. The SOHC engine, figure 9-4, is designed to keep the favorable valve placement of the OHV engine but eliminate the pushrods. With both the valves and camshafts housed in the head, the cylinder head becomes heavier, larger, and more costly than the OHV engine. To reduce engine weight, almost all SOHC engines have aluminum cylinder heads.

The SOHC engine has a light, fairly rigid valve train capable of operating efficiently at high rpm without valve float. Being able to run a SOHC engine at higher rpm improves its power output for its size. This is a great bonus for small engines and a major reason most small, 4-cylinder car engines and many V6 engines have overhead camshafts.

Most SOHC engines have two valves per cylinder, but both Honda and Mazda have a valve train system on selected models that uses rocker arms and a single camshaft to operate four valves per cylinder, figure 9-5. They employ this system to take advantage of the improved breathing provided by the extra valves, but without the complication of driving another camshaft.

The DOHC engine is the current high point in the evolution of the cylinder head, figure 9-6. Almost all manufacturers use a DOHC system on their most powerful high-performance engines. The "dual" in DOHC refers to the number of camshafts per head.

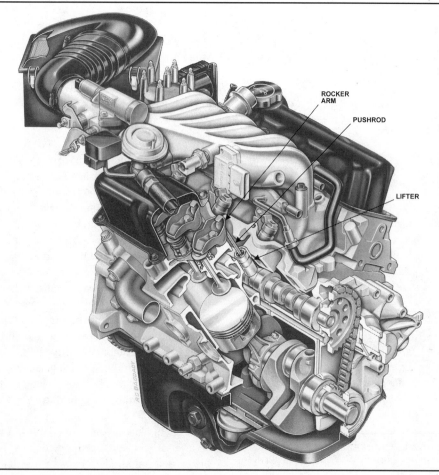

Figure 9-3. This Ford V6 uses lifters, pushrods, and rocker arms, to operate its overhead valves.

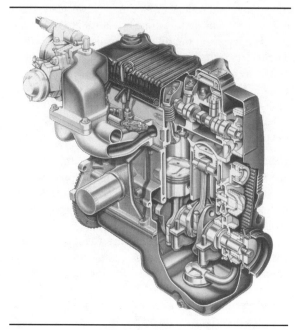

Figure 9-4. The single-overhead camshaft engine retains the overhead valves, but eliminates the pushrods.

A DOHC in-line 4-cylinder or 6-cylinder engine with a single cylinder head has a total of two camshafts. A DOHC V8 or V6 engine with two cylinder heads has a total of four camshafts — still two per head.

The use of two camshafts in the head gives the designer greater flexibity to place the intake and exhaust ports where they flow most efficiently. The designer can eliminate rocker arms and cam followers, making the valve train as light and simple as any head design, and giving the engine the highest rev potential. DOHC heads use one camshaft for the intake valves and another for the exhausts. The two camshafts also allow greater flexibility in camshaft timing, which we detail in a later section.

An additional advantage is that DOHC heads are naturals for building heads with four valves per

Cam Lobes: Egg-shaped lobes that open and close the valves. A series of cam lobes on a shaft is a camshaft.

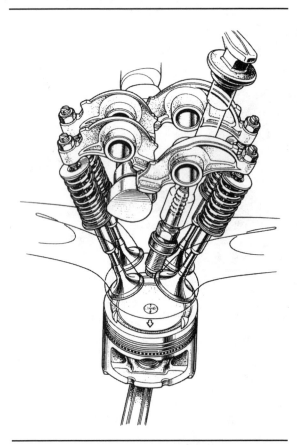

Figure 9-5. Both Mazda and Honda employ a system that operates four-valves-per-cylinder with a single camshaft.

cylinder. The breathing and power advantages of four-valve-heads were detailed in Chapter 7.

The drawbacks to the DOHC cylinder head are size, weight, and complexity. The two camshafts make the head the largest and heaviest of any design, and the most expensive to produce. Many DOHC engines use aluminum construction for both the head and block to reduce weight. The camshaft drive must also be larger and more complex to accommodate the additional valves.

CAMSHAFT DRIVES

The crankshaft drives the camshaft by chain, gears, belt or a combination of belt and chains or belt and gears. Whatever the drive method, the camshaft must revolve at one half the speed of the crankshaft. That is, for every full revolution of the crankshaft, the camshaft must turn only one-half revolution. The reason for this is that each valve opens only once for every two crankshaft revolutions.

For example, the intake valve opens on the intake stroke. Then the piston must go through the compression, power, and exhaust strokes before

getting to another intake stroke. These four strokes require two crankshaft revolutions, but if the camshaft is allowed to go through two revolutions in the same amount of time, the intake valve would open on the power stroke. Therefore, the camshaft is designed to turn only one revolution while the crankshaft goes through two, figure 9-7.

Chains

The most common form of camshaft drive is the chain. Most OHV engines use a simple, short chain drive from the crankshaft to the camshaft, figure 9-8. SOHC and DOHC engines may also use chain drive for the camshaft, figures 9-9 and 9-10. The chain drives on OHC engines must be longer and more complex than those on the OHV engines.

Chains give long, reliable service and plenty of warning before they break because they get very noisy when badly worn. Chains are made of carbon steel and are very strong for their intended use. The result is that the camshaft drive chain and sprockets can be thinner than a comparable drive belt.

Chain drive is quieter than gears but not as quiet as a belt. To reduce noise, some vehicle manufacturers use a **silent chain** in light-duty, OHV applications, figure 9-11. The silent chain is a sprocket and link-type chain that pivots around pins. Each link is a series of plates held together with pins. The number of plates determines the chain width. The word silent comes from a comparison with a **roller chain**, like those used on bicycles. The roller chain is made of links, pins, and rollers. Each link has two plates held together by pins and separated by a roller around the pin. The length of the rollers determines chain width. The roller chain is stronger, but noisier. Many OHC and heavy-duty OHV engines use the roller chain. Double roller chains, figure 9-12, are used in some high-performance applications.

Gears

Gears are the oldest type of camshaft drive, figure 9-13. Gears are also the strongest, most precise, and most reliable form of camshaft drive. Ford and GM in-line 6-cylinder light truck engines use gear drive. GM's Tech IV Code U in-line 4-cylinder used gear drive from 1985 to 1989. Volkswagen's flat, air-cooled engines also have gear-driven camshafts. Most engines designed exclusively for racing have gear-driven camshafts, figure 9-14.

A gear drive system to drive overhead camshafts requires a complex network of idler or intermediate gears to transfer the motion from the crankshaft to the camshafts. Gear drives are not usually used in OHC

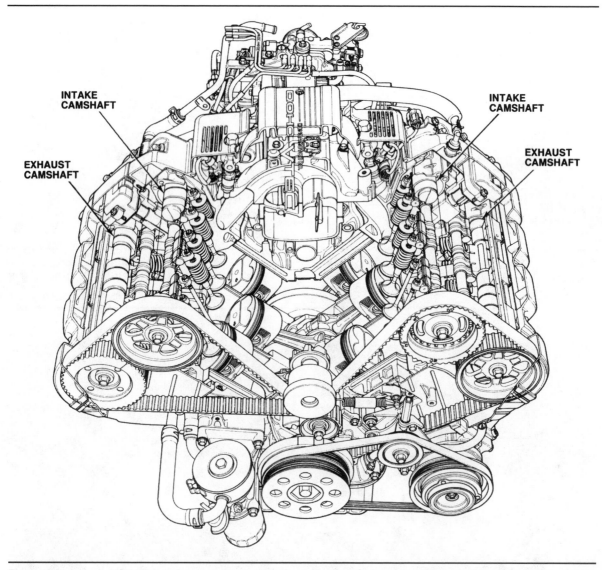

INTAKE
CAMSHAFT

EXHAUST
CAMSHAFT

INTAKE
CAMSHAFT

EXHAUST
CAMSHAFT

Figure 9-6. Acura's NSX V6 has four valves per cylinder like most dual-overhead-camshaft engines.

street engines because they are complex, expensive to produce, and noisier than other drive systems.

Engineers have used a number of designs to reduce gear noise. **Helical gears** — gear teeth cut with a slant — run quieter, figure 9-13. All manufacturers that employ gear-driven camshafts for their street engines use helical gears. Helical gears create a little more friction than straight-cut gears, so they are not used in race engines.

In 1989, Toyota introduced a special camshaft drive gear designed to reduce gear noise. It is used in selected DOHC in-line 4-cylinder, V6, and V8 engines in Toyota and Lexus models. A single, rubber timing belt drives the intake camshafts of both cylinder banks. Gears located on the intake camshafts drive the exhaust camshafts, figure 9-15. The driven gears on the exhaust camshafts are a special scissors design.

Silent Chain: A chain made of flat links that pivot around pins. Each link is a series of plates held together by the pins. The number of plates determines chain width.

Roller Chain: A chain made of links, pins, and rollers. Each link has two plates, held together by pins, and separated by a roller around the pin. The length of the rollers determines chain width.

Helical Gears: Gears with teeth cut at an angle to the shaft instead of parallel to it.

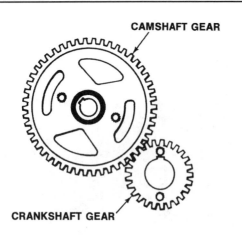

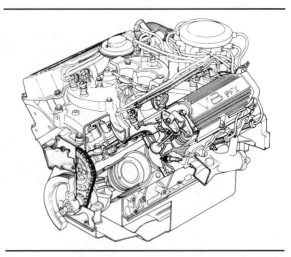

Figure 9-7. By making the camshaft gear twice as large as the crankshaft gear, the camshaft goes through one revolution for every two of the crankshaft.

Figure 9-8. Most OHV engines have a simple chain drive from the crankshaft to the camshaft. This Cadillac V8 uses a silent chain.

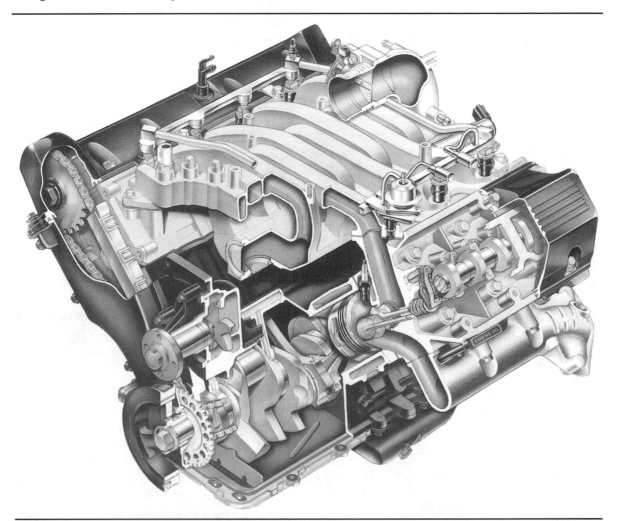

Figure 9-9. A roller chain drives the single overhead-camshaft in each cylinder head of the 4.6-liter V8 that Ford introduced in 1991.

Figure 9-10. This BMW dual-overhead-camshaft in-line 6-cylinder engine uses two single-row roller chains to drive its camshafts. The first chain runs from the crankshaft to a sprocket on the exhaust camshaft. A second chain runs from the exhaust camshaft to the intake camshaft.

Scissors gears operate more quietly than other designs when subjected to frequent changes in torque application. The scissors-gear mechanism on the exhaust camshaft uses a spring-loaded subgear with the same number of teeth as the driven gear on the exhaust camshaft, figure 9-16. Through the reaction force of the scissors spring, the subgear and driven gear teeth pinch the drive gear teeth on the intake camshaft. This pinching action reduces backlash to zero while still allowing some movement between gears. The elimination of backlash greatly reduces gear noise. Although these gears worked well, Toyota/Lexus switched back to belt drive for both camshaft sin the mid-1990s.

Belts

In the late 1960s, European designers began fitting camshaft drive belts to small-displacement OHC engines. Today, almost every manufacturer uses belt drives on at least a few of its models. These applications range from expensive luxury and high-performance cars to economy models, figures 9-17, 9-18, and 9-19. No manufacturers currently use belt drives on OHV engines, although a system is available from the aftermarket.

Cam-drive belts are made from rubber and fabric and are usually reinforced with fiberglass or kevlar. The belt and sprocket teeth are square-cut, or cogged.

Figure 9-11. The silent chain is common on OHV engines.

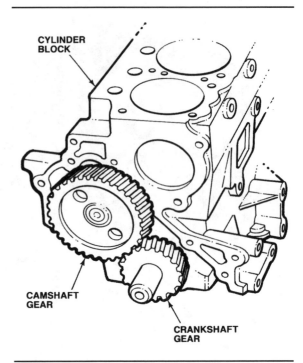

Figure 9-13. Gears are a very strong type of camshaft drive and give precise camshaft timing. The gears are usually given a helical slant, or cut, as they have on this engine, to make them operate quieter.

Figure 9-12. The roller chain is stronger than the silent chain. A double-row roller chain is often used in heavy-duty applications.

The belt-drive sprockets are made of iron or aluminum. Belt construction is unique to each engine design, varying in length, width, tooth profile, and tensioning requirements. Drive belts and sprockets reduce weight, require no lubrication, and produce less noise than the chain or gears they replace.

Belt construction and reliability have greatly improved since the first belts were introduced. However, most still require routine inspection, adjustment, and replacement. Early belts had a projected service life of about 30,000 miles (48,000 km), while late model engines call for replacement at about 60,000 miles (96,000 km). Belts give little or no warning before they break, so it is vital that they are replaced at the manufacturer's recommended interval.

Tensioners and Guides

The relatively long distance from the crankshaft to the camshaft on OHC engines requires that their drive chains or belts have tensioners and guides to prevent slap and take up slack as they wear. A slack chain can flap back and forth, making noise and increasing wear. Many chain-drive assemblies use one or two guides to control the slack side of the chain, figure 9-20. The guide usually has a metal body and a plastic or Teflon face on which the chain slides.

All OHC chain drives have a tensioner to take up slack as the chain and sprockets wear. Most tensioners consist of a plastic or fiber rubbing block. The tensioner is usually pushed out against the chain by engine oil pressure. These tensioners do not require periodic adjustment. A racheting device allows the tensioner to extend out against the chain but not to retract. A small spring keeps pressure on the tensioner until oil pressure builds up in the first few seconds when the engine is started. However, some chain tensioners are purely spring loaded and do not use engine oil pressure. They usually require periodic adjustment.

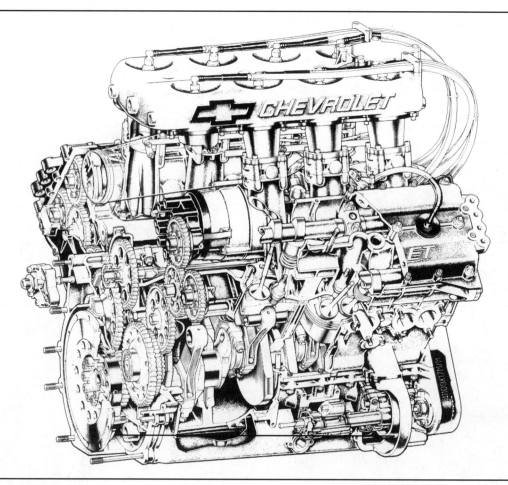

Figure 9-14. The DOHC Chevrolet Indy racing engine uses a complex system of straight-cut gears to drive its camshafts. This engine powered the winning cars in the four Indianapolis 500 races held from 1988 to 1991.

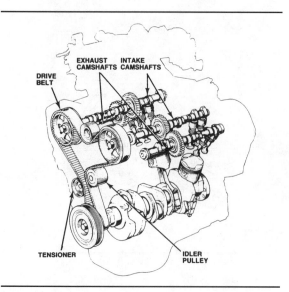

Figure 9-15. The camshaft drive for this DOHC Lexus V6 consists of a belt to drive the intake camshafts and gears to drive the exhaust camshafts.

On a belt-drive, the tensioner compensates for wear and actually determines the tension of the belt, figure 9-21. Belt-drive tensioners consist of a spring-loaded wheel that pushes inward on the back of the belt, figure 9-22. A fixed idler wheel eliminates slack on the opposite side of the belt. The wheels turn on sealed bearings.

On engines with belt-drive, you usually do not need to adjust the belt. A new belt is installed, and the tensioner is allowed to press against the belt. It maintains the proper tension until the belt's scheduled replacement.

Accessory Drives

Camshafts and camshaft drives can do more than open valves. They may also run other parts of the engine. In many OHV and OHC engines, a gear is mounted onto, or cast into, the shaft to drive the distributor, figure 9-23 and figure 9-24. The gear is usually helical to allow the distributor shaft to operate at a right angle to the camshaft. Both the camshaft gear and the distributor gear have the same number and size of teeth so

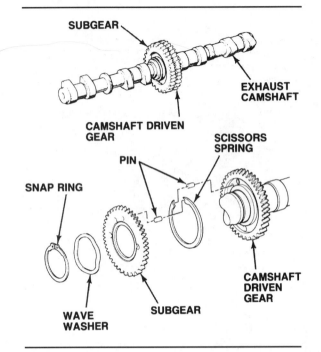

Figure 9-16. Quiet-running scissors gears drive the exhaust camshafts on some Toyota/Lexus DOHC engines.

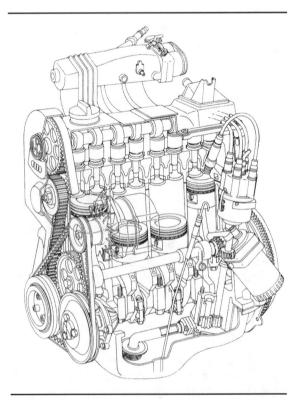

Figure 9-17. The 2.0-liter in-line 4-cylinder engine that powers the Audi 80 uses a belt to drive its single-overhead camshaft.

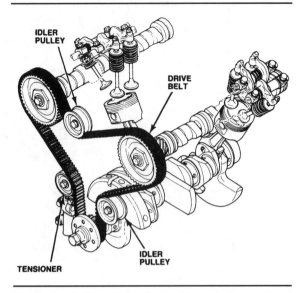

Figure 9-18. Belts effectively drive single-overhead camshafts located in separate heads, as on this Mazda 3.0-liter V6 that powered the Mazda 929 from 1988 to 1990.

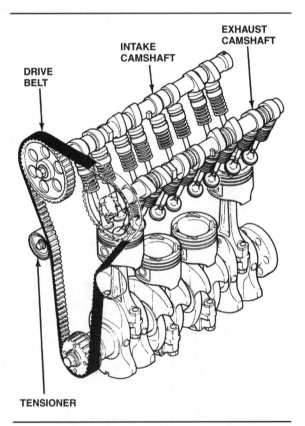

Figure 9-19. The 1.6-liter, 4-cylinder DOHC engine in the Mazda 323 used belt-drive for its camshafts.

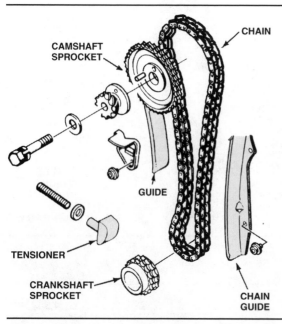

Figure 9-20. This Chrysler SOHC chain drive has chain guides on both sides of the chain and a tensioner adjusted by oil pressure.

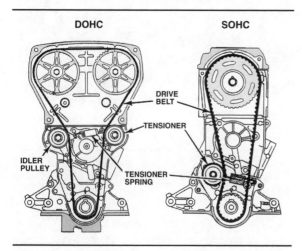

Figure 9-21. Belt-driven overhead camshafts for both SOHC and DOHC engines may use spring-loaded tensioners.

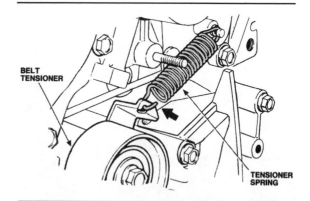

Figure 9-22. Camshaft drive belt tensioners consist of a wheel and a spring.

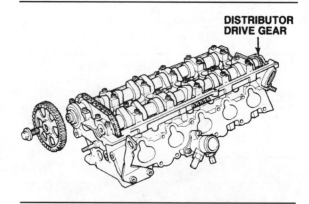

Figure 9-23. A helical gear on the end of the camshaft may drive the distributor in OHC engines.

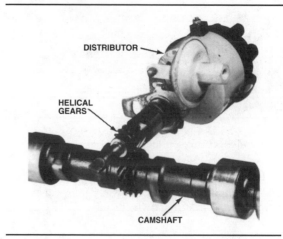

Figure 9-24. Helical gears allow the distributor shaft to operate at right angles to the camshaft.

they turn at the same speed. In most OHV engines, the same helical gear that drives the distributor also drives the oil pump, figure 9-25.

The camshaft of an OHV engine also may operate the fuel pump. This is done by an **eccentric** which is bolted to the front of the camshaft or cast into it. The eccentric changes the rotary camshaft motion to the reciprocating motion necessary to operate the fuel pump, figures 9-26 and 9-27. Since the 1980s, most cars have been produced with electric fuel pumps mounted in the fuel tank, so this fuel pump drive is not as common as it used to be.

Eccentric: A circle mounted off center on a shaft. It is used to convert rotary motion to reciprocating motion.

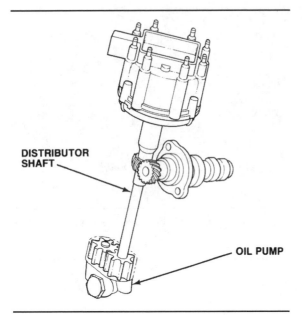

Figure 9-25. A shaft that extends for the distributor drives the oil pump in many domestic OHV engines.

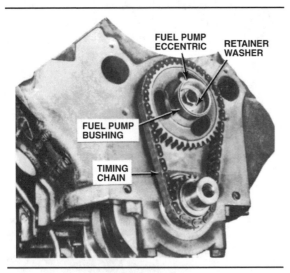

Figure 9-26. An eccentric bolted to the front of the camshaft operates the fuel pump.

In belt-driven OHC engines, some manufacturers add extra pulleys to drive the water pump or the oil pump, figures 9-28 and 9-29.

Cam Thrust

The longitudinal movement of the camshaft in its bearings is called cam thrust. The thrust can be forward or backward, depending on the loading of the cam drive or the distributor. On chain drive cams, the only thrust is from the distributor gears. The cam is pushed in the direction of the slant of the gears.

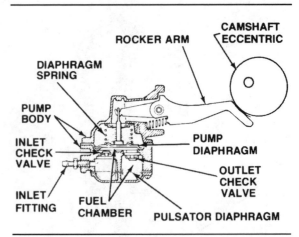

Figure 9-27. The fuel pump plunger rides on the camshaft eccentric.

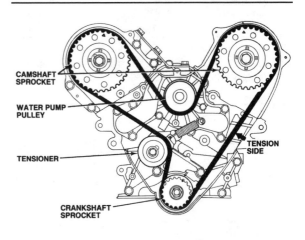

Figure 9-28. The pulley that drives the water pump also acts as an idler wheel to keep the camshaft drive belt tight.

On most OHV engines, the cam thrust is controlled by a plate that bolts to the front of the block, figure 9-30. On others, the front cover is close enough to the cam so that it can move back and forth only a few thousandths of an inch. On chain-driven camshafts, the timing chain gives a certain amount of end thrust control. When it is new, it will not bend sideways easily. But a worn chain can allow quite a bit of cam movement which sometimes becomes so great that a cam lobe actually gets under the neighboring lifter.

CAMSHAFT DESIGN

The camshaft appears to be simple, but cam design determines the important factors of how much, how long, and when the valves open. To make good power, an engine must open and close its valves rapidly. In a 4-cycle engine, at 6,000 rpm, a valve must open and

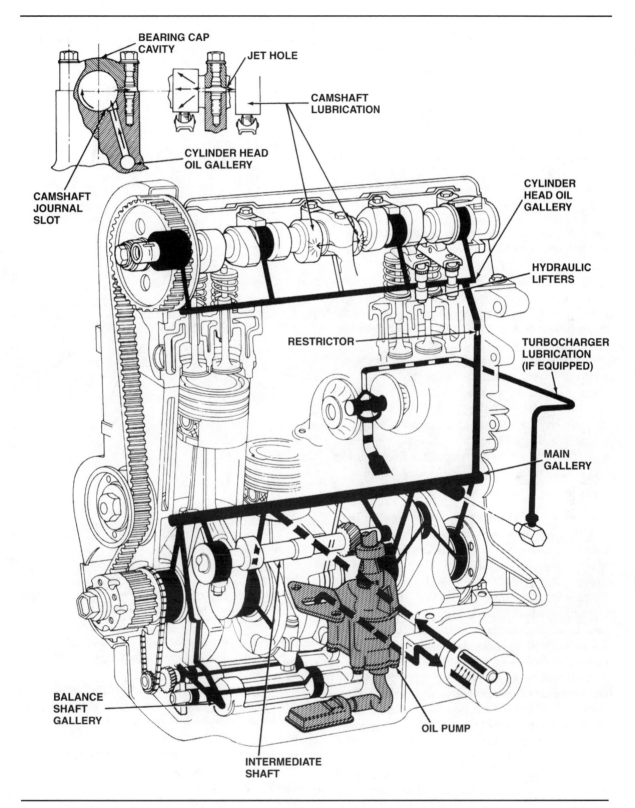

Figure 9-29. On this Chrysler 4-cylinder engine, a third pulley drives an intermediate shaft that operates both the distributor and oil pump.

Figure 9-30. On many OHV engines, cam thrust is controlled by a thrust plate bolted to the front of the engine.

close 50 times a second. Cam design must also take into account the fact that valves cannot open instantly. The components of the valve train — the lifter, pushrod, rocker arm, and valve — all have mass. If the cam lobe pushes the valve open too quickly, the force of lifting that mass at high rpm is very great, and the cam lobe will wear rapidly. To help prevent high wear, the valves must be opened gradually.

In addition, if the valve is lifted too quickly, the inertia of the valve train causes the valve to continue opening after the cam lobe has reached the full lift position, causing valve float. And, if the lobe does not allow the valve to close gradually, the lobe may drop out from under the lifter, leaving slack in the valve train. The valve spring then slams the valve against its seat causing wear and possibly breaking the valve.

Cam Lobe Construction

The cam lobe is not centered with the camshaft bearings so that as it turns, it transmits a lifting action to the valves through the valve train. Cam lobe construction consists of two imaginary circles that can be drawn on the profile of a cam lobe, the **nose circle** and the **base circle**, figure 9-31. The nose circle intersects the very top of the lobe, the highest point on the cam lobe in relation to the valve train. The base circle intersects the very bottom of the lobe, the lowest point on the cam lobe in relation to the valve train.

Each lobe on a camshaft is designed to gradually take up the clearance in the valve train, figure 9-32. As the camshaft rotates, the cam lobe **ramps** lift and lower the valve rapidly but smoothly. The **lash ramp** is the initial portion of the lobe that meets the lifter as the cam rotates off its base circle. The lash ramp compensates for small deflections and takes up slack in the valve train without actually lifting the valve. However, cams ground for hydraulic lifters do not need lash ramps because they operate at zero clearance without lash.

As the cam continues to rotate, the lifter moves up to the **opening ramp** which does the actual opening

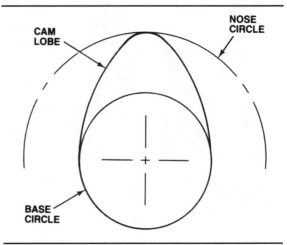

Figure 9-31. Cam lobe construction is defined by the nose circle and the base circle.

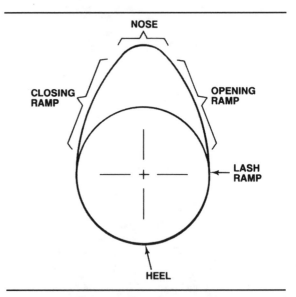

Figure 9-32. The ramps on the cam lobe open and close the valves quickly, but under control.

of the valve. The rate the valve is lifted is accelerated significantly on the opening ramp, up to a fairly constant opening speed.

Near the fully open position, the cam lobe slows its rate of lift. The idea is to slow down the rate of opening so that when the lobe reaches the fully open position, the valve will be moving slowly. Some camshaft ramp designs accelerate and slow the rate of lift several times on the opening ramp to open the valve as quickly as possible but still keep the lifter in contact with the cam lobe, avoiding valve float.

The top of the lobe, or **nose**, holds the valve in the fully open position. Then, as the top of the lobe passes out from under the lifter, the valve starts to close on the **closing ramp**. Just before the valve reaches the fully closed position, the lobe reduces its rate of letdown so the valve closes very gently against the seat. The ramps on each side of the cam lobe, between the nose and base circle, are also called the **flanks**.

While the valve is closed, the lifter stays in contact with the cam lobe on the base circle, which prevents slack in the valve train. The point directly opposite the nose is called the **heel** of the cam lobe. All valve lash adjustment is done with the lifter in contact with the heel.

Camshaft Materials

Camshafts are usually cast from nodular iron. They are brittle and must be handled with care. Some manufacturers use cast steel camshafts in high-performance applications. The high valve spring pressures in the high-performance engines require steel's greater strength to provide good wear resistance.

Roller lifter camshafts place the highest loads per square inch on cam lobes of any style lifter. To withstand the pressures, these camshafts are usually made from forged steel billets.

Cam Lobe Conditioning

Because the cam lobes receive such high loading from the valve train, manufacturers apply several processes or treatments to camshafts to reduce wear. Nodular iron camshafts are commonly induction-hardened in a process similar to that used to harden integral valve seats in cast-iron heads. The manufacturer heats the camshaft electrically and then rapidly cools it in oil.

In 1991, General Motors began nitriding the nodular iron camshafts of its 2.5-liter Tech IV engines. GM heats the camshafts by induction, then the entire camshaft is placed in a 400 to 500°F (204 to 260°C) molten salt bath for about two hours. Finally, the camshaft is removed from the bath and allowed to cool to room temperature.

Another method used to improve the wear properties of the lobes is a casehardening process called **carburizing**. In Chapter 3, we explained that the carbon content of a steel helped determine to what degree a steel could be hardened. Carburizing causes more carbon to be absorbed into the surface of the steel so that surface can be made harder with heat treatment. The core of the shaft does not absorb any extra carbon and remains softer and less brittle than the surface of the lobes.

To carburize the camshafts, the manufacturer heats them for several hours in an oven filled with a carbon-rich gas or liquid. In two to four hours, the part is carbonized to a depth of 0.020 to 0.030 inch (0.50 to 0.75 mm). To prevent carburizing parts of the camshaft that do not require it, the entire shaft is copper-plated and the lobes and bearing surfaces are ground to remove the copper. The carburizing process works only on the parts not covered by the copper. After carburizing, the camshaft is then hardened by quenching, taking advantage of the extra carbon in its surface.

Some camshafts have a hard-face overlay of tungsten-carbide, a chrome-nickel alloy electrically fused to the ramps and the noses of the lobes. The hard-face overlay provides good wear resistance with the extremely stiff valve springs used in some race engines. Hard-face overlays are seldom found in production cars.

Nose Circle: An imaginary circle that can be drawn on the profile of a cam lobe that intersects the very top of the lobe. It is the highest point on the cam lobe in relation to the valve train.

Base Circle: An imaginary circle that can be drawn on the profile of a cam lobe that intersects the very bottom of the lobe. It is the lowest point on the cam lobe in relation to the valve train.

Ramps: The part of the cam lobe that lifts and closes the valve.

Lash Ramp: The initial portion of the cam lobe that meets the lifter as the cam rotates off its base circle. The lash ramp compensates for small defections and begins to take up slack in the valve train without actually lifting the valve.

Opening Ramp: The part of the cam lobe that does the actual opening of the valve.

Nose: The highest part of the cam lobe that holds the valve in the fully open position.

Closing Ramp: The part of the cam lobe that closes the valve. The closing ramp lobe reduces its rate of letdown so the valve closes very gently against the seat.

Flanks: The ramps on each side of the cam lobe, between the nose and base circle.

Heel: The part of the cam lobe directly opposite the nose. All valve adjustment is done with the cam in this position.

Carburizing: A process that causes the cam lobe surfaces to absorb additional carbon. Then, the surfaces can be hardened to a greater degree than the parent metal.

Camshaft Lubrication

Two parts of the camshaft require lubrication, the bearing journals and the cam lobes. The bearing journals are relatively easy to lubricate; the lobes are more difficult.

A camshaft in the block of an overhead-valve engine is lubricated by oil splash, drain back, and full oil pressure, figure 9-33. The rotating crankshaft and connecting rods churn up the oil in the crankcase into a mist that lubricates the cam lobes. Oil draining back from the rocker arms also drips onto the cam lobes. This is why it is so important to keep the revs up at a moderate 2,000 rpm or so after first firing up a new or rebuilt engine. If the engine were just allowed to idle, the cam lobes would not receive sufficient oil during break-in and would be quickly damaged.

The camshaft bearings of overhead-valve engines receive pressurized oil through passageways from the oil pump. Usually, these passages are fed from the same source as the crankshaft main bearings. Although worn camshaft bearings may not make any noticeable difference in the way an engine runs, they can make a big difference in oil pressure. A worn camshaft bearing can bleed off so much oil that the oil pressure at the main bearings drops dangerously low, causing main bearing knock. This, in turn, reduces the oil pressure at the connecting rod bearings and may eventually result in a rod knock. Therefore, camshaft bearings should be inspected and replaced, if necessary, along with all other bearings when an engine is rebuilt.

OHC engines have oil fed under pressure directly to the cam bearings, internally or externally. In the internal feed, figure 9-34, oil is pumped through a passage inside the camshaft and comes out at a hole in each cam bearing journal. In the external feed, oil squirts or drips on each journal from a passageway that runs through the cylinder head near the camshaft, figure 9-35. In both cases, excess oil that bleeds out of the journals also lubricates the cam lobes before draining back to the crankcase.

Camshaft scuffing

A common camshaft problem is scuffing of the lobes and lifters when a rebuilt engine is first started. Since the camshaft may be installed several hours or days before the engine is started, the lubricant can drain off the cam lobes in that time. In some cases, an engine may be put into storage for several months before being run. When it is started, a dry cam lobe running against a dry lifter will break through the hard outer surfaces and cause rapid wear.

Many camshaft manufacturers apply a manganese-phosphate coating on the cam. This slightly rough coating holds a film of oil and prevents scuffing

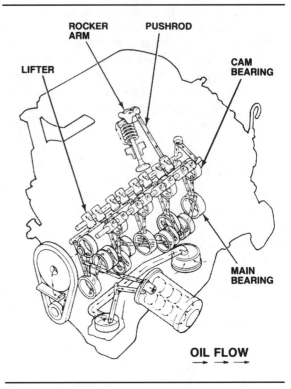

Figure 9-33. Full oil pressure reaches most internal parts of a domestic, OHV engine. The camshaft receives further lubrication from oil splash and drain back.

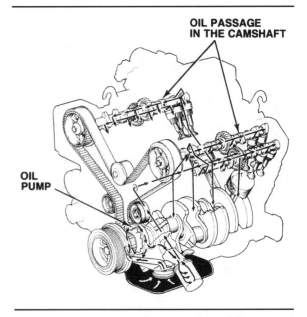

Figure 9-34. In this Lexus 400 engine, oil is pumped through passages in the camshafts. The oil squirts out a hole in each camshaft bearing journal.

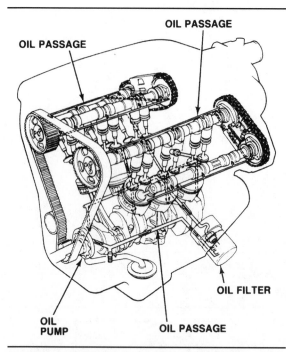

OIL PASSAGE

OIL PASSAGE

OIL PASSAGE

OIL FILTER

OIL PUMP

OIL PASSAGE

Figure 9-35. Oil passages in the cylinder heads deliver oil to the camshafts in this Ford SHO V6.

during dry starts. Cams that are coated with this material have a dark gray appearance.

In spite of precautions taken by cam makers, there is still danger of wear from dry starts. For this reason, assembly lubricant is a must. Most camshafts are sold with an assembly lubricant, usually a heavy grease with additives, that sticks to the lobes and works under extreme pressure.

VALVE TIMING

Engine designers rate camshafts by the valve timing they provide. As we stated earlier, valve timing determines how much, how long, and when the valves open. Since the valves allow the air-fuel mixture into the engine, these factors affect horsepower, torque, fuel economy, exhaust emissions, and idle quality.

Engine designers select valve timing that works best in the rpm range that they want an engine to operate in. To a large part, the timing *determines* the engine's operating range. The timing optimizes airflow and cylinder pressure in a certain range. The optimum valve timing is found through experimentation on an engine dynamometer. A change of a few degrees one way or the other can have a big effect on performance, fuel mileage, and exhaust emissions. No one camshaft grind works best at all engine speeds, so valve timing is a compromise. Some engines use variable valve timing systems to eliminate valve timing compromise, and we detail them in a later section.

■ **Donnell A. Sullivan — One of Ford's Better Idea Men**

The career of Don Sullivan was a long one marked with success in the automotive engine design field. Sullivan was born in 1904 in Port Huron, Michigan. At the age of 18, he built a steam turbocharger made out of parts from his mother's vacuum cleaner. He attached the turbo to a Model T roadster which made that vehicle the talk of Port Huron. A few years later, Sullivan went to work for the Ford Motor Company where his list of achievements is impressive:

- Principal design engineer for Ford's 1932 flat-head V8
- Developed the first over and under intake manifold for Ford's 1934 V8
- Developed Ford's first 2-barrel intake manifold
- Held the original patent for a 2-carburetor intake manifold for a V8 engine, using two 2-V carburetors. His design employed engine vacuum to operate the second carburetor's linkage, a principle that led to the development of the 4-barrel carburetor.
- Developed a new water pump system with the pump near the bottom of the cylinder block. Colder water was used in the system. This system was used on the 1933 and later model Ford V8s.
- Developed the Ford V8 engines used in the Miller-Ford Indy cars for the 1935 Indianapolis 500 race
- Qualified a flathead Ford V8-powered car for the 1940 Indianapolis 500
- Designed the Hutton-Sullivan high-performance, finned-aluminum cylinder heads in 1946 for Ford flathead V8s
- Designed the high-performance camshafts and aluminum intake manifolds for the 1956-57 Ford 292- and 312-cid engines, as well as the later 289-, 302-, 352-, 390-, 406-, and 427-cid engines
- Developed the engines for the LeMans-winning Ford GT-40 race cars of the 1960s

After he retired from Ford in 1969, Sullivan continued his engine design work as an independent engineer and Ford Motor Company consultant. Since then, he designed camshafts for A.J. Foyt's Indianapolis 500 winning race cars, the Oldsmobile diesel engine, Mercruiser marine engines, a national champion motorcycle engine, and others.

In the early 1990s, Sullivan turned his talents to designing camshafts that improve fuel efficiency and reduce exhaust emissions. One of his successful emission camshafts had a very small "bump" on the base circle of the exhaust cam lobe. The bump let in a small amount of exhaust gases during the intake stroke, eliminating the need for an EGR valve.

Sullivan worked as a consultant for Ford's Special Vehicle Operations (SVO) since its beginning, and worked on 32-valve modular engines prior to his death in 1994. When he worked on Ford's NASCAR race engines, he reworked the 351 V8 engines prior to Bill Elliott's successful era with the NASCAR Ford Thunderbird in the mid-1980s. Sullivan contributed most of the design expertise for the NASCAR Ford SVO V6.

We first describe the important valve timing events, then explain how they affect engine performance. The important ways to rate or measure valve timing events are:

- Lift
- Duration
- Overlap
- Lobe separation angle
- Camshaft installation position.

Lift

Valve lift is measured in thousandths of an inch or hundredths of a millimeter. Each cam lobe has a specific amount of lift, called the **lobe lift**. On most cams, you can measure lobe lift with a micrometer, subtracting the smallest measurement (across the base circle) from the largest (from the heel to the nose), figure 9-36. However, some cam lobes have measurable lift for more than 180 degrees of rotation, so the smallest measurement is not the true base circle, and micrometer measurement will not give an accurate number, figure 9-37. The only way to measure the lobe lift of these cams, and the most accurate for all cams, is to rotate the camshaft between centers and measure the rise and fall with a dial indicator.

Gross valve lift is the distance that the valve lifts off its seat. Gross valve lift is a theoretical number because it does not take into account lift lost from deflections or clearances in the valve train. Net valve lift is the actual amount the valve lifts off its seat in an assembled engine. **Net valve lift** is slightly less than gross valve lift because of the bending of pushrods, rocker arms, and flex in the camshaft itself. You can measure net valve lift at the lifter or valve stem tip with a dial indicator by rotating the engine.

In OHV and OHC engines with rocker arms or followers, gross and net valve lift are greater than lobe lift because of the **rocker arm ratio**. One side of the rocker arm or follower is longer than the other so that the valve lift is multiplied. Rocker arm ratio is explained in detail in the section on rocker arms.

In OHC engines and flathead engines without rocker arms or followers, gross valve lift is the same as lobe lift. On these engines, net valve lift is usually the gross lift minus any clearance present with mechanical tappets.

Valve lift is important to the breathing of an engine because, up to a certain point, the valve head is a restriction to airflow. Once a valve in a street engine reaches approximately 0.300-inch (8-mm) lift, the port flows as if the valve were not there. However, it is still helpful to lift the valve higher than this point because the valve then spends more cam duration time

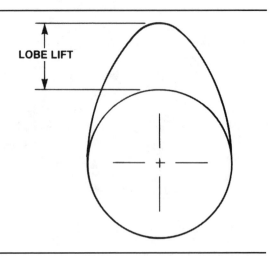

Figure 9-36. Lobe lift is the amount the cam lobe lifts the lifter.

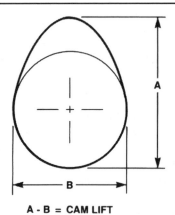

A - B = CAM LIFT

A - B DOES NOT EQUAL CAM LIFT

Figure 9-37. Some cam lobes provide lift for more than 180 degrees of camshaft rotation, so it is not possible to get an accurate base circle measurement with a micrometer.

at or above 0.300-inch (8-mm) lift. For best airflow, it is important to keep the valves open as long as practical. There are no driveability, fuel economy, or emission drawbacks to high valve lifts, so some manufacturers keep duration and overlap moderate and use high lift to gain airflow.

However, high lift does place greater loads on the valve train. This is especially true in pushrod engines because of the extreme changes in rocker arm angle high lift causes. High valve lifts may also cause valve to piston interference. If a valve touches a piston, major engine damage results. Most street engines keep valve lifts moderate to avoid interference and reduce loads on the valve train.

Duration

The length of time the valve is open, expressed in degrees of crankshaft rotation, is called the camshaft **duration**. At the end of each stroke, the piston is either at top dead center (TDC) or bottom dead center (BDC). A stroke requires one-half of a crankshaft rotation, or 180 degrees.

If a valve is described as opening at TDC and closing at BDC, the total time it is open is 180 degrees of crankshaft rotation. Since intake valves always open before TDC and close after BDC, the additional degrees must be added to 180 to get the total duration. For example, figure 9-38, if an intake valve opens 5 degrees before top dead center, and closes 15 degrees after bottom dead center, the duration of the valve opening is the total of 180, 5, and 15 degrees, or 200 degrees. The same is true of an exhaust valve that opens at 15 degrees before bottom dead center, and closes at 5 degrees after top dead center. 15 plus 180 plus 5 equals 200 degrees of duration.

This early opening and late closing of the valves is necessary because intake air and exhaust gas have inertia. Neither air nor gas begins entering or leaving the cylinder as soon as the valve opens, so some of the time a valve is open is actually wasted. All camshafts open the intake valve just before TDC and before the piston begins its intake stroke, thus, giving the intake charge some "lead" time to get moving into the cylinder. Since exhaust gases also have inertia, all camshafts open the exhaust valve before BDC on the power stroke, prior to the piston starting its rise on the exhaust stroke. This also gives the exhaust gases some lead time.

Similarly, all camshafts keep the intake valve open after the piston reaches BDC. As the piston starts upward after it passes BDC and begins its compression stroke, it actually takes some time before pressure builds to the point that it would push some of the intake charge out of the cylinder. It is possible to keep the intake valve open well after BDC to more fully fill

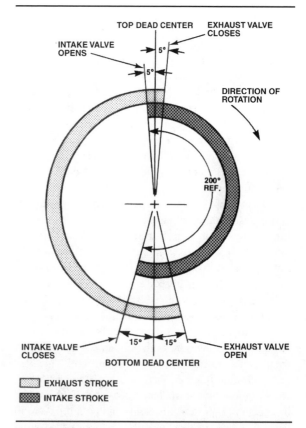

Figure 9-38. A typical valve timing diagram, like this one, shows the periods in which the intake and exhaust valves are open and closed.

the cylinder. All camshafts keep the exhaust valve open after TDC at the beginning of the intake stroke. Inertia continues to keep the exhaust moving out the exhaust valve despite the intake valve having opened.

Lobe Lift: The lift provided by the cam lobe. Best measured with a dial indicator.

Gross Valve Lift: The distance that the valve lifts off its seat. It is a theoretical number because it does not take into account lift lost from deflections or clearances in the valve train.

Net Valve Lift: The actual amount the valve lifts off its seat. It is slightly less than gross lift because of the bending of pushrods, rocker arms, and flex in the camshaft itself.

Rocker Arm Ratio: The ratio of the valve side of the rocker arm to the cam side, measured from the pivot.

Duration: The valve timing period that a valve is open. Measured in degrees of crankshaft rotation.

Duration measurement would seem to be very simple, but there is some disagreement on how it is measured. In the past, duration was commonly measured from 0 lift just before valve opening to 0 lift just after valve closure. Some cam makers still rate duration this way, and it is called the "advertised duration." However, since it is important to take up slack in the valve train as well as lift and close the valve gently, the lobe moves the valve very slowly at the beginning and ending of its cycle. This also makes it very difficult to determine or measure *exactly* when the valve begins to open or close.

To get around this inaccuracy, most cam makers now measure duration at specified points in the valve lift. Some common points are at 0.006 inch (0.15 mm), 0.020 inch (0.51 mm), and 0.050 inch (1.30 mm) valve lift. 0.050 is the most popular. Lift is easy to measure, and this gives the cam maker or engineer an accurate point at which to start and stop measuring duration. When you compare cam timing specifications, remember to find out between what lift points the duration is measured. The higher the lift at which the duration measurement begins and ends, the less rated duration that camshaft will have. For example, the duration of a cam lobe measured between the 0.050 inch (1.30 mm) lift points can be more than 50 degrees less than that measured between the actual opening and closing points, figure 9-39. The former is a more meaningful measurement of duration.

Intake duration and exhaust duration are usually very close. What works well to get the intake mixture usually works well to get the exhaust gases out, but this is not always the case. Some engine designs have intake tracks that are more efficient than their exhaust tracks, or vice versa. These engines need "dual pattern camshafts," with different timing specifications for intake and exhaust.

Camshaft duration probably has more effect on how an engine runs than any other single valve timing specification. A camshaft with a lot of duration is an excellent performer at high rpm. A low-speed engine may have a valve duration of 200 degrees, but the same engine designed for high speed may have more than 260 degrees of duration (both measured at 0.050-inch lift), figure 9-40. Longer durations help overcome the inertia of the incoming or outgoing air. A long duration cam allows more time to fill the cylinder with the air-fuel mixture and it also allows more time to exhaust a larger amount of spent gases. The result is more power at high rpm. At low rpm, a long duration camshaft causes an engine to run roughly and return poor fuel economy.

On the other hand, a camshaft with less duration will idle smoothly, produce good low rpm power, and

CAMSHAFT INTAKE DURATION			
CAMSHAFT	ADVERTISED DURATION	DURATION AT 0.050	DURATION AT 0.200
A	242	194	101
B	255	203	112
C	250	205	120

Figure 9-39. The lift at which a cam manufacturer checks camshaft duration can greatly affect the duration measurement.

return favorable fuel economy. Yet, at high rpm the engine cannot get enough air and fuel and power falls off. Most street engines have camshafts with a moderate amount of duration to provide the best compromise between power, emissions, and fuel economy.

Overlap

Another term used to describe valve timing is **overlap**. The intake and exhaust valves are usually designed to both be open for a short time when the piston is near TDC at the beginning of the intake stroke. If an intake valve opens 12 degrees before TDC, and an exhaust valve closes 12 degrees after TDC, add the two figures to get the overlap. In this case, 24 degrees.

During overlap, the intake valve opens before the exhaust valve has closed. All 4-stroke engines have at least a short overlap period, but long duration camshafts usually extend the overlap period, which helps make large amounts of power at high rpm. The overlap period packs more mixture into the cylinder by opening the intake valve near the end of the exhaust stroke while the exhaust valve is still open. The rush of exhaust gases out of the cylinder helps to pull in the intake mixture, even though the piston has not yet reached top dead center. The exhaust valve is still open when the piston starts down on the intake stroke. This does not allow exhaust gases to be sucked back into the cylinder at high rpm, because air velocity is high and the outward rush of exhaust gases is still pulling the intake mixture into the cylinder.

However, an extended overlap period of a long duration cam can cause the engine to idle roughly and produce very little power at low rpm. Some of the fresh, incoming air-fuel mixture can get sucked out of the exhaust valve, increasing HC emissions and greatly reducing fuel economy and low rpm power. Also, at low rpm airflow velocity is slow and some exhaust gases can get sucked back into the cylinder. The exhaust gases dilute the intake charge, reducing power and causing the engine to run rough.

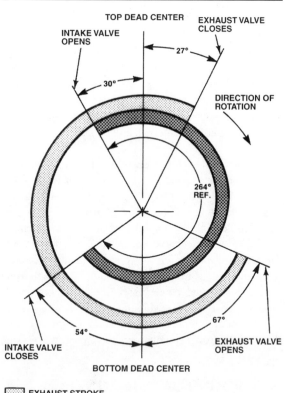

EXHAUST STROKE
INTAKE STROKE

Figure 9-40. The valve timing diagram of this high-performance camshaft shows that it has more duration and overlap than a typical camshaft.

Lobe Separation Angle

Camshafts for OHV and SOHC engines have both the intake and exhaust valve lobes on the same shaft. The **lobe separation angle** is the angle between the centerline of the intake lobe and the centerline of the exhaust lobe for a single cylinder. While all other camshaft timing specifications are measured in degrees of crankshaft rotation, lobe separation angle is measured on the camshaft, figure 9-41. An average lobe separation angle is 110 degrees. The range is from approximately 106 degrees to about 114 degrees.

Reducing the lobe separation angle on any given cam lobe profile causes the exhaust valve to open later and the intake valve to close sooner in terms of crankshaft rotation. This may be easier to visualize if you view figure 9-41 upside down, with the lobe noses pointed down and the exhaust valve lobe on the left and the intake lobe on the right. The camshaft turns clockwise (in most engines), so now the exhaust lobe will not begin to touch its lifter until later in the rotation and the intake lobe will be closing its valve sooner.

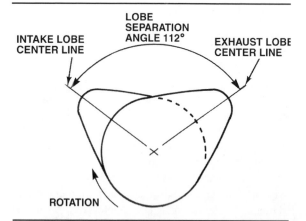

Figure 9-41. Changing the intake and exhaust cam lobe separation angle can alter the engine's performance characteristics.

Camshafts for OHV and SOHC engines, which have both the intake and exhaust valve lobes on the same shaft, have fixed lobe separation angles. Once the camshaft is ground, the angle is set. DOHC engines have separate camshafts for the intake valves and for the exhaust valves, so a machinist can easily change the lobe separation angle when he assembles the engine.

Lobe separation angle affects engine performance because it affects the opening and closing points of the valves. For a given duration, a relatively narrow lobe separation angle (106 to 110 degrees) not only increases overlap, the time both valves are open, but it also increases the time both valves are closed. These are critical valve timing events because they determine the time the cylinder is sealed and cylinder pressure can push on the piston. The result is more power at high rpm. Low-speed torque will also improve unless the increase in overlap is excessive. The longer the cam duration, the greater the overlap problem. To maintain idle quality and driveability, the lobe separation angle must be wider the longer the cam duration.

Most street engines have lobe separation angles of 110 to 114 degrees to avoid excessive overlap. Street engines must have smooth idle quality and high vacuum to run accessories.

Overlap: A relatively short valve timing period during which both the intake and exhaust valves are open. Measured in degrees of crankshaft rotation.

Lobe Separation Angle: On a camshaft, the angle between the centerline of the intake lobe and the centerline of the exhaust lobe of a single cylinder.

Camshaft Installation Position

The camshaft is timed to the engine in relation to the crankshaft. When installing a camshaft in an engine, you line up the timing marks for the camshaft and crankshaft drive. Assuming the marks are accurate, this usually places the number one cylinder at TDC and the camshaft near the middle of overlap for that cylinder. This is called installing the camshaft "straight up."

This is not the only position in which you can install the camshaft. By turning the camshaft clockwise or counterclockwise and leaving the crankshaft stationary, you can change an engine's power band. If you move the camshaft ahead in relation to crank position, we say the camshaft is advanced; if you move the camshaft back, we say it is retarded.

Changing the camshaft installation position alters the opening and closing events of the valves without affecting lift and duration. Advancing the camshaft opens the valves sooner and closes them sooner, which generally improves low rpm power. The famous engine designer and camshaft expert, Ed Winfield, was fond of saying that the closing point of the intake valve is the most important valve timing event. The real benefit of advancing the cam to low rpm power is the earlier intake valve closing. Air velocity is slow at low rpm, air inertia is low, and cylinder packing is less efficient. Closing the intake valve earlier keeps the intake mixture from flowing back out the intake port, an unwanted occurrence called reverse pumping.

Retarding the camshaft opens and closes the valves later, which generally increases high rpm power. At high rpm, where the cylinder must fill in a few thousandths of a second, keeping the intake valve open as long as possible greatly improves power output.

Changing camshaft installation position on SOHC and OHV engines changes both the intake and exhaust timing because these engines have both intake and exhaust cam lobes on a single camshaft. Therefore, camshaft installation position changes to these engines must be limited to a few degrees. DOHC engines allow much more freedom in changing camshaft position because they have separate shafts for intake and exhaust. You can change the intake timing without affecting the exhaust timing, for example, or vice versa.

Valve to piston interference

All engines can be divided into two categories, free-running and interference. In a free-running engine, valve lift or angle prevent the valve heads from entering the cylinder volume swept by the piston. As a result, the camshaft and crankshaft can rotate independently because there is always clearance between

a piston near TDC and a fully opened valve. With an interference design, valves and pistons sweep the same space at different times in the engine cycle. When timing is out of phase, open valves can strike the pistons as the pistons rock over top dead center. DOHC engines, depending upon the angle and lift of the valves, may also experience valve-to-valve interference between intake and exhaust. The end result is a tangled mass of internal engine parts and a large repair bill.

Because interference designs offer improved efficiency and increased power, the majority of OHC engines and many OHV engines are designed this way. When engineers are not constrained to a free-running design, they can alter the angle and position of the valves to achieve improved flow and combustion, raising power without compromising economy. Unfortunately, a timing belt or chain failure will bring this efficient and powerful engine to an immediate and expensive halt. Manufacturers' recommended replacement intervals are intended to protect both their high-strung engines and their customers' bank accounts.

REGROUND CAMSHAFTS

A worn camshaft is usually replaced during a rebuild. However, some machine shops regrind worn camshafts to save their customers money. A number of machine tool companies sell a cam regrinder for this purpose, figure 9-42. The cam regrinder grinds away the worn portions of the cam lobes, providing a smooth surface for the lifter. By grinding away equal amounts of material from both the nose circle radius and the base circle radius, the reground camshaft ends up with the stock contour and valve timing is unchanged, figure 9-43. The reground camshaft is then checked for straightness before being installed in an engine.

Some aftermarket camshaft manufacturers regrind camshafts to provide greater engine performance. The camshaft maker can grind material from the lobe nose to increase duration, and as long as more material (heightwise) is removed from the base circle than the nose, valve lift increases, figure 9-44. The new camshaft timing profile can also be advanced or retarded slightly on the shaft. Regrinding is best for small timing changes. Radical regrinding could leave the base circle too small and grind through all the casehardening on the original lobe surface.

VARIABLE VALVE TIMING

As we have already stated, valve timing is a compromise. No single camshaft profile for an engine is

Figure 9-42. Some machine shops regrind camshafts to restore them to serviceable condition.

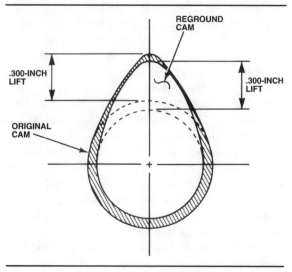

Figure 9-43. This cam lobe has been reground to provide the same profile as before.

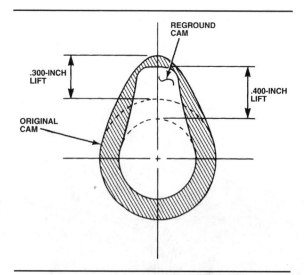

Figure 9-44. This cam lobe has been reground to provide increased lift and duration.

correct for all engine speeds. A major breakthrough in avoiding the compromise is variable camshaft timing. The first systems, used by Alfa Romeo, Nissan, and Mercedes-Benz on some of their DOHC engines could be called on-off devices. They either advance or retard the installation position of the intake camshaft, without affecting lift or duration.

Alfa Romeo introduced the first variable valve timing system on its 2.0-liter 4-cylinder engine in 1980. Its purpose was to smooth a rough idle caused by ignition and camshaft timing changes made to meet emission standards. The system employed a hydraulic-mechanical control valve mounted on the intake camshaft. At idle, the system routes engine oil pressure through a series of check valves to the control valve, which rotates and advances the intake camshaft. In 1982, Alfa switched from mechanical to electronic fuel injection with Motronic engine control. The Motronic system took over control of a revised, solenoid-activated, electro-hydraulic control valve, figure 9-45, for cam timing adjustment.

In 1990, Nissan introduced its Nissan Valve Timing Control System (NVTCS) on the 300 ZX V6 engine and the Infiniti Q45 V8 engine, figure 9-46. Each cylinder head has a separate single-row timing chain to drive its dual camshafts. However, the drive gears are not directly splined to the intake valve camshafts. A spring-loaded, movable helical gear is placed between the drive gear and the intake camshaft. A control valve and solenoid, figure 9-47, mounted on the end of each intake camshaft, governs the entry of engine oil pressure which can move the helical gears. Moving the helical gear rotates the camshaft relative to the drive gear, advancing intake valve timing 20 degrees. A computer, acting on information about engine speed, intake flow rate, cylinder wall temperature, and throttle plate opening signals the valve to change valve timing.

The intake camshafts are ground to provide optimum high-rpm efficiency. At high-load conditions (wide throttle opening, low manifold vacuum) and low to medium rpm, the camshafts are rotated to

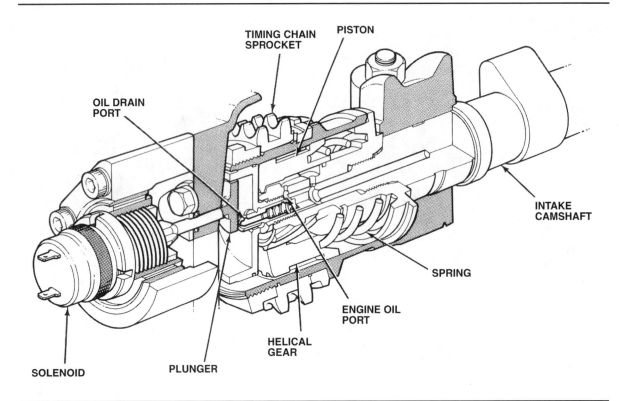

Figure 9-45. Alfa Romeo's electro-hydraulic cam timing control valve, introduced in 1982.

advance them. At higher rpm and low-load conditions, the computer shuts off oil flow to the control valve. A spring returns the camshafts to their original position. Infiniti claims the two-position camshaft timing allows good high-rpm power and boosts torque in the lower rpm ranges, increases fuel economy, and smoothes the idle.

Mercedes-Benz introduced a system similar to Nissan's for its 1990 500SL V8 engine and 300SL in-line 6-cylinder. M-B's system also uses oil pressure, a computer-controlled valve, and a helical gear to rotate the intake camshafts. It is different in that it both retards and advances the camshafts according to rpm and other input data. At idle, and up to 2,000 rpm, the camshaft is retarded to reduce overlap. From 2,000 to 4,700 rpm in the six and 2,000 to 5,000 rpm in the V8 the cams are advanced, closing the intake valves earlier to improve low rpm torque. Over those speeds, the camshafts are retarded again, delaying the intake valve closing to improve cylinder filling at high-rpm.

Honda took variable valve timing a step past the on-off devices that simply change the intake camshaft's installation position. The 1991 Acura NSX V6 uses Variable Valve Timing and Lift Electronic Control (VTEC), the first valve timing mechanism of its kind on a production car in the United States. VTEC is also used on many Honda and Acura

production vehicles. VTEC operates the valves on totally different cam profiles for low and high rpm.

The NSX V6 has DOHC heads and four valves per cylinder. The VTEC system requires three cam lobes and three followers for each pair of intake valves and each pair of exhaust valves, figure 9-48. The two outer lobes have lift and timing profiles that favor low-speed torque. The center lobe has a more radical profile that allows good high-rpm power.

The engine normally runs on the two outer followers and cam lobes. When signaled by the VTEC computer, a solenoid opens a spool valve. When the spool valve opens, engine oil pressure pushes against pins that locks the three intake rocker arms together. With the rocker arms locked, the valves must follow the profile of the high rpm cam lobe in the center. This entire process takes 0.1 second.

When the VTEC computer signals the solenoid to close the spool valve, it reduces pressure against the pins. A return spring pushes the pins back and unlocks the rocker arms. This allows the two outer arms to follow the low rpm cam lobes again.

The VTEC computer judges running conditions for activating the spool valve based on engine rpm, water temperature, and engine load. Generally, the cam timing switchover takes place at between 5,800 to 6,000 rpm.

Figure 9-46. The Infiniti Q45 V8. One of its valve timing control valves is shown on the left-side intake camshaft.

SUMMARY

The camshaft has a series of cams, or lobes, and rotates to open the engine valves. The camshaft may be located in the engine block or the cylinder head. Engines with camshafts in the block include L-head and overhead-valve (OHV) engines. Engines with camshafts mounted in the head include the single-overhead-camshaft (SOHC) and dual-overhead-camshaft (DOHC) engines.

The camshafts are driven by gears, chain, or cog belt from the crankshaft. The camshaft must revolve at one-half the speed of the crankshaft and must be timed to the crankshaft rotation so that the valves open and close in time with piston movements.

The cam lobe is eccentric to the camshaft bearings and transmits a lifting action to the valves through the valve train. Cam lobe construction consists of two imaginary circles that can be drawn on the profile of a cam lobe, the nose circle and the base circle. Each lobe on a camshaft is designed to gradually take up the clearance in the valve train. The cam lobe ramps lift and close the valve rapidly but smoothly. Camshafts are made of nodular iron, cast steel, or steel billets.

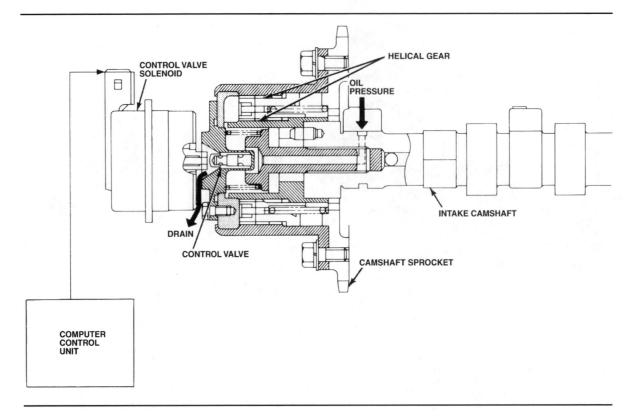

Figure 9-47. On the Q45 V8, the solenoid on each intake camshaft governs the flow of oil into the control valve, which advances the camshafts.

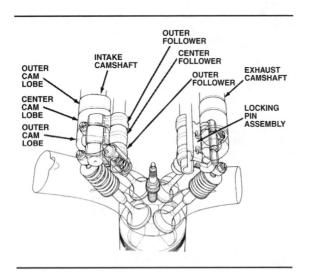

Figure 9-48. The NSX's computer-controlled hydraulic system switches cam lobes to provide optimum high-speed and low-speed performance.

The camshaft bearing journals and the cam lobes require lubrication. The bearing journals receive full pressure lubrication from the oil pump through oil passageways. The lobes get their lubrication from splash, drainoff, or excess oil that bleeds away from the cam bearings.

A camshaft designer uses lift, duration, overlap, lobe separation angle, and camshaft installation position to affect engine performance, emissions, and fuel economy.

Valve timing is a compromise. No single camshaft profile for an engine is correct for all engine speeds. A major breakthrough in avoiding the compromise is variable camshaft timing. The first systems used by Alfa Romeo, Nissan, and Mercedes-Benz either advance or retard the installation position of the intake camshaft, without affecting lift or duration. The 1991 Acura NSX V6 uses a Variable Valve Timing and Lift Electronic Control (VTEC) system that operates the valves on totally different cam profiles for low and high rpm.

Review Questions
Choose the single most correct answer.
Compare your answers with the correct answers on Page 328.

1. The camshaft opens valves:
   a. Gradually, with speed changes depending upon at which point the lifter is in contact with the cam lobe.
   b. With an instantaneous thrust at exactly the right moment
   c. As fast as this can be done by a mechanical device
   d. None of the above

2. Most camshafts are made of:
   a. Aluminum-beryllium alloy
   b. Stainless steel
   c. Forged steel
   d. Cast iron or steel

3. Which of the following treatments reduces cam wear?
   a. Hard-face overlays
   b. Carburizing
   c. Nitriding
   d. All of the above

4. If the intake valve opens 18 degrees before top dead center and closes 12 degrees after bottom dead center, the duration is:
   a. 186 degrees
   b. 210 degrees
   c. 222 degrees
   d. 310 degrees

5. Camshafts designed for high-speed operation:
   a. Usually provide comparatively long valve duration
   b. Make inlet valve duration much longer than exhaust valve duration
   c. Decrease valve overlap
   d. Have special gear ratios, to rotate faster than a regular camshaft

6. Technician A says that chain-drives for OHC engine require a chain tensioner. Technician B says that these drives require a chain guide. Who is right?
   a. A only
   b. B only
   c. Both A and B
   d. Neither A nor B

7. Technician A says that during assembly of an OHV engine you can alter valve timing by changing the camshaft lobe separation angle. Technician B says that during assembly of an OHV engine you can alter valve timing by changing the camshaft installation position. Who is right?
   a. A only
   b. B only
   c. Both A and B
   d. Neither A nor B

8. To make a reground camshaft with the stock contour or valve timing:
   a. Grind more material from the nose circle than the base circle
   b. Grind more material from the base circle than the nose circle
   c. Grind equal amounts of material from both the nose and base circles
   d. None of the above

9. Noise reduction is a major function of which of the following parts?
   a. Cam drive belts
   b. Helical gears
   c. Silent chains
   d. All of the above

10. Overlap is when:
    a. Two intake valves of the same cylinder are open at the same time during the intake stroke
    b. Two exhaust valves are open at the same time while the piston is at TDC on the exhaust stroke
    c. An intake and an exhaust valve of the same cylinder are open at the same time while the piston is near TDC on the beginning of the intake stroke
    d. None of the above

11. The lobe separation angle:
    a. Is easiest to change in DOHC engines
    b. Is set when the camshaft is ground in OHV and SOHC engines
    c. Has an effect on engine performance
    d. All of the above

12. The camshaft operates the valves, and:
    a. Has no other function
    b. May drive the distributor
    c. Always drives the oil pump
    d. Never drives the fuel pump

13. Camshafts may be driven by:
    a. Gears
    b. A chain
    c. A belt
    d. All of the above

14. On some engines, the camshaft operates the fuel pump:
    a. By a helical reduction gear
    b. Off one of the intake valve lobes
    c. By an eccentric
    d. By a concentric

15. In an overhead-valve engine, net valve lift is equal to:
    a. Lobe lift
    b. Lobe lift plus rocker arm ratio
    c. Lobe lift plus lifter clearance minus rocker arm ratio
    d. Lobe lift times rocker arm ratio minus lifter clearance and valve train deflection.

16. Technician A says that when starting a fully rebuilt engine, just let it idle for a half hour. Higher rpm will cause the engine to seize. Technician B says to bring the rpm of the rebuilt engine to a moderate rpm above the idle speed, approximately 2,000 rpm. This will provide necessary lubrication for the camshaft. Who is right?
    a. A only
    b. B only
    c. Both A and B
    d. Neither A nor B

17. Technician A says that a DOHC engine has the lightest, simplest valve train. Technician B says that an OHV engine has the lightest, simplest valve train. Who is right?
    a. A only
    b. B only
    c. Both A and B
    d. Neither A nor B

18. Technician A says that some camshafts are made of titanium billets. Technician B says that some camshafts are made of heat-treated and carburized aluminum. Who is right?
    a. A only
    b. B only
    c. Both A and B
    d. Neither A nor B

19. Technician A says that most variable valve timing systems alter the intake camshaft's installation position. Technician B says that most variable valve timing systems alter the exhaust camshaft's installation position. Who is right?
    a. A only
    b. B only
    c. Both A and B
    d. Neither A nor B

20. Variable valve timing systems:
    a. Change the lobe separation angle
    b. Provide precisely the same cam timing for all engine speeds
    c. Help eliminate the valve timing compromise
    d. All of the above

10
Lifters, Followers, Pushrods, and Rocker Arms

Lifters, followers, pushrods, and rocker arms are the valve train components that transmit the motion of the camshaft to the valves, figures 10-1 and 10-2. Like the camshaft, detailed in Chapter 9, machinists do not perform much work on these parts, but they must know how to inspect the valve train and assemble it correctly.

VALVE LIFTERS AND CAM FOLLOWERS

Lifters and cam followers have the same purpose. They ride directly on the camshaft and assist in transmitting the motion or the push of the cam lobe to the stem of the valve.

Valve Lifters

Lifters are also called tappets, and they are divided into two types: **flat tappets** and **roller tappets**, figure 10-3. Flat tappets are so called because the part of the lifter that rides against the cam lobe appears flat to the naked eye (actually, most are slightly crowned, which we explain later). Both solid and hydraulic lifters are called flat tappets. Roller tappets have a steel wheel, or roller, to ride on the cam lobe.

The lifter body is made from cast iron, and is hardened by heat treating so it will be long wearing. The outside of the lifter body is ground to very close tolerances and slides up and down in a machined lifter bore. In overhead-valve (OHV) and flathead engines, the lifter rides in a bore in the cylinder block. In overhead-camshaft (OHC) engines, the lifter bore is in the cylinder head.

Valve clearance

Having the proper clearance between the parts in the valve train is important for engine performance and valve train life. Heat is the initial reason parts require clearance. Valve stems and pushrods expand considerably from heat. If the clearance between the cam lobe and the valve stem is not sufficient, the valve will be held open when the engine warms up and the valve stem expands. If the valve does not fully seat, it cannot properly cool, and soon burns. If clearance is too great, the cam cannot transmit its total lift and duration, reducing performance.

Engine designers specify a proper clearance for the valve train, and on many engines the technician or engine builder must set the clearance. Unfortunately, wear can change the clearance as the engine runs.

Wear between the valve stem and rocker arm or cam lobe wear may increase clearance between parts in the valve train. Extra clearance allows the parts to slam together, increasing wear and valve train noise.

Figure 10-1. The Chrysler 3.8-liter overhead-valve V6 has hydraulic lifters, pushrods, and rocker arms to operate its two valves per cylinder.

Wear between the valve face and seat may reduce clearance. As the valve wears, it slowly sinks into the head, bringing the stem closer to the rocker arm, reducing clearance.

Designers must make some provision in the valve train for the increase or reduction of clearance. It is always a part of the design of either the lifter, follower, pushrod, or rocker arm that automatically compensates or allows you to adjust for this change in the valve train. As we detail these parts in the following paragraphs, we explain this feature.

Solid lifters

Solid lifters are the earliest type of lifters, figure 10-4. However, they reman in use today because they have high-performance characteristics that are still desirable. They are also called mechanical lifters to distinguish them from hydraulic lifters. Solid lifters are inexpensive, simple, one-piece iron machinings that function reliably from idle speed up to very high rpm. Most solid lifters for OHV engines have oil holes to help transmit oil to the pushrods and on to the rocker arms.

Solid lifters are also lighter than other types of lifters, so all other factors being equal, valve springs with lower pressure are required to resist valve float. As we explained in Chapter 8, the lighter the components of the valve train, the less inertia they have, and the less likely valve float is to occur.

The major drawback of solid lifters is that they cannot compensate for the wear. You must check and adjust the clearance at regular intervals. Many engines with solid lifters allow for the adjustment in the rocker arm. We detail the various methods of adjustment in a later section in this chapter on rocker arms.

Most solid lifters have a body that is basically of constant diameter over their entire length, but the **mushroom lifter** has a foot which is a great deal larger than the body of the tappet, figure 10-5. The wide foot of the mushroom lifter will follow a steeper cam lobe profile without digging into the lobe. Mushroom lifters are used in OHV racing engines when roller lifters are prohibited by the rules. Because of the large foot, you must machine the block to accept mushroom lifters. You can then only remove the mushroom lifter

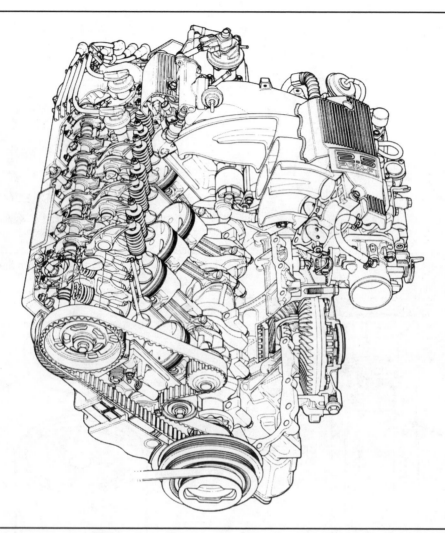

Figure 10-2. The Acura Vigor 2.5-liter in-line 5-cylinder engine has two rows of rocker arms on a single-overhead camshaft to operate four valves per cylinder.

from the bottom of the lifter bore. Some passenger-car engines, such as air-cooled Volkswagen engines, have used mushroom lifters.

Many DOHC engines have solid lifters that operate directly off the cam lobe, figure 10-6. These lifters are also called "buckets" because of their shape. You adjust the valve clearance with these lifters by exchanging shims placed either above or below the lifter, figure 10-7. Adjustment is made by replacing the shim with one of a different thickness. Changing shims makes valve adjustment more difficult than other designs, but once set these hardened parts wear slowly and stay in adjustment longer. For example, the 1990-97 Toyota Previa, which uses this type of solid lifter, has a factory recommended valve adjustment interval every 60,000 miles (96,600 km).

Flat Tappet: A valve lifter or tappet where the base of the lifter that rides against the cam lobe appears flat. Either solid or hydraulic lifters can be flat tappets.

Roller Tappet: A valve lifter or tappet with a roller that rides on the camshaft lobe. Either solid or hydraulic lifters can be roller tappets.

Solid Lifter: A valve lifter or tappet that is one-piece, lightweight, and usually made of cast iron. Solid lifters cannot compensate for wear so the valve clearance must be checked and adjusted at regular intervals.

Mushroom Lifter: A valve lifter or tappet with a foot or base larger in diameter than its body.

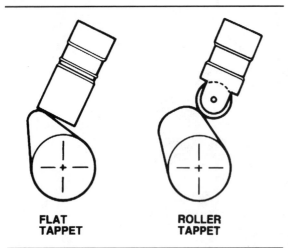

Figure 10-3. Lifters, or tappets, are made in two styles: flat and roller.

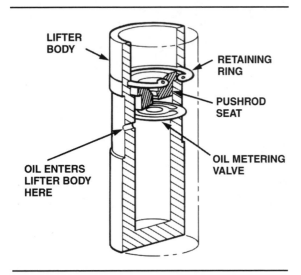

Figure 10-4. Oil enters most solid lifters through a hole in the side of the lifter body. A metering disc restricts the flow of oil to the pushrod and the top end of the engine.

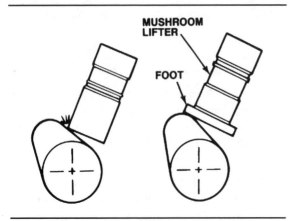

Figure 10-5. Mushroom lifters will follow a steeper cam lobe profile than a standard flat tappet.

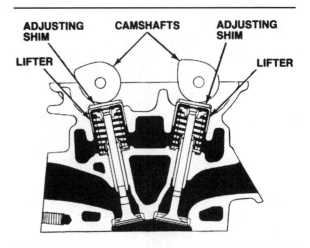

Figure 10-6. Bucket-type solid lifters fit over the valve spring and retainer.

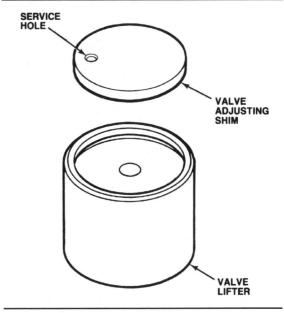

Figure 10-7. Valve clearance adjustment shims usually have holes to ease their removal. After using a tool to depress the lifter, you direct compressed air to the hole in the shim, floating it above the lifter. You then remove the shim with a magnet.

Hydraulic lifters

After many years of using adjustable tappets and adjustable rocker arms, manufacturers began replacing them in the 1950s with the **hydraulic lifter**, figure 10-8. It consists of an outer body, a plunger, and a check valve. Hydraulic lifters do not need adjustment because they automatically adjust for wear or play in the valve train. They always maintain zero clearance in the valve train without holding the valve open. Some engines require you to set the initial clearance when the valve train is assembled, but thereafter

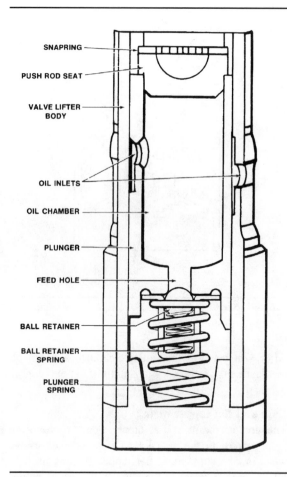

SNAPRING

PUSH ROD SEAT

VALVE LIFTER
BODY

OIL INLETS

OIL CHAMBER

PLUNGER

FEED HOLE

BALL RETAINER

BALL RETAINER
SPRING

PLUNGER
SPRING

Figure 10-8. Hydraulic lifters are more complex than solid designs, but they are maintenance free.

the hydraulic lifter needs no adjustment. Another advantage of maintaining zero clearance is that the valve train operates quietly.

The hydraulic lifter rests in a bore. Oil passages in the cylinder head and block supply oil under pressure from the engine oil pump. The cam lobe moves the body of the lifter, which slides up and down in the lifter bore. Inside the lifter body is a plunger. Oil is forced through a hole in the side of the body, and through a second hole in the side of the plunger to the inside of the plunger. A check valve in the bottom of the plunger allows oil to pass out of the plunger into the lifter body, but not back into the plunger.

A light spring under the plunger pushes the lifter body against the cam and the plunger against the pushrod or cam follower. This removes all the play from the valve train, but is not strong enough to lift the valve off its seat.

The space below the bottom of the plunger fills with oil under pressure through the check valve. Because of the check valve, this oil can pass out of the

lifter only through the very small clearance between the plunger and the lifter body.

When the cam lobe raises the lifter, the check valve closes and the plunger tries to compress the oil below it. But since oil cannot be compressed, as the lifter rises, the plunger rises with it, opening the valve. Although some leakage occurs, the amount is practically nothing. The leakage is small because the clearance between the body and the plunger is extremely small and the valve opens and closes in a fraction of a second. At an idle speed of 500 rpm, the valve is open less than nine hundredths of a second.

The pushrod seat is usually at the top of the plunger and has a hole in it in most designs. The hole allows oil to go up through the hollow pushrod to lubricate the valve stems and rockers. There may be a metering device under the pushrod seat, such as a small disc. This limits the oil pressure to the top end so that the valve guides are not flooded with oil.

Designers use the advantages of hydraulic lifters in OHC as well as OHV engines. In SOHC engines, many manufacturers install the hydraulic lifters in the tips of the rockers arms, figure 10-9, in place of a screw-type adjuster. In many DOHC and some SOHC engines, a hydraulic lifter is built into the bucket-type lifter, figure 10-10.

Hydraulic lifter pump-up

Valve float creates a sudden slack in the valve train. The hydraulic lifter quickly takes up this slack because the hydraulic pressure in the lifter causes it to expand. Since the lifter is now longer than it needs to be, it holds the valve off of its seat and the engine loses power. Pump-up occurs when an engine is run at a higher rpm than the designers of its camshaft and valve train intended.

Hydraulic lifter leak down

A small amount of leakage between the lifter plunger and body helps lubricate the lifter body and is beneficial to performance. **Hydraulic lifter leak down** allows for some heat expansion in the valve train parts and helps reduce lifter pump-up during valve float.

Hydraulic Lifter: A valve lifter or tappet that uses oil pressure to automatically adjust for wear or play in the valve train. Hydraulic lifters always maintain zero clearance in the valve train so that they need no periodic adjustment and operate quietly.

Hydraulic Lifter Leak Down: Hydraulic leakage between the lifter plunger and body. It allows for heat expansion in the valve train parts and helps reduce lifter pump-up during valve float.

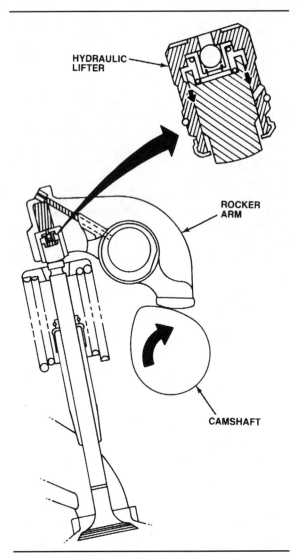

Figure 10-9. Hydraulic lifters may be built into rocker arms on OHC engines.

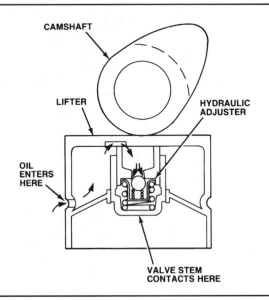

Figure 10-10. Hydraulic lifters may be built into bucket-type lifters on OHC engines.

You can check hydraulic lifters with a leak down test before installing them in an engine. This leak down rate is critical because too much leakage reduces the lifter's ability to transmit the total valve lift from the cam lobe.

Installing hydraulic lifters

When you install hydraulic lifters after performing engine work, the length of the valve train at each valve must be checked. If there is too much play, the lifter will not be compressed and will not operate. If there is too little play, the lifter may be completely collapsed, preventing the valve from closing. If the engine does not have adjustable rocker arms, adjustable or different-length pushrods must be used, or the valve tips ground to adjust the lifter clearance.

After you perform the initial clearance adjustment, readjustment is unnecessary until the next engine overhaul.

Cam and flat tappet wear

The greatest point of flat tappet wear is where the tappet rides on the cam lobe. To prevent excessive wear in one spot, most flat tappet designs used in OHV engines rotate in their bores. Camshaft manufacturers grind the cam lobe with a slight taper (approximately 0.0007 inch to 0.002 inch or 0.0178 mm to 0.051 mm) from one side of the lobe to the other, figure 10-11. When the taper is combined with lifters that have a crown on the bottom, the pressure is applied off center and the lifter rotates. You can see this rotation when you run an engine with the rocker covers off. The pushrods actually spin in their sockets from lifter rotation.

Another way to achieve lifter rotation is by having the lifter bore slightly offset from the cam lobe. Instead of rubbing across the center of the lifter, the lobe rubs to one side. This makes the lifter rotate.

Whenever you replace a camshaft with either a reground or new cam, you must also replace the lifters. Using worn lifters can damage a new cam very quickly. The surface of a worn lifter contacts the cam lobe only at the high points between the wear grooves. This increases the pressure on the cam lobe face and quickly wears away the smooth lobe surface.

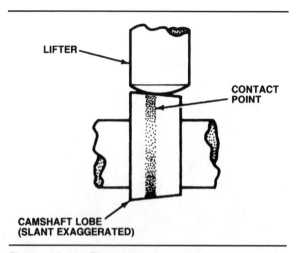

Figure 10-11. The tapered cam lobe rotates the crowned lifter. Both taper and crown are exaggerated in this illustration.

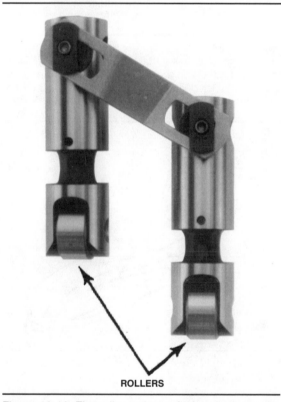

Figure 10-12. The roller reduces friction between the lobe and the lifter.

Roller lifters

The roller tappet or lifter was once an expensive high-performance lifter, but vehicle manufacturers now use it extensively on street cars. The roller reduces friction between the lobe and the lifter, figure 10-12. It also allows grinding of more radical cam lobe shapes, because the roller will follow almost any shape lobe. Consequently, cam lobe shapes designed for roller

lifters, called "roller cams," usually open the valve much faster than solid or hydraulic "flat tappet" cams. Quicker opening allows designers to specify a timing profile that keeps the valves closed longer in the cycle. Since the roller cam can open the valves faster, it still allows adequate intake duration for airflow in through the valve. The result is more horsepower with less loss in low-speed torque output, idle quality, or engine vacuum for running accessories. For more information on camshaft timing, see Chapter 9.

In the 1980s, fuel mileage and engine efficiency became so important that domestic automakers adopted roller lifters in all of their overhead-valve engines. In these engines, the rollers are built into hydraulic lifters, figure 10-13.

Roller lifters require a special device to keep the roller square with the lobe, figure 10-14. This device is usually a link connecting two tappets together, or a guide bolted to the block. Also, blocks manufactured for roller lifters usually have larger lifter bores to accommodate large-diameter roller lifters. The rollers always provide a smaller contact area on the cam lobe than flat tappets. Even if the valve spring pressure remains the same, a switch to roller lifters places greater pressure per square inch on the camshaft lobe. To withstand the extra stress, manufacturers always use a higher grade metal or heat treatment process for roller camshafts. Camshafts designed for use with flat tappets will quickly wear out if used with roller lifters.

Followers

Some overhead-camshaft engines use **cam followers** or finger followers, figure 10-15. One end of the follower rests on a pivot, while the other end rests on the valve stem. The rotating cam lobe contacts a pad on the top of the follower. The clearance in the valve train is between this pad and the cam lobe. The follower has a leverage ratio similar to a rocker arm. You multiply the lift of the cam lobe by this ratio to compute the valve lift.

Hydraulic lifters have many advantages, but they do usually add unwanted weight to the valve train. Using followers on an OHC engine allows a vehicle manufacturer to install hydraulic lifters, but keep the extra weight and inertia out of the valve train. The hydraulic lifters sit in a bore in the head, offset from the valve train instead of depositing their weight right

Cam Follower: A one-piece valve train part that follows the motion of the cam lobe and transfers it to the valve stem. A follower has a leverage ratio similar to a rocker arm. Also called a finger follower.

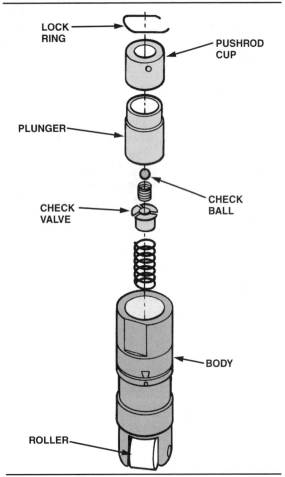

Figure 10-13. A roller is incorporated into this Ford hydraulic lifter.

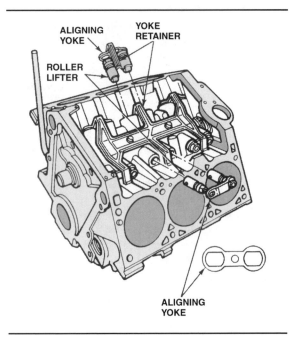

Figure 10-14. Roller lifters, such as this Chrysler design, have a yoke that ties each pair of lifters together and keeps the rollers aligned with the cam lobe. A retainer bolts to the block and keeps the cam lobes from pushing the lifters out of the block.

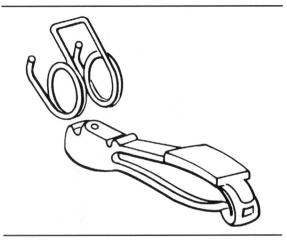

Figure 10-15. Cam followers transmit the cam lobe profile to the valve.

over the valve. In this position, they are still easily able to hydraulically preload the cam follower and keep the valve adjusted automatically, figure 10-16.

Some engines have rollers built into the followers to reduce friction, figure 10-17. In the late 1980s, Chrysler adopted roller followers for its SOHC 2.2- and 2.5-liter in-line 4-cylinder engines, and credited the change with providing a three to four percent increase in city fuel economy. To withstand the extra stress the rollers place on the cam lobes, Chrysler changed the camshafts to a higher grade of heat-treated nodular iron.

Most followers operate a single valve. However, the 4-cylinder SR20DE engine, used in some 1991-97 Infiniti and Nissan models, has Y-shaped followers that operate two valves each, figure 10-18. The engine features DOHC and 16 valves. Each follower operates a pair of intake or a pair of exhaust valves. Hydraulic lifters maintain valve lash. The engine requires only eight cam lobès to operate the 16 valves, reducing internal friction.

PUSHRODS

Pushrods, found only in OHV engines, transmit the lifting motion of the valve train from the lifters to the rocker arms, figure 10-19. Pushrods are built to be stiff, light, and flex free. Designers also try to make the pushrod as short as the engine design allows to further reduce flex. The less the pushrod flexes, the more accurately it transmits the cam lobe profile to the valve.

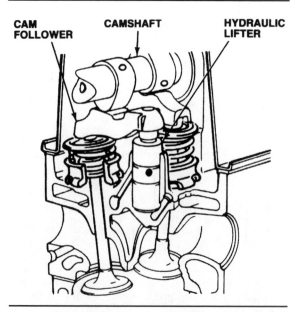

Figure 10-16. The use of cam followers allows designers to fit overhead-camshaft engines with hydraulic lifters.

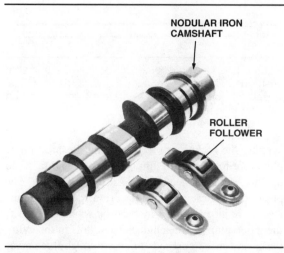

Figure 10-17. Chrysler replaced followers with sliding pads to roller followers to reduce friction and improve fuel economy.

Most pushrods are made of seamless carbon steel tubing, although some in the past have been solid steel. High-performance engines may use pushrods made of chromium-molybdenum tubular steel because of its greater strength. Chrome-moly pushrods may also have a thinner wall thickness to reduce weight. A few high-performance pushrods are made of aluminum.

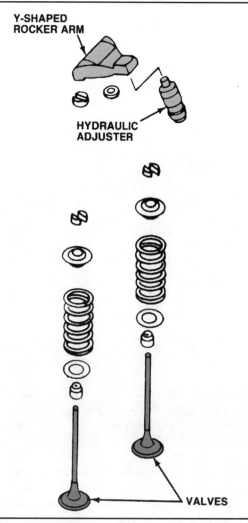

Figure 10-18. Each Y-shaped follower operates a pair of valves.

The bottom of the pushrod fits into a socket in the lifter that is deep enough that the pushrod cannot fall out. At the top, the pushrod usually fits into a cup-shaped recess in the rocker arm. To reduce wear, the pushrods have hardened tip inserts. The ends of the pushrods come in many different designs, figure 10-20.

Pushrods and Valve Clearance

Some engines do not have any provision for valve clearance adjustment. When the valves are refaced and the seats ground, the valves move into the head. This compresses the hydraulic lifters. If too many valve jobs have been done, the lifters will be compressed so much that after assembly they bottom out. The lifters will not adjust and will probably hold the valves off their seats when the parts get hot and expand. The results can be no compression and bent pushrods if the interference is major, or lowered

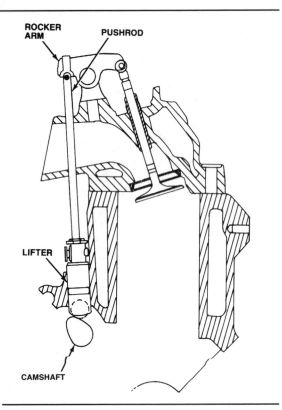

Figure 10-19. Overhead-valve engines are also known as pushrod engines because of the large pushrod that reaches from lifter to rocker arm.

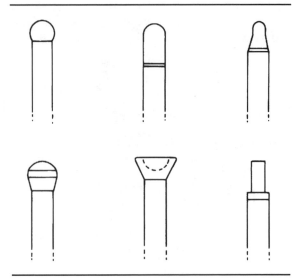

Figure 10-20. Designers have tried a number of different pushrod end designs to reduce wear and ensure retention in the lifter.

Figure 10-21. Some older cars have adjustable pushrods. To make the necessary adjustment, just loosen the locknut, turn the threaded end, and then tighten the locknut.

compression and possibly burned valves after the engine is run awhile if the interference is minor.

For engines without adjustable valves, manufacturers specify a valve stem length measurement. If the measurement is not within specifications, you have the option of grinding the stem tip or the valve seat to change the measurement. If neither produces the desired measurement, or the amount of grinding required would be severe, the machinist installs a new valve seat or valve.

Other manufacturers solve the problem by offering pushrods in different lengths. They are usually available in 0.030 or 0.060 inch (0.76 mm or 1.5 mm) under and over standard size. You may also see a few older cars with adjustable pushrods. These have an adjustment screw at the top of the rod to change the length, figure 10-21.

Pushrod Guide Plates

Stud-mounted rocker arms that pivot on a ball, as on most General Motors OHV engines, can tip from side to side when the engine is running. This movement causes wear and inaccurate transfer of the cam profile to the valve. Some of these engines have a guide plate that keeps the pushrod aligned with the rocker arm,

keeping the rocker arm from moving. Pushrod guide plates are formed from thin metal plates and fastened to the cylinder head casting, figure 10-22. The pushrods in these applications are made of hardened steel to reduce wear on the pushrod where it contacts the guide plate. Some engines have slots in the cylinder head for the pushrods to pass through that serves the same purpose of aligning the pushrods with the rockers.

Rocker arms that mount on a shaft do not need pushrod guide plates. The shaft bushing keeps the rocker from twisting.

Pushrod Lubrication

Most hollow pushrods also carry oil to the top end of the engine. Oil flows from the lifter through the hollow pushrod, figure 10-23. This oil lubricates the rocker arms, valve stems, pushrod seat, and drips over the valve springs, helping to keep the springs cool. If the pushrod does not carry oil, then the pushrod's lifter end gets oil from the lifter and its rocker arm end gets oil from the rocker.

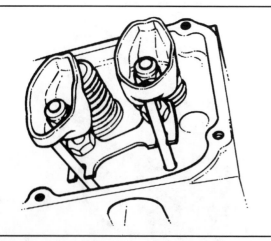

Figure 10-22. Pushrod guide plates fasten to the head by the rocker arm studs.

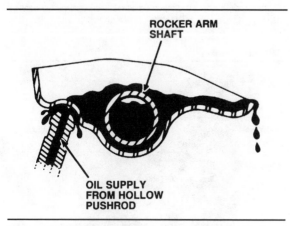

Figure 10-23. Hollow pushrods supply oil to pushrod ends and rocker arms and shafts. Oil then drips from the rocker arm onto valve springs, stems, and guides.

ROCKER ARMS

Rocker arms are used in all OHV engines and many OHC engines. In the OHV engines, a pushrod transfers the lift of the cam lobe to the top of the cylinder head, figure 10-24. The rocker arm acts as a reversing lever and transfers the force to the valve. In SOHC engines, the camshaft operates the rocker arms directly, figure 10-25.

Rocker arms are usually cast iron, stamped steel, or aluminum, figure 10-26. Cast-iron rockers are strong but add weight to the valve train. Stamped-steel rockers are lighter, inexpensive to manufacture, but not as durable. Aluminum rockers are lightweight but usually require the extra expense of special heat treatment to be durable.

Cast-iron rockers are often used on overhead-camshaft engines and imported overhead-valve engines. Domestic OHV engines typically use

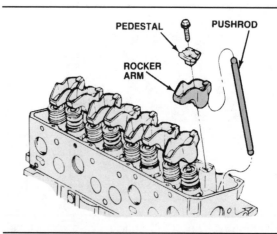

Figure 10-24. Pedestal-mounted rocker arms are common on Ford OHV engines.

stamped-steel rockers. Some import OHC engines have aluminum rocker arms that use hardened steel inserts at the valve wear point.

Like roller lifters, some rocker arms have rollers to reduce friction. Manufacturers use roller rocker arms on OHC engines and OHV engines, figures 10-27 and 10-28.

■ The Six-Stroke Engine

If two-stroke and four-stroke engines work so well, why not add a few more? Charles Linford built the first six-stroke engine in England in 1880, and several other inventors built and patented versions of this device during the next 15 years, including one in America.

In a six-stroke engine cycle, three complete revolutions of the crankshaft are necessary for one power stroke. Beginning with a four-stroke cycle, an extra crank revolution is placed between the exhaust stroke and the next intake stroke. During these extra two strokes, the piston first descends and fills the cylinder with air, and then rises and purges the cylinder of the air and any remaining exhaust gases. There is no spark, compression, or fuel intake during the extra strokes. The necessary valve action for introduction and expulsion of the scavenge air is supplied by extra cams.

The advantage of a six-stroke engine is that the cylinder is almost completely emptied of all exhaust gases before beginning the intake stroke. This increases the efficiency of combustion enough to offset the disadvantage of having a power stroke only once for every six strokes of the piston.

The disadvantage is that the engine provides no appreciable gains in power or economy over that of a standard four-stroke engine, but requires considerably more complex valve gear in order to run. Perhaps as a result, no new six-stroke engines have been introduced since 1895.

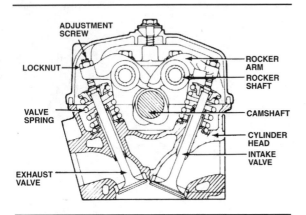

Figure 10-25. An SOHC engine with shaft-mounted rocker arms. This crossflow head requires two rocker shafts.

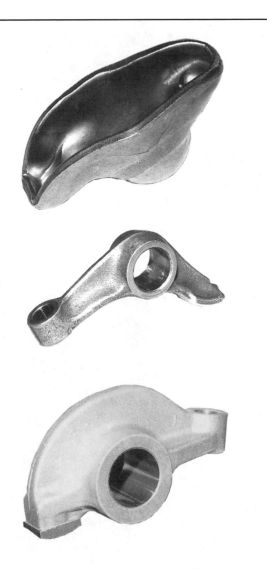

Figure 10-26. Rocker arms may be stamped steel, cast iron, or aluminum.

Figure 10-27. A Chrysler roller rocker arm for the 3.0-liter V6.

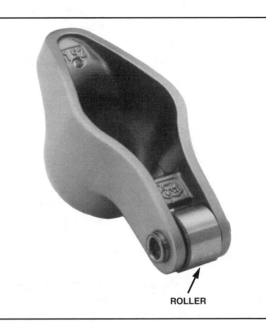

Figure 10-28. An after-market, high-performance roller rocker arm for an OHV V8 engine.

Shaft-Mounted Rockers

Some OHV engines and most SOHC engines mount their rocker arms on a heavy shaft that runs the full length of the cylinder head, figure 10-29. Stands that bolt to the head provide support for the shaft. The stands and rockers are arranged so that each rocker is against one side of a stand. This aligns the rocker arm with the pushrod and the valve stem. Some designs use a washer on the shaft to put light pressure against the rocker arm and keep it against the stand, figure 10-30.

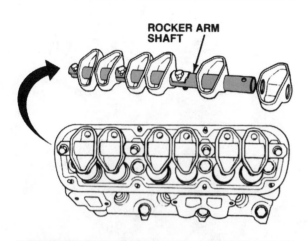

Figure 10-29. The rocker arm shaft on this Chrysler OHV V6 runs the length of the cylinder head.

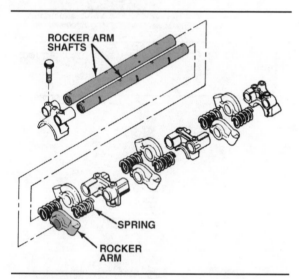

Figure 10-30. This Mitsubishi V6 uses springs to keep its rocker arms aligned.

Figure 10-31. Check valve clearance with a feeler gauge. You adjust the clearance by turning a screw, which is in place with a locknut.

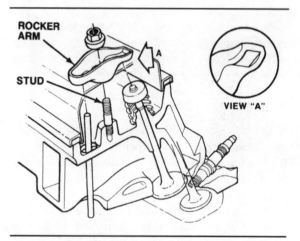

Figure 10-32. Stud-mounted rocker arms are very common on General Motors OHV engines.

Because the shaft provides a strong and stable platform for the rocker arms, shaft-mounted rockers work very well up to high rpm. They resist flex and accurately transmit the cam lobe profile to the valve. On the negative side, they add weight and expense to the engine.

On OHV engines with solid lifters or SOHC engines without lifters, you typically make the valve lash adjustment by turning a screw in the tip of the rocker, figure 10-31. The adjusting screw can be in the pushrod side of the rocker arm or the valve stem side. Shaft-mounted rockers on OHV engines with hydraulic lifters usually have no provision for adjustment. You make the clearance by grinding the tip of the valve or installing different length pushrods when you build the engine.

Stud-Mounted Rockers

In this rocker mounting system, found only in OHV engines, the designers mounted each rocker arm on a stud, figure 10-32. The stud is pressed or threaded into the head. A split ball acts as the pivot for the rocker. The split ball is held in place with a nut that you screw up or down on the stud to adjust the valve clearance. The rocker arms themselves are inexpensive, lightweight stamped-steel.

For many years all OHV engines used shaft-mounted rockers. The 1955 small-block Chevrolet V8 engine introduced stud-mounted rockers. Some engineers doubted that the stud-mounted rockers would work. Their reservations were mainly that the studs and stamped-steel rockers would flex too much and transmit the cam lobe profile poorly. But in years of street use and testing, this has not proved to be the case. They work so well that this design and variations of this design have been used by many other engine makers, and are still in use today. Only in racing engines with stiff valve springs that operate at over 7,000 rpm have stud-mounted rockers shown any weakness. On these race engines, builders fit forged-aluminum or forged-steel rocker arms, larger screw-in studs, and a girdle that clamps all the studs together, providing greater rigidity.

Pedestal-Mounted Rockers

Pedestal-mounted rockers are similar to stud-mounted rocker arms, figures 10-33 and 10-34, and are used only in OHV engines. Instead of pivoting on a split ball, the rocker pivots on a split shaft. There is no stud. The split shaft fastens to the cylinder head with one or two capscrews. This is not quite as simple as the stud-mounted rocker, but it still eliminates the long, heavy rocker shaft. The rockers are also the lightweight, stamped-steel type used on stud-mounted rockers. Pedestal-mounted rockers do not allow adjustment for valve lash. If adjustment is needed, you must grind the valve tips or fit different length pushrods.

Rocker Arm Ratio

If you look closely at a rocker arm, you will see that the distance from the pivot point to the pushrod seat (or cam lobe pad) is shorter than the distance from the pivot point to the valve stem pad, figure 10-35. This lever action gives the rocker a mechanical advantage in the ratio of the long side to the short side. This **rocker arm ratio** can be anywhere from 1.1 to 1.75 on street engines, but is usually around 1.5.

The rocker arm ratio directly affects valve lift. If the lift of the cam lobe is 0.35 inch (8.9 mm) and the rocker ratio is 1.5, the actual lift at the valve will be 0.525 inch (13.35 mm), since 0.35 x 1.5 = .0525 (8.9 mm x 1.5 = 13.35 mm).

The rocker arm ratio is designed to give the desired valve lift while keeping cam lobe size to a minimum. The positive rocker ratio does put more pressure on the cam lobe and makes it harder for the lobe to open the valve. But it is easy to deal with the increased pressure. A cam lobe cannot be physically bigger than the camshaft bearing journal. Building all the lift into the cam lobe would require a larger lobe, larger bearings, and a larger overall cylinder head or block to contain them.

Rocker Arm Lubrication

Engine designers employ a variety of systems to supply the rocker arms with oil. Every type of oil delivery system described in Chapter 6 is used on the

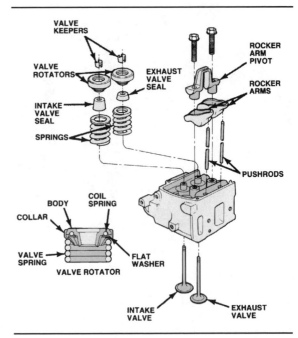

Figure 10-33. An Oldsmobile pedestal-mounted rocker arm.

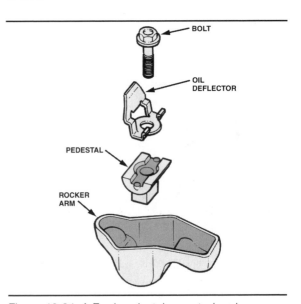

Figure 10-34. A Ford pedestal-mounted rocker arm.

rocker arms of one engine or another. The rockers in OHV engines usually get oil through hollow pushrods, as we showed earlier in figure 10-23. Rocker shafts receive oil from a gallery, figure 10-36, or a rocker arm stand. The rockers in OHC engines get oil through rocker arm stands, from an oil gallery, or an external line.

When the oil supply is through the rocker stand, it may be through a drilled hole in the stand, or the oil may come up through the rocker stand mounting bolt, figure 10-37. Since the bolt is hollow, you must

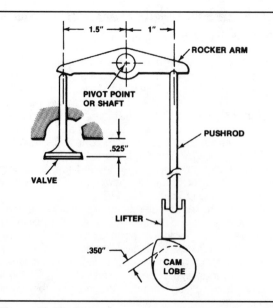

Figure 10-35. On rocker arms, the distance from the pivot to the pushrod seat is shorter than the distance from the pivot to the valve stem.

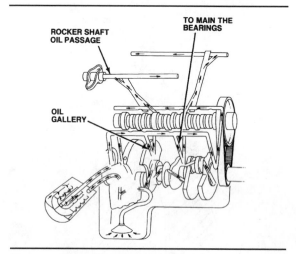

Figure 10-36. Galleries often supply oil to rocker shafts.

replace it in the stand positioned over the oil supply. In some cases, the bolt shank is a smaller diameter to allow the oil space to travel up through the stand. You must also replace the relieved bolt shank in the right stand.

Rocker Arm Geometry

Rocker arm geometry is the angle of the rocker arm in relation to the valve stem, figure 10-38. It is important because rocker arm geometry affects where the rocker arm tip contacts and pushes on the valve stem. Engineers consider the geometry correct when the center of the rocker arm tip contacts the centerline of the valve stem when the valve train is at 30 to 50 percent of its maximum lift. At this point, there should also be approximately a 90-degree angle between the centerlines of the rocker arm and the rocker contact pad or roller in relation to the valve stem, figure 10-38. If the rocker does not center on the valve stem properly, it pushes the valve stem sideways and wears out the guide prematurely.

In pushrod engines, the angle of the rocker arm in relation to the pushrod is also important. Altered rocker arm geometry may cause the pushrod to hit the side of its passage in the cylinder head or the pushrod guide plate.

Incorrect rocker arm geometry may also harm valve train lubrication in pushrod engines where hollow pushrods supply oil to the top end. In these engines, the pushrod fits into a cup-shaped recess in the rocker arm that usually has a hole in it. When the

opening in the pushrod aligns with the hole in the rocker arm, oil sprays over the rocker arms and valve springs. The alignment occurs when the rocker arm is in the lash position. As the rocker moves to open the valve, oil flow is shut off. If rocker arm geometry is incorrect, the holes may not be properly aligned and allow too little or too much to flow to the valve train.

You must maintain the correct rocker arm geometry in the engines you rebuild. Machining operations that change the distance between the camshaft and rocker arm, such as block or head resurfacing, alter the rocker arm geometry. Also, changes in valve stem height, caused by valve face or seat grinding, can change the geometry. Even installing a camshaft with a higher lift may cause rocker arm geometry problems.

Although cases of incorrect rocker arm geometry are rare, they do happen. A simple visual check while slowly turning the engine over is usually enough to see if geometry isn't correct or if anything is hitting that shouldn't be.

Rocker Arm Ratio: The ratio of the valve side of the rocker arm to the cam side, measured from the pivot.

Rocker Arm Geometry: The angle of the rocker arm in relation to the valve stem. Rocker arm geometry is correct when the center of the rocker arm tip contacts the centerline of the valve stem when the valve train is at 30 to 50 percent of its maximum lift.

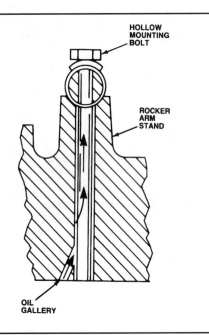

Figure 10-37. Hollow rocker arm stand mounting bolts receive oil from a gallery and direct it to the rocker shaft.

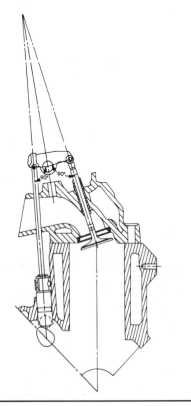

Figure 10-38. Rocker arm geometry primarily concerns the alignment of the rocker arm with the valve stem. These parts create an imaginary angle that should be approximately 90 degrees at one-half lift. On pushrod engines, the angle between the rocker arm and pushrod is important, too.

■ Cadillac's V8-6-4 Engine

The gasoline shortage of the 1970s drummed into the minds of the American public the need for good fuel economy. The domestic auto industry responded by downsizing cars and offering smaller 4- and 6-cylinder engines. Of course, this move threatened the American love affair with large cars and the big V8 engines needed to power them. Cadillac's variable displacement V8 engine, offered from 1981 to 1984, was designed to provide the needed fuel economy and still have the torque to power the heavy, luxury cars the public craved.

During cruise, deceleration, or idle, the engine computer signals solenoids in the valve train to deactivate both rocker arms in a cylinder, "shutting down" that cylinder. The system does this in two-cylinder sets, reducing the engine first to six and then to four cylinders, so Cadillac called it the V8-6-4 engine. It effectively reduced engine size for better fuel economy.

The valve selector solenoids are mounted on cylinders 1, 4, 6, and 7. When the computer signals, the solenoid unlocks a collar that holds the rocker arm in place. The rocker is then free to ride up the rocker arm stud and can no longer pivot to open the valve. The collar is spring loaded to maintain zero valve lash and proper hydraulic lifter function. When the driver steps on the accelerator pedal, the system reactivates as many cylinders as necessary to give adequate acceleration.

The V8-6-4 engine delivered its promise of fuel economy and torque, but customers complained about its less than smooth, linear operation during acceleration. At the request of customers, some mechanics tricked the computer to lock the V8-6-4 into 8-cylinder operation by grounding a wire. Today, it is difficult to find a V8-6-4 in original operating condition.

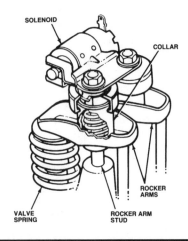

SUMMARY

Tappets, cam followers, and rocker arms transfer the motion of the cam lobes to the valve's lifters. Valve stems expand from heat, so designers make a provision in the valve train for periodic adjustment.

Lifters are divided into two types: flat tappets and roller tappets. Both solid and hydraulic lifters are flat tappets. Roller tappets have a roller wheel that rides on the cam lobe. Solid lifters, also called mechanical lifters, are inexpensive, simple, lightweight, and function reliably from idle speed up to very high rpm. Solid lifters must be checked and adjusted at regular intervals. Hydraulic lifters operate quietly and do not need adjustment because they automatically adjust for wear in the valve train. Hydraulic lifters may suffer from pump-up at high rpm or lifter leak down caused by wear.

Roller lifters reduce friction between the cam lobe and the lifter and will follow almost any shape lobe. Rollers may be built into hydraulic or solid lifters. In the 1980s, domestic automakers adopted roller lifters in all of their overhead-valve engines.

Some overhead-camshaft engines use cam followers or finger followers. One end rests on an adjustable pivot, while the other end rests on the valve stem.

Pushrods, found only in overhead-valve engines today, transmit lifting motion from the lifters to the rocker arms. Most pushrods are made of seamless, carbon steel tubing. The bottom of the pushrod fits into a socket in the lifter which is deep enough that the pushrod cannot fall out. At the top, the pushrod usually fits into a cup-shaped recess in the rocker arm.

The hollow portion of most pushrods carries oil to help lubricate the rocker arms and the valve stems. Oil flows from the lifter through the hollow pushrod.

Rocker arms are used in all OHV engines and many SOHC engines. In the OHV engines, a pushrod transfers the lift of the cam lobe to the top of the cylinder head. In SOHC engines, the camshaft operates the rocker arms directly. Rocker arms may be shaft-mounted, stud-mounted, or pedestal-mounted.

On rocker arms, the distance from the pivot point to the pushrod seat is shorter than the distance from the pivot point to the valve stem pad. This lever action gives the rocker a mechanical advantage in the ratio of the long side to the short side. The rocker arm ratio can be anywhere from 1.1 to 1.75 on street engines, but is usually around 1.5.

Rocker arm geometry is the angle of the rocker arm in relation to the valve stem. It affects where the rocker arm tip contacts and pushes on the valve stem. If the rocker does not center on the valve stem properly, it may push the valve stem sideways and wear out the guide. Machining operations that change the distance between the camshaft and rocker arm, such as block or head resurfacing, alter the rocker arm geometry. Also, changes in valve stem height caused by valve face or seat grinding can change the geometry. Installing a camshaft with a higher lift may also cause rocker arm geometry problems.

Review Questions
Choose the single most correct answer.
Compare your answers with the correct answers on Page 328.

1. Technician A says that solid lifters are made of aluminum.
   Technician B says that solid lifters are the lightest lifters available.
   Who is right?
   a. A only
   b. B only
   c. Both A and B
   d. Neither A nor B

2. Hydraulic lifters:
   a. Make it easy to set the tappet clearance
   b. Are filled with hydraulic fluid similar to brake fluid
   c. Cannot rotate; they can only move up or down
   d. Transmit steady thrust to the valves because oil cannot be compressed

3. Technician A says that hydraulic lifters must pump up to work properly.
   Technician B says that solid lifters only pump up half as much as hydraulic lifters.
   Who is right?
   a. A only
   b. B only
   c. Both A and B
   d. Neither A nor B

4. Technician A says the greatest point of flat tappet wear is where the tappet rides on the cam lobe.
   Technician B says that to prevent excessive wear in one spot, most flat tappets used in OHV engines rotate in their bores.
   Who is right?
   a. A only
   b. B only
   c. Both A and B
   d. Neither A nor B

5. Technician A says roller lifters reduce friction.
   Technician B says that roller lifters allow more radical cam lobe profiles.
   Who is right?
   a. A only
   b. B only
   c. Both A and B
   d. Neither A nor B

6. Technician A says that pushrods are used in OHV engines.
   Technician B says that pushrods are used in SOHC engines.
   Who is right?
   a. A only
   b. B only
   c. Both A and B
   d. Neither A nor B

7. Most pushrods:
   a. Are hollow
   b. Are adjustable
   c. Move freely in their guides, which are similar to valve guides
   d. Are flexible

8. Technician A says that pushrod guide plates may create some wear on the pushrod.
   Technician B says pushrod guide plates keep the pushrods aligned with the camshaft.
   Who is right?
   a. A only
   b. B only
   c. Both A and B
   d. Neither A nor B

9. Technician A says that many domestic OHV engines use stamped-steel rocker arms.
   Technician B says the rocker arm works as a reversing lever.
   Who is right?
   a. A only
   b. B only
   c. Both A and B
   d. Neither A nor B

10. Rocker arms:
    a. Are always mounted on a shaft
    b. Are unequal in length on each side of the pivot
    c. Always pivot on a ball joint
    d. Provide a mechanical advantage anywhere from 2.1 to 2.6

11. Shorter pushrods are sometimes installed because:
    a. This increases engine power
    b. The original pushrods tend to spread out
    c. The original pushrods were defective
    d. Grinding valve seats changes the position of the valves

12. Technician A says that cam followers are only used in flat-head engines.
    Technician B says that some cam followers have a ratio similar to rocker arms.
    Who is right?
    a. A only
    b. B only
    c. Both A and B
    d. Neither A nor B

13. Technician A says hollow pushrods may carry oil to the top end of the engine.
    Technician B says pushrods are hollow to reduce their weight and never carry oil.

Who is right?
a. A only
b. B only
c. Both A and B
d. Neither A nor B

14. Compared to other kinds of rocker arms, shaft-mounted rockers:
    a. Are less expensive to manufacture
    b. Add less weight to the engine
    c. Resist flex more effectively
    d. Are less reliable at high rpm

15. Technician A says that the studs for stud-mounted rocker arms are always pressed into the head.
    Technician B says that the studs for stud-mounted rocker arms may be pressed or threaded into the head.
    Who is right?
    a. A only
    b. B only
    c. Both A and B
    d. Neither A nor B

16. Technician A says that the rocker arm ratio puts less stress on the cam lobe and makes it easier to open the valve.
    Technician B says that the rocker arm ratio gives the desired valve lift but keeps the cam lobe to a manageable size.
    Who is right?
    a. A only
    b. B only
    c. Both A and B
    d. Neither A nor B

17. If the rocker arm does not center properly on the valve stem:
    a. It causes no problems
    b. The rocker arm will wear faster
    c. The pushrod will bend
    d. The valve guide will wear faster

18. Technician A says that when rocker arm geometry is correct, the center of the rocker arm tip contacts the centerline of the valve stem at 30 to 50 percent lift.
    Technician B says that when rocker arm geometry is correct, there is a 90-degree angle between the centerlines of the rocker arm and the rocker contact pad in relation to the valve stem at 30 to 50 percent lift.
    Who is right?
    a. A only
    b. B only
    c. Both A and B
    d. Neither A nor B

11

Crankshafts, Flywheels, Vibration Dampers, and Balance Shafts

For an automotive machinist, the crankshaft is second only to the cylinder block as the most important engine component. The quality of crankshaft finish, straightness, and balance are critical to engine operation and reliability. The flywheel, used on manual transmissions, and the flexplate, used on automatic transmissions, is an engine part that the machinist may or may not be concerned with, depending upon how complete an engine the customer desires. Flywheels can be machined flat or replaced with a new part. Flexplates are inspected and replaced, if necessary. Balance shafts are found on many in-line 4-cylinder and some V6 engines. As an automotive machinist, you must only inspect balance shafts for wear and reassemble them correctly. They require no special machining operations.

CRANKSHAFTS

As we said in Chapter 1, the crankshaft converts the reciprocating motions of the pistons to rotary motion. This motion makes its way through the driveline to turn the drive wheels.

Crankshafts are made of iron or steel and are either cast or forged. As with other engine parts, designers specify different materials and methods of manufacture depending on the strength required of the crankshaft and the available budget. An automotive machinist must have a full understanding of the following parts of a crankshaft, figure 11-1:

- Main bearing journals
- Rod bearing journals
- Fillets
- Crankshaft throws
- Counterweights
- Keyways and threads
- Oil Passages.

Main Bearing Journals

The crankshaft rotates in the cylinder block on main bearings, figure 11-2. The parts of the crank that rotate in the bearings are called the **main bearing journals**, which are in-line with the crankshaft's centerline. The main bearings are detailed in Chapter 13. Their purpose is to allow the crankshaft to turn easily in the block without excessive wear.

The simplest crankshaft, operating a single cylinder, requires two main bearings, one on each end of the shaft. If more cylinders are added and the crankshaft becomes longer, more main bearings are usually necessary. Two main bearings will support a multi-cylinder crankshaft, but they will not keep it from bending. As we stated in Chapter 7, the more

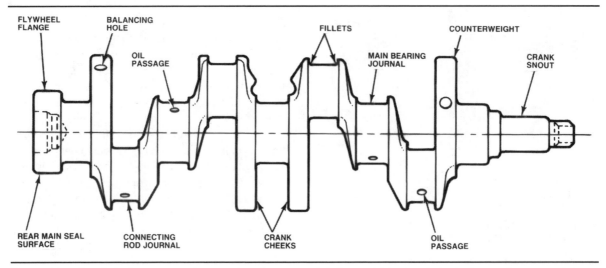

Figure 11-1. Crankshafts are one piece and perform a straightforward job, but their design and construction are complex.

main bearings an engine has, the stiffer its structure and the more durable it is. A minor drawback is that adding main bearings increases internal friction. It is now common for in-line 4-cylinder and V8 engines to have five main bearings, for V6 engines to have four main bearings, and for in-line 6-cylinder and V12 engines to have seven main bearings, figure 11-3.

Thrust surface

Clutch release and engagement forces or pressure in the automatic transmission cause thrust loads that push and pull the crankshaft back and forth in the engine block. A thrust bearing in the engine supports these loads and maintains the crankshaft position. The designer specifies a thrust bearing surface, ground on a crankshaft cheek next to one of the main bearing journals, figure 11-4. The thrust surface is usually located at the middle or one of the end main bearings. On most engines, the bearing insert for that main bearing has thrust bearing flanges that ride against the crank's thrust surface. Some engines use a separate thrust bearing insert. These bearings are detailed in Chapter 13.

Rod Bearing Journals

The **rod bearing journals**, also called the **crankpins**, are offset from the crankshaft's centerline. The big end of each connecting rod attaches to the crankshaft on the rod bearing journal. Insert-type bearings fit between the big end of the connecting rod and the crankpin of the crankshaft.

Burning the air-fuel mixture in the combustion chamber pushes the piston down on its stroke. The piston pushes on the piston pin, which is in the small end of the connecting rod. The connecting rod is the link between the piston and the crankshaft. The piston

requires leverage to turn the crankshaft and the flywheel. The amount of leverage is equal to the distance from the centerline of the crankpin to the centerline of the crankshaft, figure 11-5. The illustrations show that this distance, called the **crankshaft throw**, can be longer or shorter to change the leverage. This distance determines the stroke of the piston.

As we detailed in Chapter 1, the displacement of an engine in cubic inches, liters, or cubic centimeters is computed from its bore and stroke. The crankshaft, with throws that measure one-half of the stroke, has a direct relationship to the displacement of the engine.

Bearing journal diameter

Designers must consider bearing longevity and crankshaft strength when they determine bearing journal diameter. The result is usually a compromise. Large journals add material to the crank, strengthening it. However, the larger the diameter, the higher the surface speed at a given rpm. To understand this relationship, imagine a record turning at $33^1/_3$ rpm. In miles per hour, the outer rim of the record turns faster than a point near its center. The significance of this relationship for crankshafts is that higher bearing journal surface speeds reduce bearing life.

In the first half of this century, most bearing journals were relatively small to keep surface speeds down. Improvements in bearing materials and lubricants allowed designers to enlarge bearing journal diameter without reducing bearing life. The stronger crankshafts that resulted allowed designers to increase maximum engine rpm for greater horsepower output.

Bearing journal finished size

The bearing journals, either main or rod, are among the most critical parts of an engine. The manufacturer

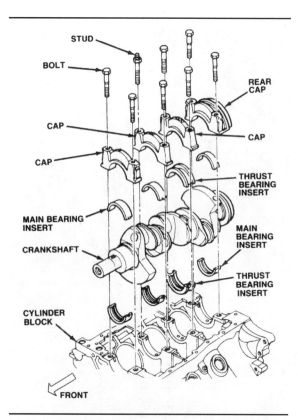

Figure 11-2. The crankshaft rotates in main bearings. It is comparatively easy to replace the bearings, so they are designed to absorb any wear and allow the crank to remain at its original size. It usually takes high mileage or a problem with the lubrication system before the crank shows any wear.

or machinist must keep the journal diameter within a tolerance range that usually varies no more than one thousandth of an inch (.025 mm). In many cases, the tolerance may be just a few ten-thousandths of an inch (hundredths of a millimeter). For its 2.3-liter Quad 4, GM specifies a finished main bearing journal size tolerance of 2.0470 to 2.0480 inches (51.996 to 52.020 mm). For its 4.0-liter V8, Lexus specifies a finished main bearing journal size tolerance of 2.6373 to 2.6378 inches (66.989 to 67.000 mm).

Main Bearing Journals: The parts of the crankshaft on which it rotates in the cylinder block.

Rod Bearing Journals: The parts of the crankshaft that the connecting rods attach to. They are also called crankpins.

Crankpins: Another name for rod bearing journals.

Crankshaft Throw: A part of the crankshaft measured from the crank centerline to the crankpin centerline. It is equal to one-half of the crankshaft stroke.

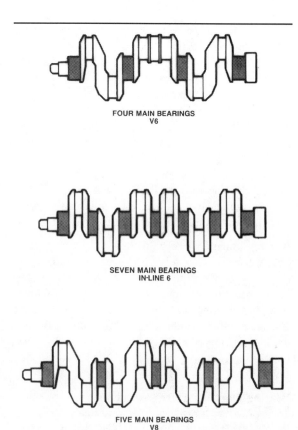

Figure 11-3. The main bearing journals also support the crank. The longer the crank is, the more main bearing journals it needs.

■ The Non-Rotating Crankshaft

Crankshafts always rotate, right? Wrong! It isn't any more necessary for a crankshaft to rotate than it is for the exhaust to exit from the side of the cylinder head. In an air-cooled engine there is no heavy cylinder block or liquid cooling system. The structure to which the cylinders are attached could be allowed to rotate, while the crankshaft remained stationary. It may sound crazy, but it has actually been done. During World War I, radial aircraft engines were built that way. They were manufactured by several companies, and powered planes from France, England, and Germany. Because the crankshaft was the only stationary part of the engine, fuel had to be pumped through the crankshaft to get to the cylinders. As odd as it was, the radial was a great success. It was the first successful air-cooled engine, which isn't surprising, considering the breeze that those whirling cylinders must have stirred up. The radial proved that air cooling would work, and paved the way for the modern air-cooled radial engine.

If the finished journal size falls outside that range, the running clearance is affected. For example, if the journal is much smaller than the specified size, the excessive clearance between the journal and bearings allows the crank to hammer against its bearings on the power stroke, quickly destroying the bearings as well as itself. With excessive clearance, lubrication is also poor because oil leakage between the journal and bearings lowers total oil pressure.

If the journal is larger than the specified size, making the clearance too small, there will not be enough room for sufficient oil between the journal and the bearing. The parts will overheat and destroy each other.

Bearing journal surface finish

The surface finish, or relative smoothness, of the bearing journals is very important for maximum bearing life. Journal surface finish is measured in microinches (µin) or micrometers (µm), which we discussed in Chapter 3. After grinding, a machinist micropolishes the journals to yield a very smooth finish. Manufacturers usually specify between a 5 and a 20 µin rms finish.

Bearing journal surface treatments

To improve wear resistance, bearing journals may be hardened by the manufacturer. The use of modern hardening processes are one of the main reasons manufacturers can substitute cast cranks where forged ones were necessary a generation ago. The most popular methods include the two we detailed in Chapter 3, nitriding and tuftriding.

Hard chroming is another surface treatment for crank bearing journals. Unlike the decorative chrome plating applied to bumpers, which uses a triple layer process of copper, nickel, and a soft layer of chrome, a hard layer of chrome is applied directly to the bearing journal, providing a very hard, durable finish. The journal is then ground to size and polished. Hard chrome plating has been used to improve journal durability, but not as a general repair for worn journals. Hard chromed journals cannot be checked for surface cracks by magnetic particle inspection. Further complicating the process, chroming can cause hydrogen embrittlement of the parent metal, thus requiring a heat treatment or baking step to remove the hydrogen from the metal.

Fillets

The joints where the main or rod bearing journal meet the crank are high stress areas. Crank failures often occur there. To strengthen that area, designers call for a curve or **fillet** of a specific radius at the joint, figure 11-6. The fillet provides additional support between the journal and the crankshaft cheek. Fillets also

Figure 11-4. A ground surface on one of the crankshaft cheeks next to a main bearing supports thrust loads on the crank.

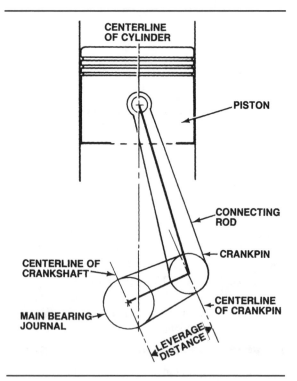

Figure 11-5. The distance from the crankpin centerline to the centerline of the crankshaft determines the leverage available to turn the crankshaft and flywheel.

eliminate any sharp corners, which are weak because they allow cracks to form more easily. The more severe duty the crank is expected to endure, the larger the fillet. High-performance, diesel, and light truck cranks have the largest fillets. Standard passenger car cranks also have fillets, but usually of a comparatively smaller radius. Fillets also provide some measure of thrust

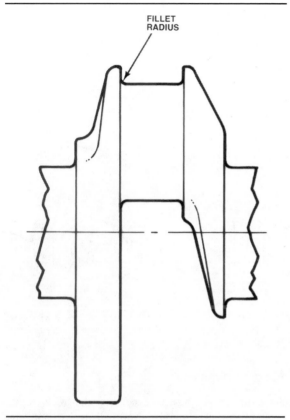

Figure 11-6. The fillets at each end of the bearing journals have a specific radius.

control on the rod because the rod and bearing cannot climb the fillet and wear against the crank cheek.

During manufacture, fillets on domestic crankshafts are usually produced by grinding the proper radius. Most experts believe that rolled fillets are superior. Rolled fillets are produced by rolling, with great pressure, hardened steel balls over the journal/crankshaft joints. This operation compresses and strengthens the metal, much like forging. Turbocharged Buick V6 crankshafts and many European car cranks have rolled fillets.

When an engine is rebuilt, don't overlook grinding the proper fillets on the crankshaft after machining the rod and main bearing journals. Some machine shops skip this step to save time, but it has a direct bearing on how long the crank will last before breaking.

Crankshaft Throws and Firing Impulses

The arrangement of rod journals or throws around a crankshaft follows a definite pattern. The throws are arranged so the firing impulses are at equal intervals, figure 11-7, which makes the engine run smoother and vibrate less. We will begin our discussion with a single cylinder engine, and add cylinders to see how they are spaced.

The crankshaft for a single cylinder engine has only one throw. Every 360 degrees of rotation, it returns to top dead center, figure 11-8. Every 720 degrees of rotation there is a firing impulse in a 4-stroke engine. Now add three cylinders to create a 4-cylinder engine with four crank throws. We then divide 720 by four to determine the crankshaft throw spacing for the engine. This gives us a firing impulse every 180 degrees.

A V8 engine also has four crank throws but has two connecting rods per crank throw. Each throw has a single crankpin that carries two connecting rods, side by side. A V8 must be designed so the angle between the cylinder banks is the same as the angle between the crankshaft throws, or a multiple thereof. If it is not, the firing impulses will not be equally spaced. Actually, this rule applies to any arrangement of V, radial, or opposed cylinder engines. Following this rule, most V8 engines have cylinder blocks with the angle between the cylinders set at 90 degrees and a crankshaft with 90 degrees between throws. Most V6s designed from a clean sheet of paper have 60-degree blocks and crankshafts.

Some engines don't follow the rules and have unequally spaced firing impulses.

The most common examples are the V6 engines built on 90-degree cylinder blocks. These engines are an economical solution to the manufacturer's need for a small displacement engine because they are built on existing V8 tooling. They can be viewed as V8s with two cylinders missing. The original Buick V6 engine, figure 11-9, used a 90-degree block with a 120-degree crankshaft and three crankpins. The two cylinders that were attached to the same crankpin fired in succession. After firing one cylinder, the crankshaft had to rotate only 90 degrees to reach top dead center on the opposite cylinder because 90 degrees was the angle between the cylinder banks. After the second cylinder fired, the engine had to rotate 150 degrees to get to the next cylinder. The engine fired from every other crankshaft throw. With 120 degrees between throws, going to the second throw took 240 degrees of rotation. The engine fired this way continuously, alternating between 90-degree and 150-degree firing intervals. It was known as the odd-firing V6.

Buick eventually decided to make the V6 fire evenly so it would run more smoothly. They did this by splitting the crankpins. Instead of a common crankpin for each pair of cylinders, the pins are splayed on the

Fillet: A radius ground on the crankshaft where a journal joins a cheek. The fillet strengthens this joint.

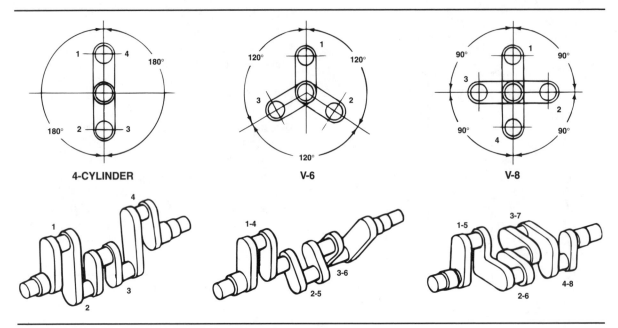

Figure 11-7. The arrangement of throws around the crankshaft makes it possible for the firing impulses to occur at equal intervals.

Firing Order

As described in Chapter 1, each cylinder in an engine is numbered. The numbers are usually cast into the top of the intake manifold. In some cases, the cylinder numbers appear only in the service literature.

Cylinder numbering sequence varies from manufacturer to manufacturer. All in-line engines are numbered in sequence from front to rear (flywheel end), figure 11-12.

When cylinder numbering is combined with the arrangement of the crankshaft throws and camshaft timing, a firing order results. The firing order is part of the design of the engine, but it can change. In the early 1980s, Ford changed the camshaft timing on some applications of its small-block V8, which also required a different firing order. The two different firing orders continue through 1997 on these Ford V8s.

Firing orders are always listed starting with cylinder number one. A typical firing order for a 4-cylinder engine is 1-3-4-2. In-line 6-cylinder engines have the firing order 1-5-3-6-2-4. V8 and V6 engines have several different firing orders, mainly because the cylinder numbering sequence varies, figure 11-13.

Counterweights

Most modern crankshafts are balanced by **counterweights** cast or forged in one piece with the shaft, figures 11-14 and 11-15. A crank is said to be fully counterweighted when it has counterweights on both sides of each rod bearing journal. A fully

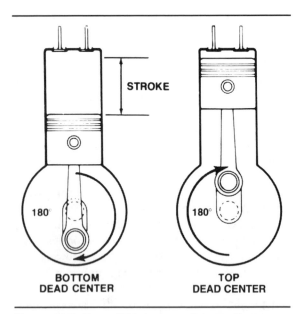

Figure 11-8. Every 360 degrees, the one throw on a single cylinder engine returns to top dead center.

even firing V6, figure 11-10. The separation between adjacent crankpins on the same throw is 30 degrees.

Another exception to the evenly spaced firing impulse engines is the V6 introduced by Chevrolet in 1978. This engine uses a 90-degree V8 block with two cylinders cut off, and has alternating 108-degree and 132-degree firing intervals. It also uses splayed crankpins, figure 11-11, like the even-firing Buick V6.

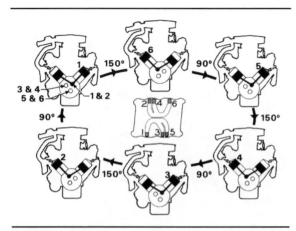

Figure 11-9. The firing impulses on this Buick V6 are unequally spaced.

V6 CRANKSHAFT

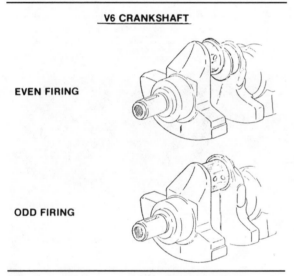

EVEN FIRING

ODD FIRING

Figure 11-10. On the even firing Buick V6, there is no common crankpin. The pins are splayed.

counterweighted crankshaft is the smoothest running and most durable crankshaft design, but also the heaviest, largest, and most expensive to manufacture. Many production cranks have fewer than this optimum amount because of space and cost constraints or to lighten the shaft. A lighter crank requires less of the engine's power to turn it, so the engine can transmit more power to the drive wheels and use less fuel in the process. An engine with a light crank also has a reduced rotating mass that allows the engine to accelerate quicker.

It might appear that a crank's counterweights are heavier than they need to be because each pair of counterweights is much heavier than the journal. However, the counterweight must also counter the weight of the bearings, connecting rods, and pistons. When the piston is at the top of its stroke, the counterweights for that journal hang below the crankshaft.

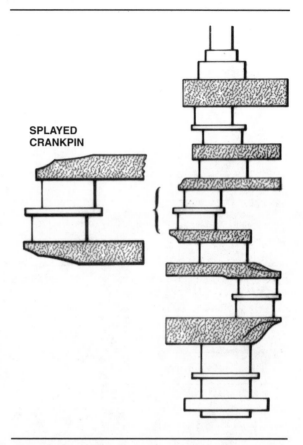

SPLAYED CRANKPIN

Figure 11-11. The Chevrolet V6 also has splayed crankpins.

When the piston is at the bottom of its stroke, the counterweights swing above that journal. Only when all the reciprocating parts are assembled is the crankshaft in anything close to perfect balance. Manufacturers drill holes in the counterweights to provide small adjustments for a final balance.

Horizontally opposed engines, such as the VW flat four-cylinder, endure lower crankshaft loads compared to V or in-line engines. Designers arrange the crank throws so that pistons in opposite banks move in opposing directions, toward and away from each other, figure 11-16. The opposing movements cancel out some of the forces acting on the crank and main bearings. Designers can specify comparatively lighter counterweights and still produce a well-balanced crankshaft.

Counterweights: The weights opposite the rod journals on a crankshaft. They balance the weight of the reciprocating mass: the journal, bearing, connecting rod, and piston assembly.

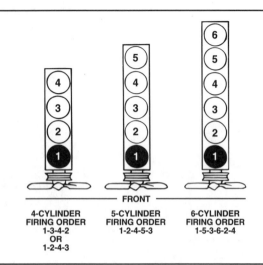

Figure 11-12. Most in-line engines are numbered front to back.

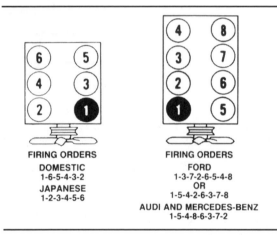

Figure 11-13. Regardless of what firing order a V engine uses, it always alternates firing impulses between cylinder banks to help maintain smooth running.

Engines that rely completely on counterweights for crankshaft balance are called **internally balanced engines**. The most common example is the Chevrolet small-block V8 in 265 to 350 cubic-inch sizes. However, the design of many engine crankcases is simply not big enough to accommodate counterweights large enough to balance the forces created by the rod and piston assembly. To correct this imbalance, designers add small amounts of extra counterweighting to the flywheel and the vibration damper to complete crank balance, figure 11-17. These engines are called **externally balanced engines**. Common examples include the 400 cubic-inch Chevrolet small-block, the 454 cubic-inch Chevrolet big-block, Ford's small-block V8 engines, Buick's V6 and V8 engines, some Mercedes-Benz in-line 6-cylinder engines, as well as the Mazda rotary engines.

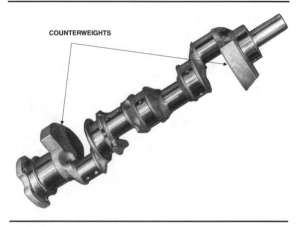

Figure 11-14. Engineers add counterweights to the crankshaft to balance it.

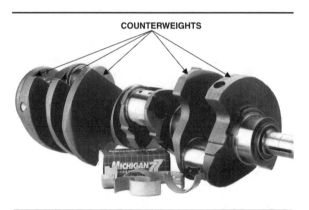

Figure 11-15. Additional counterweights improve crank balance and reduce loads on the main bearings.

Keyways and Threads

The ends of the crankshaft serve a number of important functions. These jobs are common to cranks in almost all engines, V, opposed, or in-line. The front of the crankshaft, or snout, has a keyway and a seal area. A key locks the timing gear or sprocket to the crank, figure 11-18. The seal area of the snout rides in a seal held by the front cover, which retains the oil in front of the engine. On OHV and chain-driven OHC engines, the seal is in front of the gear because the chain and gears run in the oil. On belt-driven OHC engines, the seal is behind the sprocket because belts run dry. The snout usually has internal threads for a large bolt, figure 11-19. The bolt retains the vibration damper, if the engine is so equipped, or a large flange, if it is not. Tension from this bolt also helps retain the timing gear or sprocket in place. The main accessory drive pulley bolts to the damper or flange.

The rear of the crankshaft has a built-in flange and a seal area. The flywheel or flex plate bolts to the

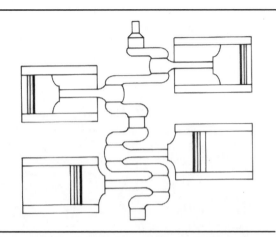

Figure 11-16. The flat-four engine has crank throws every 180 degrees so that its pistons move in opposing directions, canceling vibration forces.

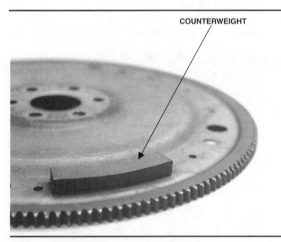

Figure 11-17. Manufacturers may also add weight to flexplates. The balance weight on this flexplate is much larger than necessary to simply balance the flywheel. When bolted to the crankshaft, the weight provides extra counterbalance for the crank.

flange. The rear seal area is for the rear main seal that holds the oil in the bottom end of the engine. On engines with manual transmissions, the rear of the crank is drilled for a pilot bearing or bushing, which locates and supports the input shaft from the transmission. Engines with automatic transmissions do not need the bearing, and have a simple locating hole.

Lubrication

As explained in Chapter 6, the crankshaft bearings are lubricated by oil under pressure. The oil pump forces oil through galleries and drillways in the block to the main bearing saddles. Outlet holes in the main bearing saddles align with holes in the main bearing inserts, allowing oil to flow to the main bearings and crank journals. Drillways in the crankshaft's main

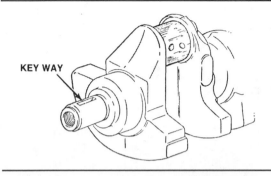

Figure 11-18. A key in the keyway locks the timing gear or sprocket to the snout so that the crank can drive the camshaft.

Figure 11-19. This is an important bolt. It helps retain the timing gear, vibration damper, and accessory drive pulley.

bearing journals route oil to the connecting rod bearings, figure 11-20. Most engines have a single oil hole in the main bearing journal. When it aligns with the hole in the bearing saddle and bearing, pressurized oil is forced through the drillways in the crank to outlet holes in the rod bearing journal, figure 11-21.

When the oil reaches the main or rod bearings surfaces, oil pressure forces it outward over the bearing

Internally Balanced Engines: Engines that rely on the crankshaft counterweights for all crankshaft balance.

Externally Balanced Engines: Engines that have crank counterweights but require additional weights on the flywheel and/or vibration damper to complete crank balance.

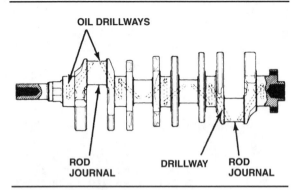

Figure 11-20. The drillways in the crank route oil to the rod bearings.

surface. A thin film, or wedge, of oil builds up between the bearings and the crankshaft journals. The oil film is strong enough to prevent the bearings and crank journals from touching each other. However, under heavy loads or during starting, the oil film may break down and allow the parts to touch. Most modern engines have long crankshaft and bearing life. A well maintained engine, disassembled after high mileage, often has no perceptible wear on the crankshaft.

Improved rod journal lubrication
Designers often extend the portion of crank rotation during which the rods are provided with oil. The most common method is grooving the upper half of each main bearing insert, figure 11-22. The groove helps route oil to the drillway that carries oil to the rods.

Another method that works on the same principle uses a lead-in groove. The manufacturer grinds a groove in the main bearing journal itself, figure 11-23. However, grooves in the bearing or journal always reduce surface area and load-supporting capacity.

A superior method is a **cross-drilled crankshaft**. A second hole, roughly 90 degrees from the first, is drilled into the main bearing journal, figure 11-24. Cross-drilling is a common racing modification machinists perform when they build a competition engine. Some engines are cross-drilled by the factory.

Cast Crankshafts

The majority of production crankshafts are cast from iron. They are inexpensive to manufacture and are strong enough for passenger vehicle use. Improvements in metallurgy make modern cast-iron cranks superior to the forged-steel cranks of the 1920s and 1930s. The best cast crankshafts of today are made of high-grade nodular cast-iron. Certain manufacturers refer to some of their crankshafts as "cast steel," but the metal used to make these cranks is probably more accurately called cast iron.

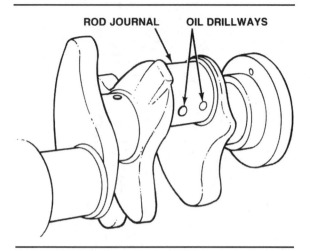

Figure 11-21. Engines with rod journals that share two connecting rods, such as V8s, have two outlet holes side-by-side, one for each rod.

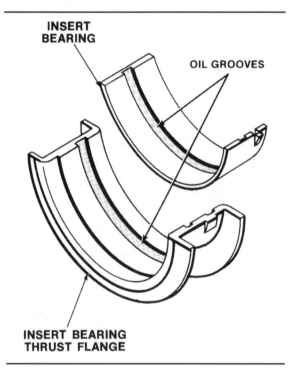

Figure 11-22. Grooves in the main bearing shells allow oil to quickly reach around the entire bearing and channel oil to the drillways that supply the rod journals.

To cast a crank, the foundry melts the iron, pours it into a mold, and allows it to solidify. The sophisticated mold produces a part very much like the finished crank. Little additional machining is required to complete the crank, making casting ideal for high-volume production. Final machining consists of finishing the areas that require a very close tolerance, such as the bearing journals, or completing the details, such as cutting the threads and keyways. A last advantage to

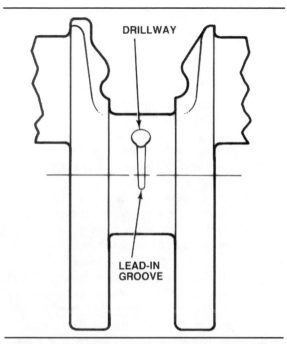

Figure 11-23. Grooves in the main bearing journals channel oil to the drillway for the rod journals.

casting crankshafts is that it is easy to cast internal portions of the crankshaft hollow, reducing weight and internal stress. You can quickly identify a cast crank by its straight cast mold parting line, figure 11-25.

Forged Crankshafts

Forged-steel cranks are the strongest and most expensive crankshafts made by production vehicle manufacturers. Designers specify these cranks where the extra strength justifies the added cost. Forged crankshafts for passenger cars are made of carbon steel such as SAE 1045 or 1053.

To forge a crank, the manufacturer heats a steel bar to about 2,000°F (1,090°C) and then places the steel between the two halves of a forging die. Two strokes of great pressure by a press make the bar conform to the imprint of the die. The result is a raw crankshaft forging that requires extensive machining to complete.

Forging compresses the grain structure of the steel, making it dense and very strong. Many high-performance, racing, light truck, and diesel crankshafts are forgings. Given equal size, a forged crank is heavier than a cast crank because the forged metal is denser. Designers compensate by making the counterweights on a forged crank smaller, and, on some crankshafts, by drilling out the rod journals to make them hollow.

While casting easily produces intricate shapes, forging usually produces parts in a single plane. This is no problem for an in-line 4-cylinder engine with its throws 180 degrees apart. Two of the throws are up and two are down, all in a single plane. For engines with two-plane crankshafts, such as the V8, which has a throw every 90 degrees, the manufacturer must twist the crankshaft to put the throws in the correct place. The in-line-six has a throw every 120 degrees and requires a three-plane crank.

To produce a two-plane crank, the manufacturer twists the crank right after forging when it is still hot. From a theoretical standpoint, the twisting probably weakens the crank because it shears the grain structure at the twist points. However, the forged crank is still stronger than it *needs* to be and much stronger than a cast crank.

■ Racing Crankshafts

Racing crankshafts are very much like their street counterparts in their design and function, although they can differ greatly in their manufacture and materials. Forging produces the strongest crankshafts, but the twist used to arrange the rod journals reduces crank strength. Many race mechanics feel the twisted forgings are more than strong enough. However, some race shops build cranks from non-twist forgings. This requires a much more complex forging die. It also adds a great deal to the cost of the crank, but it is absolutely the strongest piece available. Both twisted and non-twisted racing cranks are forged from higher grade steel, such as SAE 4340 or 5140.

Another method for producing a racing crank is machining it from a solid steel billet or blank. The billet is a round, cylindrical bar with a radius the same size as the desired stroke and counterweight configuration of the finished crank. The billet crank is not quite as strong as a forging, but a race engine builder may choose it because an unusual stroke is needed or a forging is simply not available. Machining a billet also makes it uncomplicated to arrange the counterweights exactly where they are needed for a fully counterweighted shaft. Unfortunately, the extensive lathe work by an experienced machinist required to make a billet crank can increase the cost substantially.

Hard chrome plating of bearing journals is favored for nitro-burning drag race engines. The improved hardness of the journal is a welcome upgrade, but the main reason the drag racers like the process is because the chrome prevents journal corrosion caused by substances released during the combustion of the exotic nitromethane fuel mixtures. These engines are usually supercharged, and the immense combustion pressures they develop cause higher than normal blowby, filling the crankcases with caustic substances.

Cross-drilled Crankshaft: A crankshaft with an additional oil drillway in the main bearing journal that improves lubrication to the rod journals.

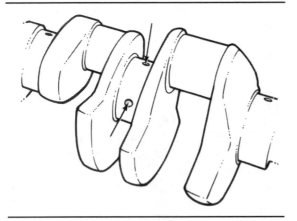

Figure 11-24. A cross-drilled crankshaft has two outlet holes for oil to reach the drillway that supplies the rod journal.

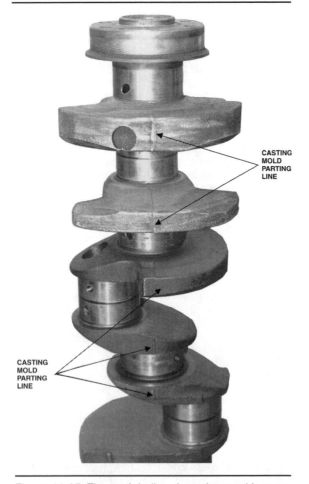

Figure 11-25. The straight line down the crank's counterweights is the parting line. It is created by a small seepage of metal between the parting line of the casting mold. The extra metal trimmed off of this area is called casting flash.

It is easy to identify a forged crank because of the twist. The separation lines, where the manufacturer removed excess flash, will be offset, figure 11-26.

FLYWHEELS

Engines with manual transmissions use a flywheel. The flywheel must be heavy so its momentum will keep the crankshaft rotating until the engine is in position for the next firing impulse. It also adds mass and inertia to the crankshaft, stringing together the power strokes into a smooth flow.

Flywheels are made of nodular cast iron or steel. Steel flywheels are much stronger but they are more susceptible to distortion and warpage.

The flywheel bolts to the flange on the end of the crankshaft, figure 11-27. The rear surface of the flywheel is machined smooth to accept the clutch friction disc. A clutch pressure plate is bolted over the clutch disc onto the flywheel. The pressure plate squeezes the disc against the flywheel, engaging the engine with the transmission. When the clutch pedal is depressed, the pressure plate is forced back from the disc, allowing the disc to turn freely and disengage the engine from the transmission.

The flywheel is a convenient location for the starter to connect to the engine. A ring gear is mounted on the circumference of the flywheel, to allow starter engagement. The large diameter of the flywheel provides adequate leverage for a small starter to turn the engine against its compression.

Dual-mass flywheels

A conventional, single-mass flywheel, like that described above, generally does an acceptable job. However, at low speeds the power strokes are far enough apart to cause speed fluctuations at the flywheel. All but an excessively heavy flywheel might allow these fluctuations to reach the driveline as **torsional vibrations**. These oscillations can produce rattling and other noises in the gearbox and driveline, due to gear backlash and necessary running clearances. They may even cause the driveline to **resonate**. A flywheel heavy enough to eliminate all low-speed oscillations would add too much mass to the crankshaft, hindering the engine's ability to accelerate.

A dual-mass flywheel, used on the Chevrolet Corvette since 1989, some Ford trucks and some Porsche models, absorbs most of the irregular oscillations before they reach the transmission. A dual-mass flywheel can also change the resonant speed of the driveline. The car maker tries to design the dual-mass flywheel so that it reduces the resonant speed below any speed at which the car will ever operate, such as below the idle speed.

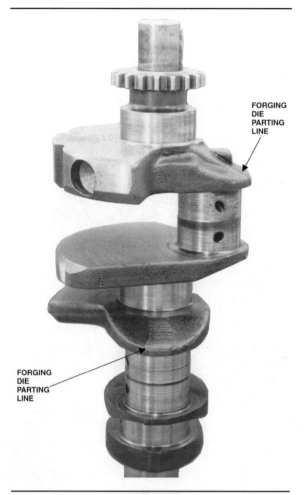

FORGING
DIE
PARTING
LINE

FORGING
DIE
PARTING
LINE

Figure 11-26. Forged cranks also have a parting line created by metal squeezed out between the halves of the forging die. The parting does not line up because the crank is twisted after forging.

A dual-mass flywheel may weigh no more than a conventional flywheel, but distributes its total weight in two parts, the primary and secondary flywheels, figure 11-28. The primary flywheel bolts to the crankshaft. The secondary flywheel, which provides the friction surface for the clutch disc, rides against the primary flywheel on ball bearings. The two flywheels can twist out of phase with each other by up to about 60 degrees to absorb torsional oscillations. The oscillations are damped by a system of inner and outer springs.

FLEXPLATES

On an engine equipped with an automatic transmission, the heavy torque converter acts as a rotating mass for the engine, so a flywheel is not needed. Instead, a flexplate is used in place of the flywheel, figure 11-29. It is very light and its only functions are to provide a mount for the ring gear and to take up any thrust

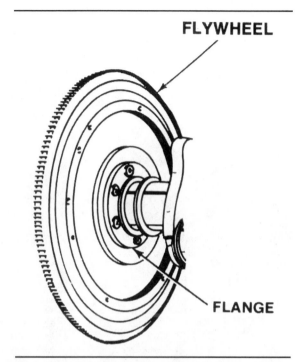

FLYWHEEL

FLANGE

Figure 11-27. The flywheel bolts onto a flange on the end of the crankshaft.

movement from the crankshaft. The flexplate is named for its ability to absorb this thrust. Instead of the torque converter being thrust in and out of the transmission, the flexplate cushions this movement by flexing.

VIBRATION DAMPERS

The crankshaft receives tremendous force, easily as much as two tons, on each power stroke. These forces can cause the crank to bend, twist, or even break. The vibration damper, or harmonic balancer, helps the engine deal with these forces and run more smoothly, figure 11-30. It must be thought of as an integral part of the crankshaft.

When the crankshaft throw receives the force of the piston on the power stroke, the crankshaft twists slightly. This twisting may be so minute that it is scarcely measurable. It may only stress in the metal without causing exterior movement. The crankshaft resists the force and after the initial push from the

Torsional Vibration: Vibration forces that place a twisting stress on a part.

Resonate: An effect produced when the natural vibration frequency of a part is greatly amplified by reinforcing vibrations at the same or nearly the same frequency by another part.

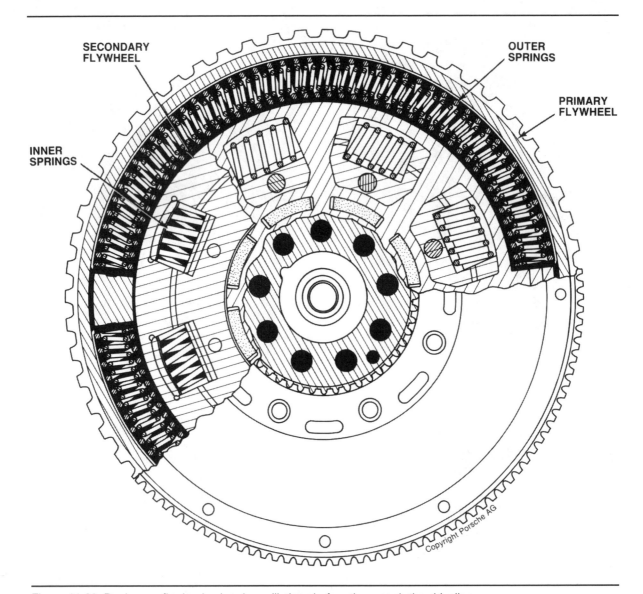

Figure 11-28. Dual-mass flywheels absorb oscillations before they reach the driveline.

piston and connecting rod, it pushes back. In effect, the piston tries to bend the crankshaft, and the crankshaft tries to straighten itself.

When the crankshaft pushes back, it overreacts and then tries to straighten itself in the other direction. These oscillations take place for several cycles and finally die out, similar to the way a tuning fork vibrates.

Every time there is a firing impulse in the engine, the crankshaft gets a push and starts vibrating again. At certain speeds, the pushes received by the crankshaft occur simultaneously with the efforts of the crankshaft to straighten itself. This increases the magnitude of the oscillations. At certain speeds these torsional vibrations can become so great that the shaft breaks. It can also cause heavy wear on timing chains.

The sprocket at the front of the crankshaft constantly oscillates against the chain. Torsional vibration rarely causes damage at the flywheel end of the engine because the flywheel and driveline resist the oscillations, transferring all the movement to the front.

To absorb torsional vibration, manufacturers fit a vibration damper or harmonic balancer. It is usually the front pulley on the crankshaft or has the front pulley bolted to it. The damper is made of a heavy outer inertia ring, an elastomer (a rubber-like material) ring, and an inner hub that attaches to the crankshaft, figure 11-31. When the crankshaft oscillates, the balancer oscillates with it. When it gets to the end of the oscillation, the inertia ring keeps going and twists the elastomer ring slightly. This happens at the end or

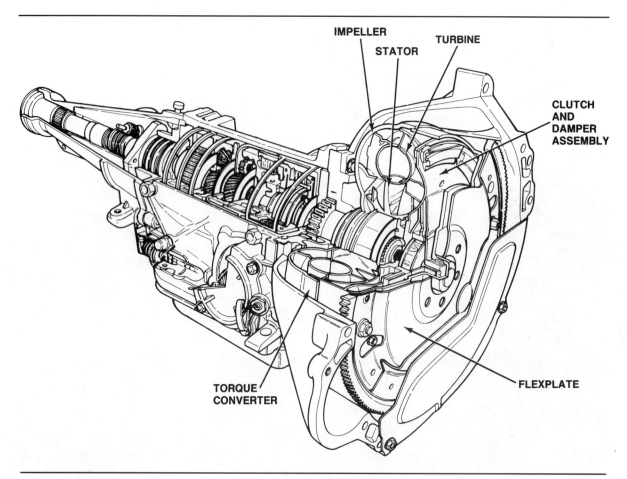

Figure 11-29. A flexplate takes the place of the flywheel when the engine mates to an automatic transmission.

beginning of each oscillation. The outer inertia ring stretches the elastomer, damping the torsional vibration in the crankshaft, by absorbing the vibration. Dampers absorb vibration and give off the energy as heat — making dampers too hot to touch.

Vibration dampers must be tuned to the crankshaft. A 4-cylinder engine usually does not require a damper. The crankshaft is short, so it resists twisting well, and the number of impulses is less than an engine with more cylinders. An in-line 6-cylinder engine always requires some kind of damper. A V8 engine with a timing chain ordinarily uses a damper. When a V8 engine uses timing gears instead of a sprocket and chain, the gears act as a damper.

Dampers can wear out. The counterweight constantly stretches the elastomer and will sometimes lose its bond to the ring. Inspect the damper and replace it before it comes loose. If it gets too loose, it can explode.

ENGINE BALANCE

Vibration is inherent in internal combustion engines because they continually rotate, reciprocate, and fire a combustible mixture. Designers have found that the number of cylinders and the cylinder arrangement has a lot of influence on the amount of vibration an engine generates.

Let's begin our explanation with a single cylinder engine. As the piston slows as it reaches TDC in the cylinder, it tends to pull the engine up. When the piston slows as it reaches BDC, it tends to pull the engine down. This up-and-down movement is called **primary vibration**. A designer can add a counterweight equal to the piston or reciprocating weight that arrives at the opposite dead center position simultaneously when the piston reaches TDC or BDC. The counterweight cancels out any up or down vibration force.

Unfortunately, the engine is now only perfectly balanced when the piston is at either TDC or BDC. At mid-stroke the rod is at an angle and the

Primary Vibration: Strong, low-frequency vibration produced in a single plane. It is caused by the up-and-down movement of the piston and connecting rod.

Figure 11-30. This Ford vibration damper also has the ignition timing marks stamped on it.

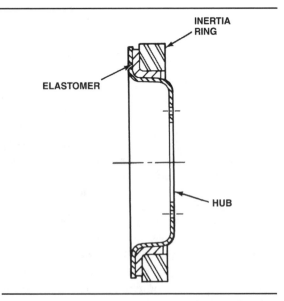

Figure 11-31. Most engines require a vibration damper. Omitting the damper from an engine designed to use one can lead to a broken crankshaft.

counterweight swings out to the side but is not counterbalanced by the entire reciprocating mass, figure 11- 32. The engine doesn't vibrate vertically, but it vibrates almost as violently horizontally. As a compromise, the designer makes the counterweight approximately 50 percent the weight of the reciprocating mass, so the engine is not perfectly balanced when the engine is at TDC or BDC, but it is not as far out of balance when the piston is at mid-stroke. Thus, the engine vibrates only half as much in four directions instead of twice as violently in only two.

Early designers saw that to provide an engine with perfect primary balance, they needed at least two cylinders with crank throws 180 degrees apart, figure 11-33. The twin pistons, rods, crank throws, and counterweights counter balance each other.

However, the alternating rising and falling masses of the 180 degree twin set up another form of vibration, called a **rocking couple**. An engine with a rocking couple rocks itself from end to end, even if it is in perfect primary balance. The solution to eliminating the rocking couple is a four-cylinder engine, which is essentially two 180-degree parallel twins joined together. The two end pistons reach TDC as the two center pistons reach BDC, figure 11-34. The rocking couples set up between cylinder pair 1 and 2 and cylinder pair 3 and 4 cancel each other out. This is why the 4-cylinder engine is generally the minimum cylinder arrangement designers use for cars.

The in-line 4-cylinder engine has good primary balance and eliminates the rocking couple, but it is not perfect. In-line four-cylinder engines suffer another form of vibration forces, called **secondary vibration**. Since the connecting rods of engines are always longer than the crankshaft stroke, all engines have a rod-length to stroke-length ratio. Depending on the engine, it varies from about 1.5 to 1 to more than 4 to 1. Because of this ratio, the piston has less time to slow down, stop, and change directions at TDC than at BDC, figure 11-35. Piston acceleration and inertia are different at TDC than at BDC. Therefore, the piston creates more upward force than downward force, causing secondary vibration. The intensity of secondary vibrations is about $1/4$ of primary vibrations. Also, since the change in inertia occurs twice per revolution, secondary vibrations have twice the frequency of primary vibration. In other words, they occur with less force but have a more "buzzy" nature.

The answer to eliminating secondary vibrations is to add more cylinders. In-line 6- and 8-cylinder engines as well as V6, V8, and V12 (or more) engines have enough cylinders to cancel out secondary vibrations.

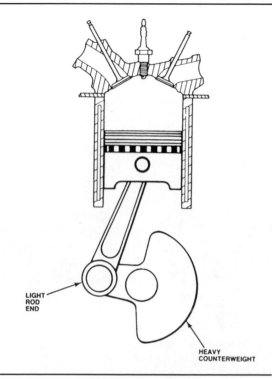

Figure 11-32. At mid-stroke, the piston does not counterbalance the crank's counterweight.

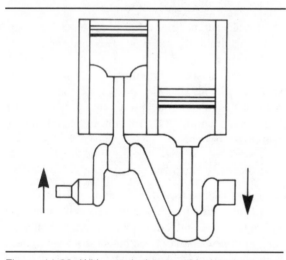

Figure 11-33. With crank throws 180 degrees apart, one piston reaches the TDC when the other piston reaches BDC.

Balance Shafts

It is not always a good idea to add more cylinders to an engine to eliminate vibration. Besides requiring an extensive redesign, the resulting engine has more internal friction as well as increased bulk and weight. In-line 4-cylinder engines are small, lightweight, and compact — perfect for the small, economical cars they usually power. When a buyer wants an economical car

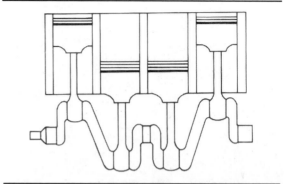

Figure 11-34. An in-line 4-cylinder engine is, essentially, two 180-degree parallel twins joined together. One pair of cylinders cancels out the rocking couple set up by the other pair.

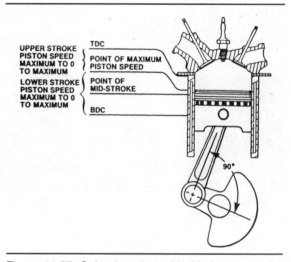

Figure 11-35. Only when the rod is 90 degrees to the crankshaft is the rod/piston assembly moving at the same speed as the crank. At any other position, the rod is moving sideways as well as either up or down.

possessing the smoothness and refinement of a larger car with more cylinders, a designer's best choice may be a small engine with balance shafts.

The Italian manufacturer Lancia was probably the first to employ balance shafts to smooth smaller engines. Ford also used balance shafts on the German Ford

Rocking Couple: Vibration or unbalance set up by an alternating rising and falling of two masses that are out of phase.

Secondary Vibration: Relatively weak, high-frequency vibration. It is caused by the difference in upward and downward force of the piston due to the rod-to-stroke ratio.

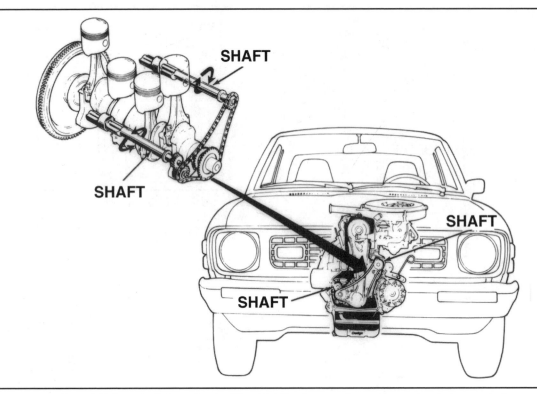

Figure 11-36. The 1976 Dodge Colt engine by Mitsubishi had a set of counter-rotating balance shafts.

Tannus V4 in the early 1960s. Mitsubishi started the current automotive trend in balance shaft use on its in-line 4-cylinder engines in the late 1970s, figure 11-36.

Balance shafts add an imbalanced weight to the reciprocating mass of the engine. The imbalance is out of phase with the engine's imbalanced period, so the two cancel each other out. The result is a smoother, longer-lasting engine and a vehicle that is more pleasant to drive.

In-line 4-cylinder engines are the most common users of balance shafts, figure 11-37. The shafts spin in opposite directions at twice crankshaft speed to eliminate secondary vibrations. Most balance shafts for in-line 4-cylinder engines are driven by chains or belts off the crankshaft, figure 11-38.

Some V6 engines built on 90-degree cylinder blocks also use balance shafts. The split crank journals the designers usually employ to even out the firing intervals on these engines improve smoothness a great deal. Special motor mounts help isolate the driver from most remaining vibration. However, the 90-degree V6s still suffer a rocking couple that causes the engines to rock front to back. To provide the smooth refinement many buyers require in a luxury car, these engines need a balance shaft.

The Ford 3.8-liter V6 and General Motors 3800 V6 have used a balance shaft since 1988. General Motors added a balance shaft to the Vortec 4.3-liter V6 in mid-1992. These engines employ a single balance shaft located in the lifter valley of the block below the intake manifold, figure 11-39. The cast-iron balance shaft spins in a bearing at each end of the block. A gear on the camshaft drives the balance shaft at crankshaft speed, but opposite crankshaft rotation. The balance shaft starts a rocking motion in the opposite direction to cancel out the rocking couple inherent in this engine design.

Crankshaft Balance

Designers take pains to design engines that are naturally "balanced"; however, all the parts that make up the engine are slightly out of balance. After manufacture, even new engines must be balanced — in much the same way a technician must balance a new wheel and tire assembly.

A rotating part is in balance when its mass axis and its geometric axis are the same. In other words, when its weight is evenly distributed around its center of rotation, figure 11-40. Outside of this ideal condition, there are three ways that a rotating part can be imbalanced:

- Static imbalance
- Couple imbalance
- Dynamic imbalance.

When a part has static imbalance, one side of the part is heavier than the other. The unevenly distributed

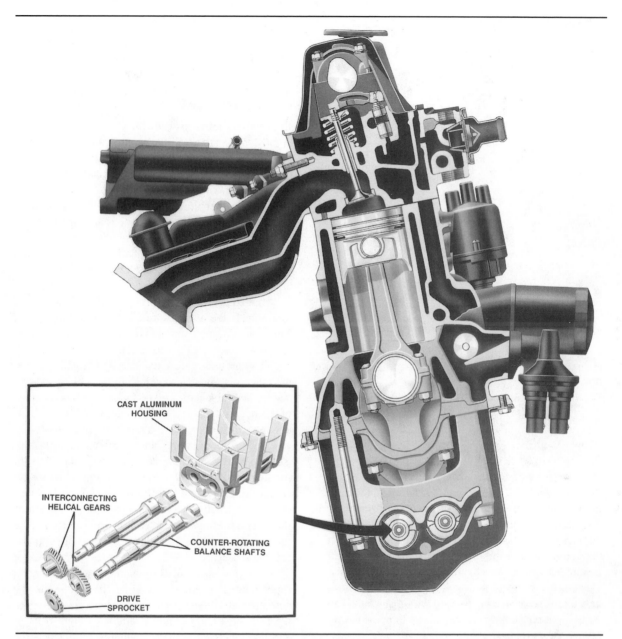

Figure 11-37. In 1986, Chrysler introduced a set of balance shafts on its 2.5-liter, in-line 4-cylinder engine. The shafts are located in the crankcase and driven by a chain.

weight displaces the mass axis from the geometric axis, figure 11-41. However, the mass axis is still parallel to the geometric axis. To restore static balance you must remove weight from the heavy side or add an equal amount of weight to the light side. Static imbalance is found primarily in narrow, disc-shaped parts, such as clutches and flywheels.

A part has couple imbalance when two equal weights exist at opposite ends of the part 180 degrees from each other, figure 11-42. The mass axis intersects the geometric axis at the center of mass, but the two are not parallel. To correct a couple imbalance

you must add or remove weight from two places on the part. Couple imbalance is found primarily in long, cylindrical parts, such as crankshafts and driveshafts.

Dynamic imbalance is a combination of static and couple imbalance. The unevenly distributed weight causes the mass axis to neither be parallel with the geometric axis nor intersect it at the center of mass, figure 11-43. Dynamic imbalance affects any shape of part. Correcting dynamic imbalance is similar to correcting couple imbalance because you must add or remove weight from two locations.

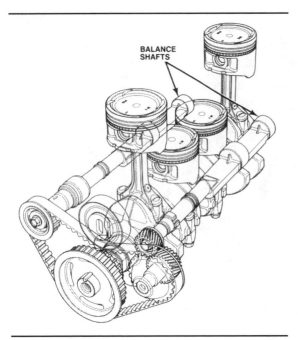

Figure 11-38. The balance shafts on Honda's 2.2-liter, in-line 4-cylinder engine are belt driven. Both shafts are in the same horizontal plane, 3 inches (75 mm) above the crankshaft centerline. Honda claims this location helps reduce vibration at mid- as well as high-rpm ranges.

Custom Engine Balancing

A factory only balances an engine closely enough to prevent objectional vibration. Some manufacturers allow as much as a 20-gram difference in weights of reciprocating parts. Few factories balance their engines closer than 3 grams. Many engine rebuilders use sophisticated equipment to provide custom engine balancing that is much closer to perfect balance. Custom balancing is an exacting process that usually gets all the reciprocating parts to within ½ gram. This is too time-consuming and expensive to perform on a large production basis.

Why balance?

While it's clear that an imbalanced engine will vibrate, why worry about a difference as small as a few grams? The reason is the effect of centrifugal force greatly multiplies any imbalance as rpm rises. The force of the imbalance increases by the square of the rpm. In other words, if the rpm doubles, the force quadruples. If the rpm triples, the force is multiplied by nine. At 500 rpm, an imbalance of one ounce that is one inch (25 mm) from the center of rotation is multiplied to a force of 7 ounces. At 5,000 rpm, the force would increase to 44 lbs (20 kg). By 8,000 rpm, that same 1 ounce would have a force of 114 lbs (52 kg).

When you rebuild, you usually change the weights in the rotating or reciprocating mass. Reboring and adding new pistons, replacing a rod, or even regrinding the crankshaft, imbalances the engine. Imbalanced parts create vibrations that not only annoy the driver, they increase wear on the bearings. Vibration also decreases performance because energy that could be used to turn the crank is being diverted into vibration.

Engine balancing procedure

A machinist gathers all the rotating and reciprocating parts of the engine to balance as a unit. For all types of engines this includes: the crankshaft, connecting rods, pistons, rings, and wrist pins. For balancing V engines the machinist also requires the rod bearing inserts. Finally, the machinist needs the parts that bolt to and rotate with the crankshaft: the flywheel or flexplate, clutch, pressure plate, harmonic balancer, and any pulley that bolts to the end of the crank, including all fasteners.

When machinists perform a balancing job, they first match the weights of the reciprocating parts. They remove small amounts of metal from the pistons and the connecting rods to equalize their weights.

On V engines, before balancing the rotating mass, the machinist must also calculate the **bob weight** to add to the crankshaft. The bob weight is a percentage of the mass on a single rod journal, which consists of a connecting rod/piston assembly with wrist pin, locks, rings, and rod bearings. Experienced machinists also add an arbitrary amount of weight to account for the oil that clings to these components in a running engine. For a typical V8, the bob weight is 100 percent of the rotating weight and usually 50 percent of the reciprocating weight on a rod journal. The machinist assembles the bob weight and bolts it to the crankshaft, figure 11-44.

Then, the machinist spins the crankshaft on a sophisticated balancing machine that pinpoints the amount and location of any imbalance in the shaft, figure 11-45. The machinist removes weight from or adds weight to the counterweights to balance the shaft. On externally balanced engines, the flywheel or flexplate and the harmonic balancer are assembled on the crankshaft and the machinist balances the whole assembly as a unit. However, only the counterweights have weight added to or removed from them to provide a final balance. On internally balanced engines, each of the rotating parts, such as the flywheel, is balanced separately on the balancing machine. We detail the custom engine balancing procedure in Chapter 13 of the Shop manual.

SUMMARY

The crankshaft converts reciprocating motion to rotary motion. Main bearings support the crank, while the connecting rods turn on the crankshaft journals of

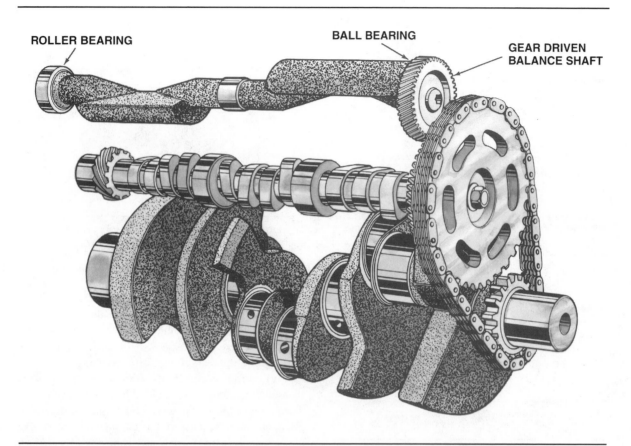

Figure 11-39. The balance shaft on this General Motors 4.3-liter V6 helps cancel a rocking couple.

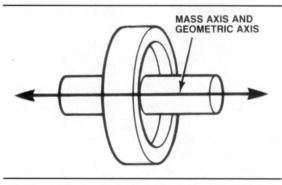

Figure 11-40. A part is balanced when its weight is equally distributed around its center of rotation.

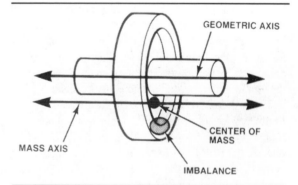

Figure 11-41. Static imbalance occurs when one side of the part is heavier than the other.

each crank throw. Bearing journal diameter, finished size, and surface finish are of primary importance to the life of the crank. The point where the journals join to the crank have fillets to improve crankshaft strength. The arrangement of rod journals or throws around a crankshaft follows a definite pattern which makes the engine run smoother and vibrate less. When cylinder numbering is combined with the arrangement of the crankshaft throws, a firing order results. The firing order is part of the engine design.

Bob Weight: A weight added to the crankshaft of a V engine prior to spinning the crank on a balancing machine. The bob weight simulates a percentage of the reciprocating mass.

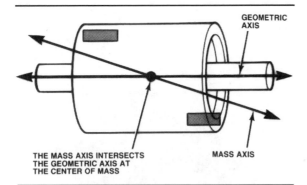

Figure 11-42. Couple imbalance occurs when two equal weights exist at opposite ends of the part, but they are 180 degrees out of phase with each other.

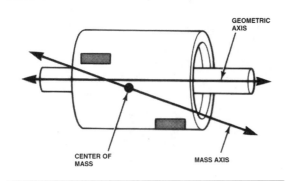

Figure 11-43. Dynamic imbalance is a combination of static imbalance and couple imbalance.

Counterweights balance the weight of the rods and pistons. Engines that rely completely on counterweights for crankshaft balance are called internally balanced engines. Externally balanced engines have small amounts of extra counterweighting on the flywheel and the vibration damper to complete crank balance.

Crankshafts are lubricated by oil under pressure. Drillways in the crankshaft's main bearing journals route oil to the connecting rod bearings. When the oil reaches the bearing surfaces, it forms a wedge strong enough to prevent the bearings and crank journals from touching each other.

Most production crankshafts are cast from iron. They are inexpensive to manufacture and are strong enough for passenger car use. Forged-steel cranks are the strongest and most expensive crankshafts made by auto makers. Designers specify these cranks where the extra strength justifies the added cost.

Flywheels are used to smooth out the firing impulses of the engine. A dual-mass flywheel absorbs irregular oscillations before they reach the transmission. A flexplate is used in place of the flywheel. It is very

Figure 11-44. Machinists install bob weights on the cranks of V engines before balancing. The bob weights simulate a portion of the rotating and reciprocating mass, so the machinist can balance the crank a more precisely.

light and provides a mount for the ring gear and takes up any thrust movement from the crankshaft. Vibration dampers reduce torsional vibrations and help prevent the crankshaft from breaking.

Vibration is inherent in internal combustion engines because they continually rotate, reciprocate, and fire a combustible mixture. Designers arrange the amount and location of the cylinders to reduce or eliminate a rocking couple and primary and secondary vibration. The in-line 4-cylinder engine has good primary balance and eliminates the rocking couple, but it requires a properly designed engine with six or more cylinders to also eliminate secondary vibrations. Balance shafts can help eliminate secondary vibrations in a 4-cylinder engine or the rocking couple in a 90-degree V6.

A rotating part is in balance when its mass axis and its geometric axis are the same. Outside of this ideal condition, there are three ways that a rotating part can be imbalanced: static imbalance, couple imbalance, and dynamic imbalance.

A factory job only balances an engine closely enough to prevent objectional vibration. Many engine rebuilders use sophisticated equipment to provide custom engine balancing that gets all the reciprocating parts to within ½ gram.

When machinists perform a custom balancing job, they match the weights of the reciprocating parts, add a bob weight percentage to the crank on V engines, and spin the crankshaft on a sophisticated balancing machine that pinpoints the amount and location of any imbalance in the shaft.

Figure 11-45. Crankshafts are balanced on centrifugal balancing machines that spin the part with an electric motor. Suspension springs and a moving pickup coil in the balancer detect vibrations created by imbalances in the part. The balancer converts the vibrations into electric signals so that its computer can calculate the amount and location of the imbalance.

■ The Inexact Science of Engine Balancing

Racing engine expert, Smokey Yunick, believes that engine balancing is anything but an exact science. The major problem is not with the balancing method or equipment, but with the motor oil. Most machinists add a few grams to the bob weight calculation to account for oil between the bearings and the oil being pumped through the crankshaft. Unfortunately, there is no way to accurately account for the weight of the oil that clings to the outside of the crank. Smokey's observations of a running test engine fitted with a clear oil pan revealed a great deal of oil whipped up by the spinning crank throws. The oil wraps around the crankshaft and since the oil has weight, he insists it must affect the crankshaft balance. The oil clings to the crank in a totally random pattern, making it impossible to accurately account for. Sometimes the oil wraps around the front of the crank, or at the back, or in the middle, or any combination. The amount of oil wrapping around the crank changes randomly as well.

While Smokey agrees that engine balancing is important, he regards it as an approximation at best. If he has to change pistons in one of his racing engines with a previously balanced reciprocating assembly, he doesn't rebalance if he can get the piston end of the rod within about two grams of the original weight. If the difference is as much as three grams, he recommends rebalancing to be on the safe side.

Review Questions
Choose the single most correct answer.
Compare your answers with the correct answers on Page 328.

1. The crankshaft converts recipro-cating motion into _____ motion.
   a. Vertical
   b. Rotating
   c. Horizontal
   d. Directional

2. The crankshaft rotates in:
   a. Rod bearings
   b. Main bearings
   c. Ball bearings
   d. Journal bearings

3. Technician A says that the crank throw supports the crankshaft against forces that push and pull it in the block.
   Technician B says that the thrust surface supports the crankshaft against forces that push and pull it in the block.
   Who is right?
   a. A only
   b. B only
   c. Both A and B
   d. Neither A nor B

4. Technician A says that the rod bearing journals are in-line with the crankshaft's centerline.
   Technician B says that the main bearing journals are also called crankpins.
   Who is right?
   a. A only
   b. B only
   c. Both A and B
   d. Neither A nor B

5. Crankshaft fillets:
   a. Strengthen the joint between the journal and crank cheek
   b. Are only found on high-performance cranks
   c. Reduce crankshaft vibration
   d. Improve oil flow

6. In a modern crankshaft, the counterweights are much heav-ier than the journal in order to also balance the:
   a. Rod bearings
   b. Connecting rods
   c. Piston and rings
   d. All of the above

7. Most four-cylinder engines with four crank throws have a firing impulse every _____ degrees.
   a. 90
   b. 120
   c. 180
   d. 360

8. How did General Motors modify the 90-degree Buick V6 engine to even the firing impulses?
   a. Evenly spaced the crank journals
   b. Eliminated the crank pins
   c. Split the crank pins
   d. Alternated the crank throws

9. The spacing of the throws on the crankshaft is determined by:
   a. The number of cylinders
   b. The arrangement of cylinders
   c. Whether the firing intervals are designed to be equal or unequal
   d. All of the above

10. Technician A says that an inter-nally balanced engine relies completely on the counter-weights for crankshaft balance.
    Technician B says that an exter-nally balanced engine has extra counterweighting on the flywheel and the vibration damper to complete crank balance.
    Who is right?
    a. A only
    b. B only
    c. Both A and B
    d. Neither A nor B

11. The purpose of a flywheel is to:
    a. Engage the engine and transmission
    b. Smooth out the firing impulses
    c. Disengage the transmission
    d. None of the above

12. Technician A says that a dual-mass flywheel lightens the rotat-ing mass of the driveline.
    Technician B says that a dual-mass flywheel absorbs torsional vibrations.
    Who is right?
    a. A only
    b. B only
    c. Both A and B
    d. Neither A nor B

13. Technician A says that a flex-plate is used only with an auto-matic transmission.
    Technician B says that a flex-plate absorbs crankshaft thrust.
    Who is right?
    a. A only
    b. B only
    c. Both A and B
    d. Neither A nor B

14. A vibration damper or harmonic balancer is used to:
    a. Improve crank lubrication
    b. Make the engine accelerate quicker
    c. Prevent the crankshaft from wearing out
    d. Reduce torsional vibrations

15. A vibration damper is made of:
    a. A heavy outer weight, a rub-ber ring, and an inner hub
    b. Cast iron outer ring, an alu-minum core, and an inner hub
    c. Billet aluminum outer weight, a silicone core, and an inner hub
    d. A rubber core, a titanium outer ring, and an inner hub

16. The starter motor ring gear is mounted on the:
    a. Clutch or harmonic balancer
    b. Engine or clutch
    c. Flywheel or flexplate
    d. Crankshaft or connecting rod

17. Technician A says that some in-line 4-cylinder engines use bal-ance shafts to eliminate secondary vibrations.
    Technician B says that 90-degree V6 engines use balance shafts to eliminate a rocking couple.
    Who is right?
    a. A only
    b. B only
    c. Both A and B
    d. Neither A nor B

18. Dynamic imbalance affects:
    a. Only cylindrical parts
    b. Only disc-shaped parts
    c. Only cam-shaped parts
    d. Either cylindrical or disc-shaped parts

19. Technician A says it is important to balance a rebuilt engine if the weights of the reciprocating parts are significantly changed.
    Technician B says it is important to balance a rebuilt engine only if the engine's crank is external-ly balanced.
    Who is right?
    a. A only
    b. B only
    c. Both A and B
    d. Neither A nor B

20. Technician A says it is neces-sary to add bob weights to the crankshaft of a V engine before balancing.
    Technician B says it is neces-sary to add bob weights to the crankshaft of an in-line engine before balancing.
    Who is right?
    a. A only
    b. B only
    c. Both A and B
    d. Neither A nor B

12

Pistons, Rings, and Connecting Rods

The pistons, rings, and rods make up the reciprocating parts of an engine's powertrain. A great deal of machine work involves reconditioning and repairing these pieces. Automotive machinists must also know the correct assembly methods.

PISTONS

Pistons receive the full force of the burning air-fuel mixture. An engine with an 8:1 compression ratio develops about 700 psi (5,000 kPa) of combustion pressure. This amounts to about 8,800 lbs (4,000 kg) of force on top of the piston, so the piston must be very strong. Pistons transmit this force to the connecting rod, which pushes on the crank throw to turn the crankshaft. A piston must also be very light. Pistons accelerate to top speed and come to a dead stop twice per revolution, creating very high inertia loads. A light piston reduces the loads on the bottom end parts and helps an engine achieve better power and fuel economy. An automotive machinist must have a full understanding of the following parts of a piston, figure 12-1:

- Piston crown
- Pin boss
- Ring grooves and lands
- Piston skirts.

Piston Crowns

The **piston crown** is the top of the piston and it forms the bottom portion of the combustion chamber. Because the crown receives the greatest force and heat, some crowns have reinforcement ribs on the underside of the piston, figure 12-2. The ribs also act as cooling fins to transfer some heat away from the crown.

Many pistons have small areas cut into the crown for valve head clearance, called **valve reliefs**, valve pockets, or eyebrows, figure 12-3. The depth of the reliefs depends on how far the piston reaches into the combustion chamber and how extreme the camshaft timing is, particularly the lift.

The crown may contain information for the engine builder. Some pistons have a mark, usually a notch or an arrow, machined into the top edge of the piston, figure 12-4. The notch must face the front of the engine so that the valve reliefs and the piston offset, if any, are properly aligned.

Piston crowns and compression ratio

Designers shape or size piston crowns to fit the combustion chamber, which has a direct effect on the compression ratio. As we stated in Chapter 1, the compression ratio compares the total cylinder volume when the piston is at bottom dead center (BDC) to the

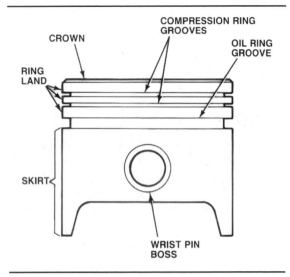

Figure 12-1. All pistons share these parts in common.

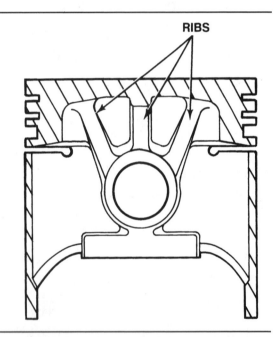

Figure 12-2. The ribs, usually found inside cast pistons, help strengthen the piston and draw off heat.

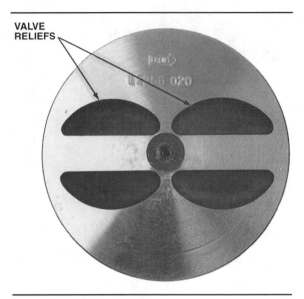

Figure 12-3. Valve reliefs allow extra room for the valve head when the camshaft is at full lift.

Figure 12-4. The arrow on this piston tells you which end must face the front of the engine.

volume when the piston is at top dead center (TDC). The larger the volume at BDC and the smaller the volume at TDC, the higher the compression ratio.

The portion of cylinder volume due to the cylinder is fixed by engine displacement. The portion of cylinder volume provided by the combustion chamber is easy for engine designers to change by specifying larger or smaller combustion chambers or different pistons. You can also easily change the compression ratio by installing pistons with different crown shapes when you rebuild an engine.

Before strict exhaust emission standards were introduced, pistons for many OHV engines had crowns that projected into the combustion chamber. These are called domed or pop-up pistons, figure 12-5. Domed pistons help fill the combustion chamber, raising the compression ratio. The domes were needed in the large hemi and wedge combustion chambers in use at the time. High-performance engines with the highest compression ratios had the largest or tallest domes.

Pistons in today's street engines most often have flat or nearly flat crowns, figure 12-6. We refer to these as flat-top pistons. The trend in modern combustion chambers is to use small chamber volumes that provide rapid burn time. With these designs, flat-top pistons easily give adequate compression ratios for passenger car engines. Even high-compression racing engines with small, modern chambers have pistons with modest domes or none at all.

Figure 12-5. The large combustion chambers of older engines required domed pistons to provide higher compression ratios.

Figure 12-7. In a dished piston, most of the combustion chamber is in the piston crown.

Ring Grooves and Lands

Below the crown and above the pin bosses are the **ring grooves** which hold the piston rings. The areas of the piston between the grooves are called **ring lands**. Lands are always about 0.020 to 0.040 inch (0.5 to 1.0 mm) smaller in diameter than the piston skirt to keep them from touching the cylinder wall.

Piston manufacturers machine the ring grooves very carefully. The top and bottom of the groove must be as smooth as possible to provide a sealing surface for the rings.

Rings on most modern pistons are above the pin. Usually there are three rings: two compression rings

Figure 12-6. Flat-top pistons can yield adequate compression ratios when combined with small, modern combustion chambers.

Some piston crowns are shaped like a bowl or dished. A large part of the combustion chamber is actually in the bowl, figure 12-7. Dishing pistons is one way of lowering the compression ratio.

Pin Bosses

The area where the wrist pin or piston pin fits into the piston is called the **pin boss**. Sometimes strengthening webs are cast into the walls of the piston to connect them with the pin bosses. On many pistons, designers add extra material just beneath the pin bosses, figure 12-8. These balance pads provide areas where a machinist can remove material when balancing the pistons.

Piston Crown: The top of the piston. It forms the bottom of the combustion chamber.

Valve Reliefs: Small pockets cut into the piston crown to provide clearance for the valves.

Pin Boss: The area surrounding the hole in the piston for the piston pin.

Ring Grooves: Grooves in the upper part of the piston that hold the piston rings.

Ring Lands: The part of the piston between the ring grooves. The lands strengthen and support the ring grooves.

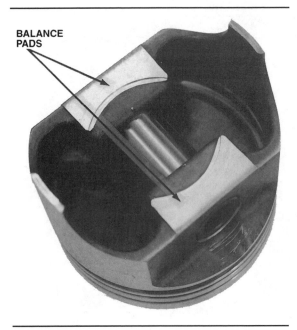

Figure 12-8. These balance pads allow you to remove the metal from a non-structural part of the piston to balance piston weights.

that are uppermost and one oil control ring below. The two compression rings prevent gases from getting past the piston. The oil control ring scrapes off oil splashed on the cylinder walls so the oil will not get into the combustion chamber.

Older engines were often made with four piston rings. The fourth ring was below the piston pin and was strictly for oil control. Some modern industrial engines that run at slow speeds still use four piston rings. Because of their high compression ratios, most diesel engines have four or more rings to prevent blowby and to give added oil control.

Piston Skirts

The lower portion of the piston is called the **piston skirt**. It stabilizes the piston in the cylinder to prevent excessive rocking, which causes wear. Fully skirted pistons are usually found in truck or commercial engines, figure 12-9. Most engines use slipper skirts where the portion of the skirt below the piston pin is cut away. This design lightens the piston and provides more clearance for the crankshaft counterweights. The piston can slide down farther without hitting the crankshaft. The slipper design retains enough of the skirt to provide adequate piston stability.

Piston Materials

Pistons in early cars were iron or steel. They were heavy, but it did not matter because the engine did not

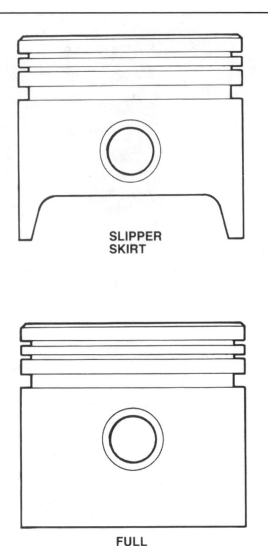

Figure 12-9. Pistons with slipper skirts are much lighter than pistons with full skirts.

turn very fast. As engines began turning faster and producing more horsepower, the inertia of the heavy pistons put added strain on other parts, which began to break. Since the piston must stop and reverse direction at the end of each stroke, a heavy piston puts a greater load on the bearings and the connecting rods. As engine speeds increased, the need for a lighter piston became more evident.

Iron pistons were made from thinner metal, but they were not durable and frequently broke. Pistons for modern gasoline engines are made of aluminum alloy, which is about one-third the weight of iron. Aluminum also conducts heat more than four times faster than iron, so the heat of combustion will be conducted into the cylinder walls and into the oil much faster.

Aluminum alloy pistons are approximately 80 to 90 percent aluminum with the other 10 to 20 percent being metals such as copper, zinc, and chromium, as well as materials such as silicon.

Piston Construction

Pistons are either cast or forged. Most passenger cars use cast pistons. As we learned in Chapter 3, casting is less expensive, but does not produce as strong a part. However, casting does provide some advantages for automotive pistons. Cast pistons have a less dense grain structure. When cast pistons heat up inside a running engine, they do not expand as much. So, generally speaking, machinists can fit them to the cylinders with tighter tolerances. They rattle less in the bore during startup, causing less wear when the engine is cold.

Forged pistons have a dense grain structure and are immensely strong, stronger than is usually necessary for a passenger car engine. However, they are often used in racing engines and some severe service passenger car engines, such as turbocharged and other high-performance engines.

Because forged pistons are less porous than cast pistons they conduct heat more quickly. Forged pistons generally run 18 to 20 percent cooler than cast pistons, figure 12-10. A cooler-running piston helps an engine resist detonation better.

Because of their dense nature, forged pistons usually expand more from heat and require more piston-to-bore clearance — but it depends on the alloy. Some aluminum alloys used to make forged pistons actually have a lower coefficient of thermal expansion than the common alloys used in cast pistons.

Hypereutectic cast pistons

The main difference between hypereutectic cast pistons and other cast pistons is the amount of silicon in each. Hypereutectic cast pistons have as much as 16 to 20 percent silicon. Typical cast pistons have about 8 to 11 percent silicon.

Hypereutectic pistons get their name for the way they are manufactured to hold the silicon. Aluminum can dissolve only so much silicon and hold it in suspension. It is like adding sugar to coffee. A small to moderate amount easily dissolves in the coffee. Add still more and, eventually, the coffee reaches a saturation point and the sugar remains in crystal form and sinks to the bottom. Aluminum is eutectic when the maximum amount of silicon remains dissolved in the alloy when it cools. For aluminum, the eutectic point is 12 percent. Less than 12 percent and the alloy is hypoeutectic. More than 12 percent and the alloy is hypereutectic. The goal in manufacturing hypereutectic

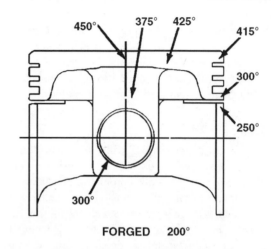

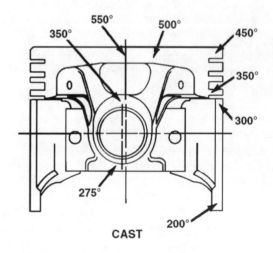

Figure 12-10. The critical crown temperature can be 100°F (37.8°C) cooler on a forged piston compared to a cast piston.

pistons is keeping the hard, undissolved silicon particles dispersed throughout the alloy rather than sinking to the bottom of the mold. Piston makers accomplish this by closely controlling the heating and cooling of the alloy during manufacture.

Hypereutectic pistons have better high-temperature fatigue resistance than other cast pistons. This strength allows manufacturers to make the pistons thinner in cross section so they can be as much as 10 percent lighter. Silicon is a very slippery substance

Piston Skirt: The part of the piston below the pin bosses. The skirt stabilizes the piston in the bore to prevent the piston from rocking.

and it gives the hypereutectic pistons better scuff resistance, better wear properties, and the capacity for tighter piston-to-bore clearances. Although they are not as strong as forged pistons, manufacturers use hypereutectic cast piston, in applications where a more common cast piston may not be strong enough but the cost and greater strength of a forging is not justified.

Piston Finish

The finish on pistons varies with the manufacturer. Many pistons have an oil-retaining finish to resist **scuffing**. Scuffing is mainly a problem when breaking in an engine. Until the high spots on the cylinder wall and on the piston are worn smooth, the piston can rub hard enough to transfer metal to the cylinder wall. Once this happens, metal continues to come off the piston until it is ruined, figure 12-11.

Distorted or overheated cylinders may cause scuffing. The overheating results from a poorly maintained cooling system. The radiator coolant need not boil because hot spots occur from rust and the clogging of the passage around the cylinder. Bent or twisted rods may also cause scuffing.

To ensure that the piston skirt will hold enough oil to prevent scuffing, some pistons come with a fairly rough finish on the skirt. The finish may be obtained by blasting with small beads or by turning the piston in a lathe. The turning cuts a very fine thread on the skirt that collects oil and lubricates the skirt, figure 12-12.

Other pistons are ground smooth and then tin plated or anodized. The plating gives a surface that can rub off to prevent scuffing.

Piston Expansion

With the advent of aluminum pistons, engineers reduced piston weight by half. Unfortunately, aluminum expands at twice the rate of cast iron. As a result, if a piston is fitted to a cast-iron cylinder with a slip fit at room temperature, it will seize when the engine reaches operating temperature.

The solution to this problem is to make the piston smaller to give it a loose fit at room temperature. Then, when it warms up, it will fit perfectly. However, a loose piston rattles in the bore when the engine is cold and it takes the piston several minutes to warm up and expand. While the piston rattles in the bore, great wear occurs. Enough wear causes **piston slap**, indicating the need for a cylinder rebore and a new piston.

Experimentation taught engineers to design pistons that did not rattle when cold but still provided the proper clearance when warm. Unlike the ring lands, all piston skirts today are cam ground or machined. Instead of being machined in a perfectly round shape, the skirts

Figure 12-11. When pistons scuff, they can rub hard enough to transfer metal to the surface of the cylinder wall.

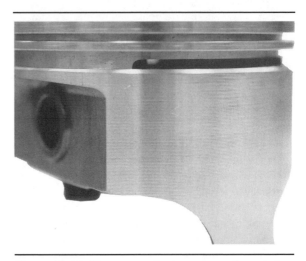

Figure 12-12. Piston manufacturers sometimes lathe-turn oil-retention grooves on the piston skirt.

are ground in an ellipse. The skirt is about 0.010 inch (0.25 mm) less in diameter across the pin boss than it is at a 90-degree angle to the pin, figure 12-13. The thrust surfaces fit closely to eliminate slap, but loosely across the pin boss to allow for expansion. When the piston warms up it expands across the pin boss until it is essentially round at operating temperature.

Engineers tried several other refinements. A slot is cut or cast between the head of the piston and the skirt, figure 12-14. The slot prevents heat transfer from the very hot crown of the piston to the skirt. If the skirt does not get as hot, it does not expand as much.

Instead of cutting the skirt, most modern manufacturers produce a tapered skirt. The top of the skirt is slightly smaller than the bottom because the top is closer to the flame. Because of its proximity to the flame, the top of the piston skirt will get hotter and expand more, evening out the diameter.

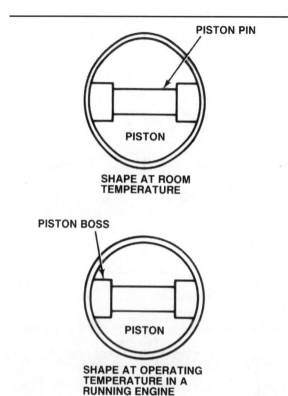

Figure 12-13. Manufacturers cam-grind pistons to produce the elliptical shape.

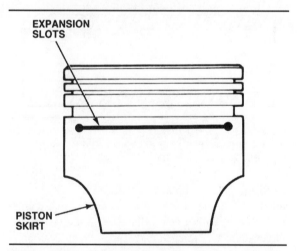

Figure 12-14. Today's automotive engines produce high power with low fuel consumption and low exhaust emissions — not an easy design task.

Some pistons are cast with steel struts inside the aluminum to control expansion, figure 12-15. Since the steel strut does not expand as readily as the aluminum, it limits the amount the piston can expand in a certain direction.

Engineers also found that adding silicon to the aluminum piston alloy greatly reduced the thermal expansion of the piston.

■ Changing the Compression Ratio

Raising the compression ratio is a time-honored method of improving engine performance. Higher compression raises horsepower and torque without sacrificing low-speed engine performance. Raised compression also makes an engine use its fuel more efficiently, improving fuel economy. Aftermarket companies help out by offering a wide variety of piston designs to choose from when rebuilding an engine.

However, beware if you select pistons that yield a higher than stock compression ratio when you rebuild an engine. Raising compression is not without its pitfalls nowadays — you could cause your customer a variety of problems. The first problem is the increased octane requirement of a high-compression engine, as we pointed out in Chapter 2. The octane of readily available pump gasoline is limited. You may force your customer to buy aviation gasoline or to doctor the fuel with "octane boosters" to avoid engine detonation. Both solutions are expensive, inconvenient, and potentially illegal.

Late-model engines with electronically controlled fuel and ignition systems provide other challenges. The mapping of the computer management systems will probably not be correct for a raised compression ratio. It will require extensive recalibration before you see the performance benefits. While this is sometimes possible with sharp tuners and aftermarket computer management equipment, it is rarely legal. Although you can often make these modifications invisible to a smog station inspector, the engine may not pass the tailpipe test that is becoming common in most states. As we pointed out in a Chapter 2 sidebar, raised compression usually results in a higher output of oxides of nitrogen (NO_x), a pollutant. You, as a professional, are subject to an assortment of state and federal penalties if you perform such modifications.

Lowering the compression ratio may be a more common modification customers will ask you to perform. Older, high-performance engines usually need a lower compression ratio these days to be used as practical street engines. As a rule of thumb, overhead-valve V8 engines run best on today's gasoline with a maximum compression ratio of about 9:1 or, in some instances, 10:1. Primarily, it varies depending on the combustion chamber design and the camshaft timing. The customer technical support people at piston manufacturers and camshaft grinders can help you select a practical compression ratio for your customer's engine.

Scuffing: The transfer of metal between two rubbing parts. It is usually caused by a lack of lubrication.

Piston Slap: A noise caused by a loose-fitting piston skirt hitting against the cylinder wall.

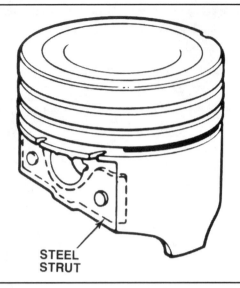

Figure 12-15. Steel struts, cast inside pistons, help control piston expansion.

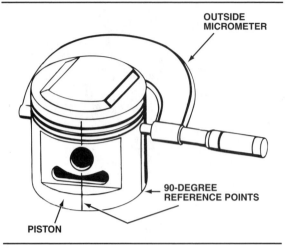

Figure 12-16. You measure the diameter of the piston skirt at the largest diameter, usually in a place perpendicular to the pin bosses.

Piston clearance

Manufacturers fit pistons to cylinders with clearances from 0.001 inch (0.025 mm) to as much as 0.008 inch (0.200 mm). For example, the small clearance could be for a cast piston with a steel strut and the large clearance for a forged piston in a high-performance engine. However, it is risky to generalize. The required clearance can vary greatly, depending as much on the piston alloy, design, and application as on the manufacturing technique. Engine machinists must always follow the piston manufacturer's clearance recommendations.

The measurement for the diameter of the piston is taken at a specific place on the skirt. The exact place is recommended by the piston manufacturer or in the factory shop manual, figure 12-16.

When rebuilding an engine, the clearance is established by boring or honing the cylinder to fit the piston. When the factory assembles engines, several different sizes of pistons are available. The pistons are selected to fit each individual bore, which produces the best match for the least machining costs.

Piston Pins

The piston pin attaches the piston to the connecting rod but still allows movement between the two. A **piston pin** is also called a wrist pin because it has somewhat the same action as the human wrist. Pins are made from high-grade carbon steel strong enough to withstand the full force of the piston pushing and pulling on the rod. The outer surface of the pin has a high polish so that it can run freely on, or be pressed into, the aluminum piston's pin boss without scoring or gouging.

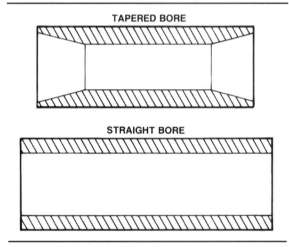

Figure 12-17. Most piston pins are hollow to reduce weight and have a straight bore. Some pins have a tapered bore to reinforce the pin.

Piston pins are hollow to reduce their weight, figure 12-17. Some pins have a tapered internal diameter, or bore, with a small diameter in the center and larger diameters at each end. Tapered construction is more expensive to manufacture, but makes the pin stronger where it bears the greatest loads.

Piston pin attachment

Through the years, designers have secured piston pins in many different ways, figure 12-18. However, most current engines use only two methods: press fit and full floating. The press fit pin is free to turn in the piston, but fits so tightly in the rod that it cannot slip out and hit the cylinder wall. Machinists use a special machine to press the pin into the rod. There is no bushing in the rod.

Pin Type & Description	Cutaway View
Type A Full Floating	
Type B Oscillating in Bushed Piston, Clamped in Rod	
Type C Oscillating in Piston (no bushing), Clamped in Rod	
Type D Oscillating in Piston — press fit in rod	
Type E Set Screw Type Piston	

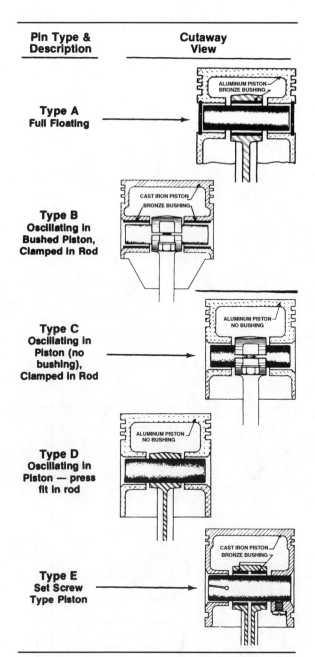

Figure 12-18. Although manufacturers have tried many different ways to attach pistons to rod, the only common methods for automotive applications are Type A, full floating, and Type D, press fit.

Full floating pins are able to turn in both the piston and the rod. Manufacturers install a bronze bushing in the small end of the rod to allow pin movement. The pin is retained with **circlips** or locks at each end of the pin hole in the piston, figure 12-19. The full floating pin requires extra lubrication, so the manufacturer drills a hole or cuts a slot in the small end of the rod to allow oil to enter, figure 12-20.

Proponents of the press fit pins claim that they are more reliable. Scoring of the cylinder wall from loose

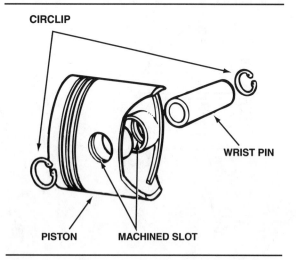

Figure 12-19. Circlips hold full-floating pins in place.

Figure 12-20. The holes drilled in this Volkswagen connecting rod allow oil to reach the full-floating pin.

pins was a problem before the adoption of press-fit pins. The small circlips that retain the pins at each end sometimes come loose, permitting the pin to slide out and contact the cylinder bore.

Many other manufacturers use full-floating pins without reliability problems. They claim the full-floating design allows better alignment (or a lack of misalignment) between the pistons, rods, and crank, which helps engine reliability and longevity.

Piston Pin: The hollow metal rod that attaches the piston to the connecting rod. Also called a wrist pin.

Circlips: Round spring-steel retainers that fit into grooves at the end of the piston pin bore to hold the pin in place.

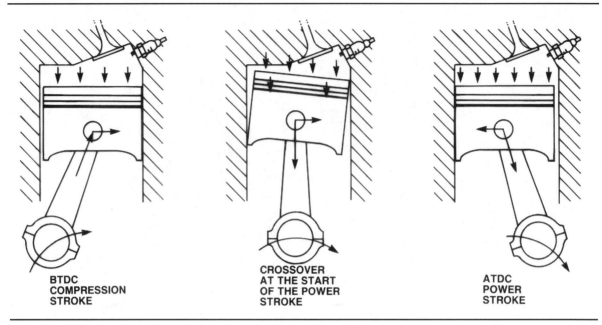

Figure 12-21. Engine rotation and rod angle during the power stroke causes the piston to press harder against one side of the cylinder, creating a major thrust surface. In this clockwise-rotating engine, the major thrust surface is on the left side.

Piston Offset

Most piston pin bores are dead center in the piston because it is cheaper to manufacture a piston that way. Some manufacturers offset the piston pin bore to reduce piston slap. The reasoning behind this design concerns the thrust forces on a piston caused by the angle of the connecting rod. When the piston goes up or down in the cylinder, the angle of the rod forces the piston against one side of the cylinder, figure 12-21. This is exactly the same as the way your weight at the top of a ladder leaning against the side of a house makes the ladder dig into the wall.

The power stroke generates much more force on the piston than the compression stroke, so the side of the piston that receives the thrust on the power stroke is called the **major thrust surface**. The side of the piston pushed against the cylinder on the compression stroke is the minor thrust surface.

This also causes the cylinders to wear unevenly around the circumference of the bore. In a clockwise-rotating engine, the major thrust surface is on the left (or passenger) side of the cylinder. Cylinder wear is greater on that side because the connecting rod angle jams the piston into the cylinder wall as gas pressure forces it down the cylinder. When the piston rises under compression, it jams against the opposite side of the cylinder (the minor thrust surface), but not nearly as forcefully. The result is that the cylinder wears out-of-round, or roughly oval.

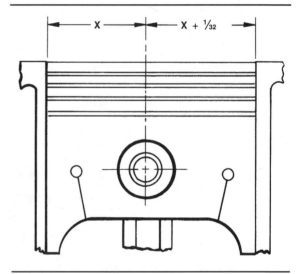

Figure 12-22. Some manufacturers offset the piston pin to prevent piston slap.

At top dead center, between the compression and power strokes, the thrust on the piston changes from the minor to the major thrust face or side. If there is any clearance between the piston and the cylinder, the piston will slap against the wall. This makes noise and can damage the piston. To prevent slap, the piston pin may be offset up to $\frac{1}{16}$ inch (1.6 mm) toward the major thrust side, figure 12-22.

Offsetting the pin causes the piston skirt to tilt as the thrust shifts from one side to the other. With a

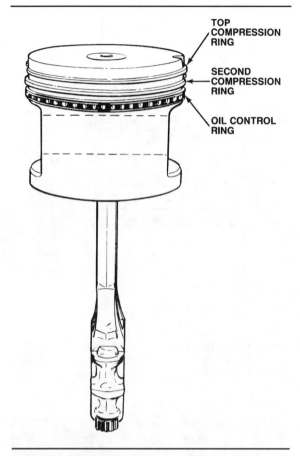

Figure 12-23. Modern pistons seal the cylinder adequately with three piston rings.

slight amount of tilt, the bottom of the skirt touches first against the cylinder wall. The bottom of the skirt flexes slightly and allows the top to come in gently without slapping. Pistons with offset pins are marked with a notch or an arrow to indicate the position in which you must install them in the engine.

PISTON RINGS

Because the piston must slide freely in the cylinder, it cannot fit tightly enough to seal the combustion from above and the oil from below. A precision seal, in the form of piston rings, is necessary. Rings do not perform perfectly when you first install them because they must seat. To seat, the rings wear so that they fit the minor irregularities of the cylinder wall. If the wall is smooth and the ring fits closely, very little material has to be worn away and seating occurs rapidly.

Pistons in modern engines normally have three rings above the piston pin, figure 12-23. It was common in older engines to have an oil ring below the pin or at the bottom of the piston skirt. In modern ring design, it is not necessary to have more than three rings in passenger car engines. The additional rings do not seal

significantly better and they increase friction, which consumes horsepower and reduces fuel economy.

There are two types of rings: compression rings and oil control rings. They perform three important functions:

- Sealing the combustion chamber
- Conducting heat from the piston to the cylinder wall
- Scraping oil from the cylinder walls.

Compression Rings

The top and second ring are the compression rings and their primary job is to seal the bottom of the combustion chamber by sealing the circumference of

---

■ Select Fitting Pistons

While the manufacturers use tolerance specifications when they decide whether or not to scrap a part on the assembly line, they usually compensate during assembly by select fitting parts. It's very expensive to manufacture pistons or crankshaft journals to extreme accuracy, but measuring them to an extra decimal place doesn't cost anything extra. The factories take advantage of this by grouping all the parts that fall within specifications into two or three even finer sizes, called tolerance groups. Then, when they assemble these parts into an individual engine, they select the parts that match most closely. The end result is an engine assembled with more precision than was used to manufacture the parts.

For instance, Porsche select-fits pistons to the cylinder bores of the 2.5-liter four-cylinder engine in the 944. Porsche bores the cylinders to a nominal 100.010 mm, but accepts any cylinder size between 100.000 and 100.020 mm, a tolerance of 0.020 mm (0.001 inch). They group the cylinders into three tolerance groups: 100.000, 100.010, and 100.020. The nominal piston size is 99.990 mm, but any piston between 99.980 and 100.000 is acceptable, again a tolerance of 0.020 mm (0.001 inch). The three tolerance groups for the pistons are 99.980, 99.990, and 100.000.

When the engine goes together, the builders choose pistons from whichever tolerance group that matches the tolerance group of a particular cylinder, with large, medium, and small pistons going into large, medium, and small cylinder bores. The result is that Porsche can hold the tolerance for individual piston fits in the cylinders to as little as 0.020 mm, even though a mismatch of as great as 0.040 mm (0.002 inch) would be possible with the ranges of pistons and bores as manufactured.

Major Thrust Surface: The side of the piston that pushes against the cylinder wall during the power stroke.

the piston. The first ring is the main seal and most of the leakage from the first ring is contained by the second ring. Engineers estimate that the second compression ring accounts for about 10 to 20 percent of compression sealing.

If combustion gases leak past both rings, problems result. First, the engine loses power. The force of the expanding, burning air-fuel mixture is wasted if it escapes alongside the piston instead of pushing on the piston crown. Also, as these hot gases leak past the rings they burn away the lubrication in that area, increasing cylinder wear. As we pointed out in Chapter 6, any significant blowby into the crankcase also contaminates the oil, reducing the effectiveness of the lubrication system throughout the engine.

The compression rings, and to a lesser extent the oil rings, are responsible for conducting about 75 percent of the heat away from the piston, figure 12-24. The source of heat is on top of the piston. As the heat moves down, it contacts the rings and moves into the cylinder wall. From there it goes into the coolant and is transferred to the air at the radiator.

Compression ring operation
Compression rings have a certain amount of static tension, which is the spring tension you feel when you squeeze a ring. When the ring is inserted in the piston, and the piston is inserted into the cylinder, this **ring tension** forces the ring against the cylinder wall. Ring tension is actually very small compared to the force of the combustion it must seal, so this static tension does very little of the actual sealing during the power stroke.

When the tremendous pressure of the burning mixture passes alongside the top of the piston and hits the ring, it firmly seats the ring in the bottom of its groove. More importantly, the gas pressure also gets into the space behind the ring, pushing the ring out against the cylinder wall, figure 12-25. With this extra pressure created by combustion, the ring makes a very good seal. The space in the ring groove above the ring is called the side clearance and the space behind the ring is the back clearance, figure 12-26.

During the other engine strokes, the compression rings do not seal as tightly, which reduces friction. During the intake stroke, engine vacuum tends to force oil past the rings. However, the design of most second compression rings causes the lower edge to scrape oil from the cylinder wall as the piston completes the downward stroke, assisting the oil control rings.

During the upward-moving compression and exhaust strokes, the compression rings are forced to the bottom of their grooves. However, without gas pressure from the burning air-fuel mixture, they do

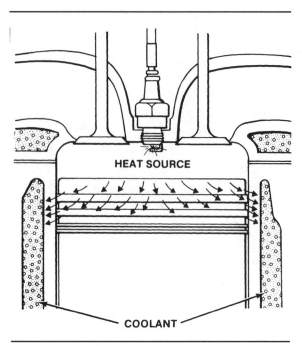

Figure 12-24. The rings conduct heat from the piston to cylinder wall where the cooling system can extract it from the engine.

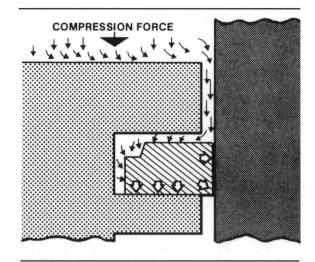

Figure 12-25. Combustion chamber pressure forces the ring against the cylinder wall and the bottom of the ring groove.

not seal tightly against the cylinder wall, allowing them to ride over the very thin oil film left behind.

The compression rings do not touch the back of the piston grooves. This allows them to ride over any imperfections in the cylinder wall without affecting the piston. The rings can follow the cylinder wall to provide the best seal.

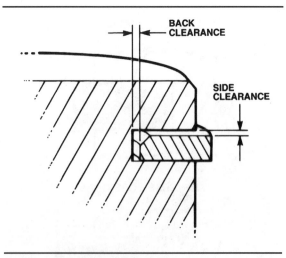

Figure 12-26. With the ring installed in the groove, the side and back clearances must be correct for the compression rings to seal properly.

Ring flutter

When the piston is moving down, the ring is against the top of the groove. When the piston reverses direction at BDC and begins traveling up, the ring is slammed against the bottom of the groove. This occurs twice every revolution. This constant hammering eventually wears out the ring grooves in the piston.

At extremely high rpm, these piston accelerations can be greater than the compression pressure holding the ring in place on the compression stroke, causing the ring to momentarily lose its seal. When the piston slows as it reaches TDC, combustion gases repeatedly slam the ring back into its groove and against the cylinder wall. Engineers call this problem **ring flutter**. The rings take a severe beating and hot combustion gases travel down the side of the piston. Lubrication is burned from the cylinder wall and the piston may overheat. The result may be ring failure or seizure. Lighter, thinner rings are better able to resist flutter than thick, heavy rings.

Compression ring design

Manufacturers make compression rings in dozens of different shapes. Most are variations of 4 or 5 basic designs. All rings are perfectly flat and smooth on their tops and bottoms, but a manufacturer will change the front or back edge depending upon the job the ring must perform. The top compression ring is usually fully rectangular, but it may have a barrel face, figure 12-27. The rectangular face provides excellent sealing and wear properties and some measure of oil scraping. The barrel face compensates for piston rocking or ring misalignment in the groove. Because its contact area with the cylinder is smaller, it seals well and seats quickly.

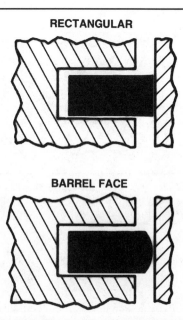

Figure 12-27. These ring shapes are common on top compression rings because they provide the best compression seal.

The second compression ring generally uses the more unusual shapes. This ring helps seal compression, but does not need to seal as well as the top ring; however, it must assist oil control. One of the first developments was the tapered-face ring, figure 12-28. It does not seal compression as well, but the sharp lower edge provides effective oil scraping on the downward piston strokes. The line contact of the sharp edge on the cylinder wall also allows the ring to seat quickly.

Engineers developed the torsional twist ring to provide better compression sealing than the taper-face ring but furnish equal or better oil control. The ring maker counterbores or chamfers a corner of the ring to change the internal stress of the ring. When the ring is compressed in the cylinder, it tilts in its groove, figure 12-29. This twist allows a line contact on the cylinder wall for good oil control and an equally good line contact on the top and bottom of the ring groove to

Ring Tension: The natural spring tension of a piston ring. It helps the ring seal against the cylinder wall during the intake and exhaust strokes and the beginning of the compression stroke, when combustion gases are not forcing the rings against the wall.

Ring Flutter: Rapid up and down movement of a piston ring that breaks the seal between the ring and the cylinder wall.

TAPER FACE

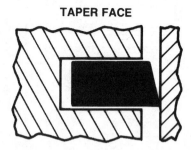

Figure 12-28. The taper face ring provides good oil control by scraping the cylinder wall. If you install the taper face ring upside down, it will pump oil into the combustion chamber.

minimize compression leakage through the groove. An additional benefit of torsional twist rings is their resistance to sticking in the piston groove because their continual twisting and untwisting is self-cleaning.

When the ring maker chamfers or counterbores the ring at its upper inside corner, it is called a positive torsional twist ring. It twists in a positive direction like a taper face ring. A chamfer or counter bore on the lower inside corner produces a reverse torsional twist ring, which twists in the opposite direction. The reverse twist provides slightly improved oil control because it has line contact at the lower outside of the ring groove. Oil is blocked from entering the groove and cannot cause the ring to float up.

Some manufacturers use torsional twist top compression rings to improve oil control. These rings usually have a rectangular face for best compression seal. When used as a second compression ring, many manufacturers also put a taper face on torsional twist rings. As we have noted, the manufacturer can get away with this because compression sealing is not of primary importance for the second ring, while good oil control is important.

Another kind of positive torsional twist ring has a counter bore at its lower outside edge, figure 12-30. This ring is also called a scraper ring because of its superior ability to scrape oil from the cylinder wall. Its compression sealing is relatively poor, so ring makers recommend you use it only as a second compression ring.

Compression ring width
A piston ring always has greater radial thickness than vertical width, figure 12-31. Therefore, combustion gas pressure bears down on a large surface to provide a large total downward force. The width, being comparatively thin, provides a high pressure per square inch (or per square meter) against the cylinder wall for a good seal.

POSITIVE TORSIONAL TWIST

REVERSE TORSIONAL TWIST

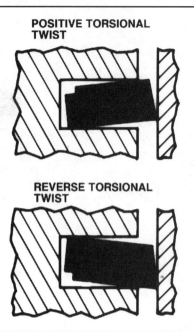

Figure 12-29. Torsional twist rings provide better compression sealing and oil control than regular taper face rings.

SCRAPER FACE

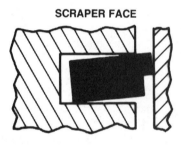

Figure 12-30. Scraper rings improve oil control.

The trend over the last few years has been toward thinner rings. Formerly, standard ring widths were ⅛-inch (3-mm) and ³⁄₃₂-inch (2.3-mm). Now, ⁵⁄₆₄-inch (2-mm) and ¹⁄₁₆-inch (1.6-mm) widths are common. Thinner widths lower ring mass and inertia, reducing groove pounding and ring flutter. Thinner widths also present less metal to the bore face and decrease scuffing. On the negative side, thinner ring grooves are more difficult to machine into the piston, and thinner rings are harder to install because they are less rigid.

Compression ring tension
Static ring tensions of 18 to 26 lbs (8 to 12 kg) used to be common. In the mid-1980s, in the search for better fuel economy, manufacturers developed low-tension rings. The new low-tension rings have about 6 to 13 lbs (3 to 6 kg) of ring tension to reduce cylinder friction.

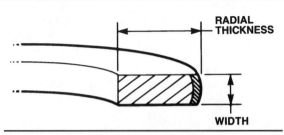

Figure 12-31. The top view of any ring shows it to be thicker than the side view.

Oil Control Rings

The third ring on modern pistons is the oil control ring, which scrapes oil from the cylinder wall and returns it to the crankcase. If oil gets into the combustion chamber, heat oxidizes the oil, turning it into carbon. The carbon can obstruct the intake and exhaust gas flow and raise compression, causing detonation. Also, super-heated, glowing particles of carbon can cause pre-ignition of the air-fuel mixture.

Oil ring operation

Unlike the compression ring, static ring tension seals the oil control ring. As the ring scrapes the oil from the cylinder wall, the oil passes through holes or slots in the back of the ring groove and drains to the crankcase, figure 12-32. Some oil also drains through holes in the top of the pin boss to lubricate the pin, figure 12-33. The end of the pin next to the cylinder wall also gets oil because it is below the oil ring.

Despite the scraping action of the oil rings, a thin layer of oil is left on the cylinder wall, which is good for several reasons, even though this oil is lost or burned. This oil film lubricates the piston and the compression rings. It traps some of the fine, gritty dust that was brought in with the air-fuel mixture. Even the most well-filtered engines allow in some contaminants. The oil also traps partially burned and carbonized oil, which, if combined with the intake dust, acts as a very abrasive compound and quickly wears out the rings, pistons, and cylinder walls.

Engineers estimate that the oil film of a perfectly set-up engine is about one ten-millionth of an inch thick. If this much oil is lost or burned, engine oil consumption would amount to about 3,000 miles per quart — a fully acceptable amount. If through engine wear or poor rebuilding practices, the film increased to just one-millionth, oil consumption would be an unacceptable 300 miles per quart. This illustrates how important cylinder preparation, ring set-up, and other rebuilding practices really are to proper engine operation.

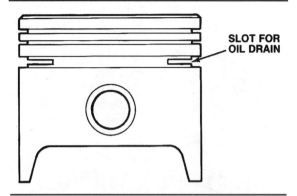

Figure 12-32. Slots or holes machined behind the oil ring allow oil to drain to the crankcase.

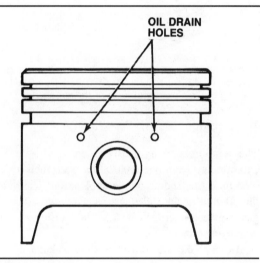

Figure 12-33. Holes drilled in the piston may help lubricate the piston pin.

Oil ring design

The modern oil control ring is usually made in two or three pieces, figure 12-34. On the three-piece ring, a spacer-expander lies between top and bottom layers called rails. The spacer-expander keeps the rails separated and pushes them out against the cylinder wall, figure 12-35. The spacer-expander between the rails does not touch the cylinder wall. The thin rails do a better job of scraping the cylinder walls than a thick, rectangular ring.

The expander is perforated and slotted. Any oil that is scraped off the wall by the top rail goes between the rails, through the holes in the expander, through the piston, and back into the crankcase. Most of the oil is scraped off by the bottom rail. The top rail only removes the oil that the bottom rail cannot handle.

The two-piece ring operates like the three-piece ring. It also uses an expander, but it pushes a single rail against the cylinder wall. Some two-piece oil rings use coil wire expanders, figure 12-36.

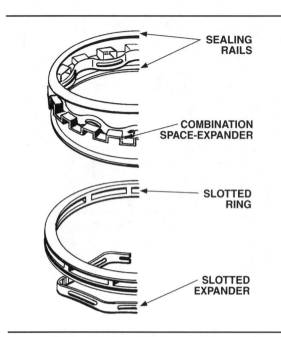

Figure 12-34. The two most common types of oil control rings are either two piece or three piece.

Ring Gap

The compression ring ends are separated by a small gap, ranging from about 0.010 to 0.025 inch (0.25 to 0.65 mm), depending on the application. The rails of the oil control rings also have square cut ends with a gap between them. Generally, larger bores require larger ring gaps.

Although the gap does allow some leakage, it is necessary to allow for the heat expansion of the ring. Also, some compression leakage past the top ring helps seal the second ring. Without any gap, the ring ends would butt together and buckle the ring when the engine got hot. The smallest gap that allows the rings to expand without butting is theoretically the best. It allows the least blowby, improving compression and raising power.

Vehicle manufacturers and ring manufacturers list a ring gap specification for the compression rings. You must measure the gap and file the ends of the ring, if necessary, to achieve the proper clearance when you rebuild an engine. The gap for the oil ring is not as critical. Even gaps as large as 0.030 inch (0.75 mm) are acceptable.

Gapless rings
A recent development is a second compression ring that operates without a conventional ring gap. Total Seal, Inc., which makes the new Gapless® rings, claims they extend ring life and reduce blowby. Gapless® rings are a little more expensive and no vehicle manufacturer currently uses them as original equipment.

The important feature of the Gapless® ring is how it provides a means for heat expansion without a conventional ring gap. The Gapless® ring is a two-piece, interlocking design, which is stepped so it overlaps, figure 12-37. The ends can slide closer together as the ring heats and expands, but the stepped end does not allow a straight path for blowby.

Ring friction and leakage
Friction is a major problem with piston rings. It is possible to make a ring that seals so completely that it will not allow any gases or oil to pass. However, such a ring would fit so tightly against the cylinder bore that the rings would get no lubrication or even so tight that the piston could not move. Modern rings are a compromise. They must leak small amounts of oil to lubricate themselves. By doing so, rings also allow small amounts of combustion gas to leak into the crankcase and they allow some oil into the combustion chamber. From a theoretical standpoint, neither is good, but as long as the leakage is below the acceptable limit, the ring performance is adequate and friction is low.

Piston Ring Materials

Manufacturers make most piston rings out of gray cast iron. For heavy-duty and high-performance applications, they choose nodular cast iron for the compression rings because it is stronger and more ductile.

Cast iron is not normally flexible, but when made with a small cross section such as in a piston ring, it flexes easily. However, it will also break if not handled with care.

Some manufacturers use steel compression rings. Steel rings are easier to manufacture because they are not as brittle. Other manufacturers are considering using a steel top compression ring because it holds up well against detonation.

The rails of oil control rings are made of cast iron or steel. The expanders are usually made of steel, often stainless steel.

Ring face coatings
Ring manufacturers usually add a face coating to the top compression ring, and sometimes to the second compression ring and to the rails of the oil control rings. These coatings improve ring performance and extend ring life.

The standard gray cast-iron ring has a soft coating, such as phosphate, iron oxide, or graphite, to prevent scuffing during the engine break-in period. These rings seat quickly, but do not have a long service life.

During World War II, manufacturers introduced a chrome-facing for improved abrasion resistance. It increased the life of engines used in desert warfare.

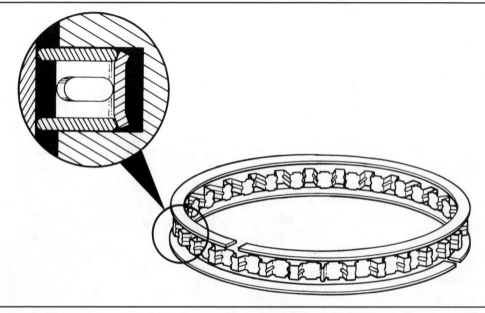

Figure 12-35. This typical three-piece oil control ring uses a hump-type stainless-steel spacer-expander. The expander separates the two steel rails and presses them against the cylinder wall.

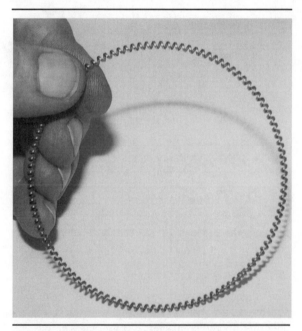

Figure 12-36. Coil wire expanders look radically different than hump-type expanders but they perform the same job in much the same way.

Manufacturers make these rings by electroplating chrome to the face of a nodular iron ring, figure 12-38.

Even though only the facing has chrome, engine builders refer to these as "chrome rings." Chrome rings are very long-lasting, but you must bore and hone the cylinder to the proper fit and finish or the rings will not seat properly. A process that helps chrome rings seat more easily consists of reversing

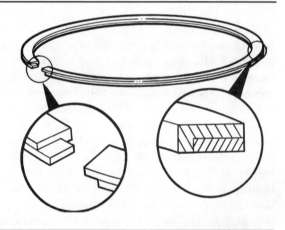

Figure 12-37. The gapless ring overlaps, while the conventional gap could allow some blowby.

the electroplating polarity during the last few moments of the plating process. This causes the outer layer of the chrome to become porous, which allows the ring to retain oil as well as seat more readily during the break-in process.

Chrome rings are so hard that vehicle manufacturers have to improve the strength of their cylinder blocks so the cylinder bores will last. They do this by increasing the amount of nickel added to the iron that is used to make the block casting. These blocks are very tough and a little more difficult for the rebuilder to hone.

In the early 1960s, manufacturers introduced molybdenum oxide-faced rings for greater resistance to heat-related scuffing. Engine rebuilders refer to

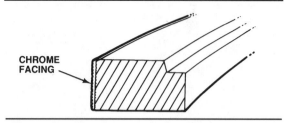

Figure 12-38. The chrome facing on this compression ring is about 0.004 inch (0.10 mm) thick.

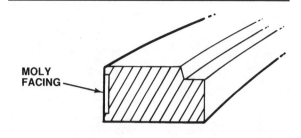

Figure 12-39. The moly facing on this compression ring is about 0.005 inch (mm) deep.

these as "moly rings." The moly surface is highly porous, allowing good oil retention for better lubrication. They work even better than chrome rings under peak loads with hot spots on the cylinder walls or when lubrication is marginal. The disadvantage of moly rings is a shorter life than chrome rings. Because of their heat resistance, moly rings are very popular these days with original equipment manufacturers.

To make a moly ring, the cast-iron ring (usually nodular iron) is produced with a groove in its face, figure 12-39. A plasma arc sprays the molybdenum oxide powder into the groove at extremely high temperature, usually 20,000°F-plus (11,000°C-plus), to bond well with the parent metal.

In the 1980s, TRW introduced a ceramic face coating that consists of a combination of aluminum and titanium oxides over a nickel-aluminide bond. These "ceramic" rings also begin as nodular iron and are plasma arc sprayed with the ceramic oxides. In addition to improving upon the abrasion and scuff resistance of other facings, ceramic rings better resist damage caused by detonation. Ceramic rings are very strong, however, detractors say they create a heat barrier and do not transfer heat from the pistons to the cylinder walls as well as all-metal rings. The result, they claim, is increased chance for detonation.

Sealed Power, Inc. introduced a plasma-sprayed chromium-carbon face coating. Sealed Power claims this heavy-duty coating has the scuff resistance of moly facings and the wear rates of chrome.

Most steel compression rings have a chrome facing, although some have a titanium nitride face coating. Titanium nitride is extremely hard and long wearing but, like chrome, it requires careful bore preparation to seat properly.

CONNECTING RODS

The connecting rod is the link between the piston and the crankshaft, figure 12-40. Connecting rods are usually made in an "I-Beam" section, 12-41. This shape resists bending and is easy to manufacture. Rods may be either cast or forged. All heavy-duty rods are forged, but casting is strong enough for some applications. After

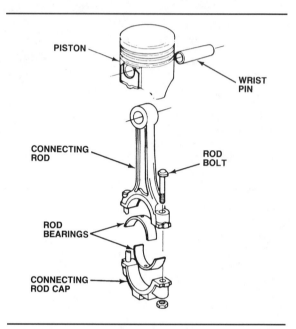

Figure 12-40. The connecting rod is probably the most highly stressed part in an automotive engine. Combustion gases try to compress it and piston inertia tries to rip it apart.

forming, the big end of the rod is usually sawn or cracked in half to separate the rod from the bearing cap.

We refer to the obvious size differences between each end of the rod as the small end and big end. We have already explained how the small end of the rod attaches to the piston. The big end of the rod has a hole that fits the crankshaft rod journal. The big end splits to simplify the installation of the bearings and the rod onto the crank. Rod bolts and nuts hold the two pieces together.

In the big end of the rod, the replaceable inserts are located by tangs that fit into notches, figure 12-42. To prevent movement of the bearing in the rod and to make sure it is tight against the rod, the bearings are slightly larger than the rod. When assembled, this mismatch provides the necessary bearing crush, figure 12-43. Bearings are covered completely in Chapter 13.

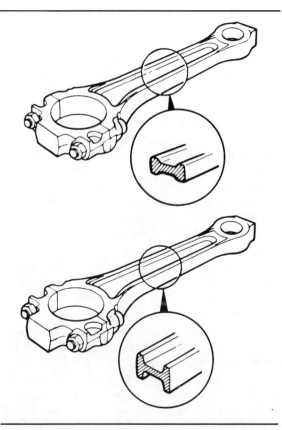

Figure 12-41. The cross section is more substantial on some rods than others. The thicker rod at the top is stronger than the rod on the bottom.

Many rods have balancing pads on the cap, and some also have a pad at the small end of the rod, figure 12-44. The pad allows you to grind material away to balance rod weights.

Connecting Rod Materials

Manufacturers usually make rods from steel. The better rods contain nickel, chrome, or other elements to make them stronger. Sometimes manufacturers make rods from titanium. They are as strong as most steel rods but about half the weight, which allows greater acceleration. Titanium is expensive, so usually only racing engines benefit from titanium rods. One exception is Acura's NSX sports car.

Some drag race engines use aluminum rods. Aluminum is much lighter than steel or titanium. However, aluminum fatigues easily, so racers change the rods frequently to prevent breakage.

High-performance rods are often shot-peened to make them stronger. As we learned in Chapter 3, shot-peening blasts the surface of a metal part with round, steel shot. Shot-peening pre-stresses the surface in compression to reduce the chance that cracks will form.

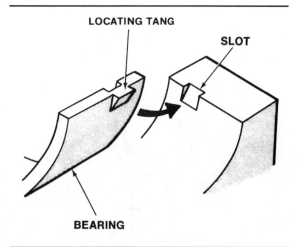

Figure 12-42. A notch in the rod cap locates the tang on the bearing insert.

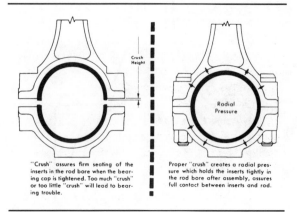

"Crush" assures firm seating of the inserts in the rod bore when the bearing cap is tightened. Too much "crush" or too little "crush" will lead to bearing trouble.

Proper "crush" creates a radial pressure which holds the inserts tightly in the rod bore after assembly, assures full contact between inserts and rod.

Figure 12-43. Rod bearing crush.

Powder-Forged Rods

Powder-forging is a new method of producing forged connecting rods, developed over the last two decades by a number of manufacturers. Ford's 4.6-liter V8, as well as some late-model Toyota, Lexus, BMW, and Porsche powerplants, use powder-forged rods. To make these rods, the foundry begins with an iron powder that is mixed with some alloying elements, such as copper and graphite, and a wax lubricant to help remove the rod from the forming die. A press shapes the powder mix into an approximation of the finished rod. This "preform" is heated to further sinter the powder together. Then, only one blow of the forging press with a finishing die forms the rod. The factory drills, taps, and deburrs the powder-forged rods like a hot-forged rod.

Ford claims the powder forgings are closer to the finished product before detailing, so there is less material waste. It also claims less weight variation between powder-forged rods, which eases balancing.

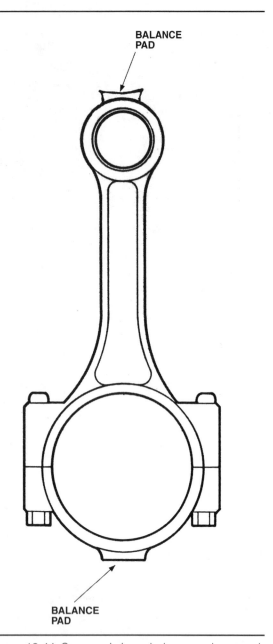

Figure 12-44. Some rods have balance pads on each end of the connecting rod.

Cracked Connecting Rods

Ford adds another manufacturing wrinkle to its powder-forged rods. Instead of sawing the rod apart, Ford scribes two notches on the inside of the big end of the rods to serve as stress points. Then, the rod is pulled apart or "cracked" with about 10,000 lbs (4,500 kg) of hydraulic tension. Ford claims the two halves of the cracked rod seat together extremely well and movement between the two parts is restricted. The procedure also simplifies manufacturing, reducing production costs.

The Effects Of Rod Length

No one agrees what the optimum rod length is, but engineers do agree that the length of the rod affects the way the piston moves and this has an effect on engine performance. When the piston is at top dead center, there are several degrees of movement of the crankshaft throw during which the piston does not move. Engineers and high-performance engine builders refer to this as piston dwell time. A long rod makes the piston dwell longer at TDC. A long rod increases the angle that the crankshaft can move without moving the piston, and decelerates the piston more slowly as it approaches top dead center. When the piston starts to leave top dead center, it also accelerates more slowly with a long rod.

With the piston at top dead center for a longer period of time, the low pressure in the exhaust manifold will pull in more intake mixture during overlap. Also, the longer rod does not tend to push the mixture back into the intake manifold because it is moving more slowly. Furthermore, the burning fuel mixture has more time to build pressure with the piston at TDC during the power stroke. Proponents of long connecting rods insist that when all other factors are equal, an engine with long rods makes more horsepower at mid- to high-rpm.

When a longer rod is at a 90-degree angle to the crankshaft throw, the angle between the side of the piston and the rod is less. This allows the piston to push more on the rod instead of against the cylinder wall, figure 12-45, resulting in less friction and wear.

The deck height of the engine is the primary limiting factor in how long a rod a designer can put in an engine. Another limit is the height of the piston. As the rod gets longer, the designer must either move the pin hole up in the piston or shorten the piston. Both of these changes can hurt piston function. The piston must be tall enough to be stable and not rock in the bore, and allow room for an adequate ring package.

A shorter rod accelerates and decelerates the piston faster near top dead center. This is an advantage in getting the piston out of the way of the valves and it can allow more radical camshaft timing. However, the more rapid piston acceleration puts more stress on the piston and rod. A rod that is too short may crack and break, or break the piston.

A short rod also increases the rod angle, so there is more side thrust on the piston against the cylinder wall, increasing friction and wear. Proponents of short connecting rods insist that the increased rod angle does have one advantage. The piston and rod spend more of their travel at angles to the crank throw. Therefore, they exert greater leverage on the crank

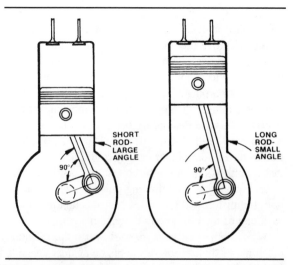

Figure 12-45. The longer connecting rod allows the piston to push more on the rod instead of against the cylinder wall.

throw during a longer portion of the power stroke. The result is an engine that produces more low-rpm torque, when all other factors are equal. Short rods also allow the designer to specify a lower block deck height, which reduces overall engine weight and allows a lower hoodline on the body for improved aerodynamics.

PISTON SPEED

During engine operation, **piston speed** is a prime factor in engine life. Piston speed is not the same thing as rpm, although rpm affects piston speed. Designers calculate piston speed as the average velocity of the piston at a given rpm. It is usually expressed in feet per minute (or meters per minute).

For example, an engine with a stroke of three inches (75 mm) will move one of its pistons three inches (75 mm) down and three inches (76 mm) up on every revolution. Therefore, piston movement on that engine would be six inches (150 mm) during each engine revolution. During ten revolutions, the total movement would be 60 inches (1500 mm) or five feet (1.5 m). If it took one minute for the engine to turn the ten revolutions, then the piston speed would be five feet (1.5 m) per minute. At 3,000 rpm, that same engine would move one piston 18,000 inches (457,000 mm), or 1,500 feet (457 m) each minute.

As the stroke gets longer, the piston must travel farther on each revolution, and piston speed increases.

Piston Speed: The average speed of the piston, in feet or meters per minute, at a specified engine rpm.

■ Oval Pistons

We learned that although pistons appear round, they are actually slightly elliptical at room temperature. We also learned that each piston requires a single connecting rod; however, Honda broke both of those rules when it built its NR750 sport bike in 1992. The reasoning behind the NR's unusual oval pistons with twin connecting rods goes back to 1978 when Honda re-entered 500cc Grand Prix motorcycle racing.

Two-stroke engines were the dominant powerplants in GP motorcycle racing then, as they are today. Despite that, for both marketing and philosophical reasons, Honda wanted to race with a four-stroke engine. Honda's designers concluded that to compete with the two-stroke, which has double the number of power strokes, their engine would need remarkably efficient breathing and turn at very high rpm. After much calculation, it appeared that only a 32-valve V8 could provide the necessary valve area and a sufficiently short stroke to reach their breathing and rpm goals. Unfortunately, the rules limited all engines to a maximum of only four cylinders.

In a brilliantly simple move, the designers merely joined each pair of pistons to make an oval-piston V4 with eight valves and two spark plugs per cylinder. Honda even retained the eight connecting rods. The biggest stumbling blocks were the cylinder bores and piston rings. The oddly shaped pistons were not easy to seal. Honda solved the problem, but has not been forthcoming about its technology.

The NR500 machines entered GP racing under a shroud of secrecy that only served to increase speculation about their unusual engines. They were fast, turning nearly 20,000 rpm, but not fast enough to overthrow the two-strokes. Honda eventually had a successful run with the oval-piston engines in endurance racing where the wide torque band the engine design allowed were great benefits. The great valve perimeter provided by the 32 valves allowed the engineers to install relatively mild camshafts, which helped torque, while the great valve area still furnished enough breathing for good high-rpm power.

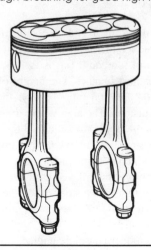

The long, oval pistons of Honda's NR750 need twin connecting rods for adequate piston stability in the bores.

The higher piston speed puts more stress on the pistons, rings, and cylinder bores. The pistons and rings also cover more distance in their bores at a given rpm with a high piston speed, so they wear out sooner. This is one of the primary reasons why long-stroke engines have lower rpm red-lines than short-stroke engines. A short-stroke engine can run at a higher rpm than a long-stroke engine and still produce a lower piston speed.

Designers try to limit the piston speeds of their engines to make them last longer. Designers have a good idea of the stresses the metals in their engines can endure, so as a rule of thumb, an average piston speed of 4,000 feet (1,220 m) per minute is considered the upper limit. Of course, racing engines with expensive, heavy-duty parts reach higher speeds. Engineers design passenger cars with transmission- and final-drive gearing to keep piston speeds around 1,500 feet (457 m) per minute for most of a car's use.

When the first automobiles were made with their heavy pistons, the bearings could not withstand much pressure and lubrication was marginal. It was considered quite an accomplishment to design an engine that would run for a reasonable length of time at a piston speed of 1,500 feet (457 m) per minute.

However, piston speed can be misleading because it is only the average piston velocity; it does not indicate maximum piston speed, which can be much greater than the average speed. There is also the acceleration of the piston to consider. Every time the piston gets to top or bottom dead center, it stops. Immediately afterwards, it accelerates to full speed. If the engine is going fast enough, this sudden acceleration will pull the piston pin right out of the piston. This is one reason why over-revving an engine can be so damaging.

SUMMARY

Pistons must be strong to endure combustion pressure and light to reduce inertia loads. The piston crown is the top of the piston and it forms the bottom portion of the combustion chamber. The height of the crown affects the compression ratio. Today, most engines have flat-top pistons. The area where the wrist pin or piston pin fits into the piston is called the pin boss.

Below the crown and above the pin bosses are the ring grooves which hold the piston rings. The areas of the piston between the grooves are called ring lands. The lower portion of the piston, called the skirt, stabilizes the piston in the cylinder, preventing excessive rocking and wear.

Pistons for modern gasoline engines are either cast or forged aluminum alloy with the addition of small amounts of copper, zinc, chromium, and silicon. Hypereutectic cast pistons have a large amount of silicon to improve their strength. Manufacturers fit pistons to cylinders with clearances from 0.001 inch (0.025 mm) to as much as 0.008 inch (0.203 mm).

The piston pin, or wrist pin, attaches the piston to the connecting rod but still allows some movement between the two. Pistons are secured on the rods with a press fit or full floating design. The press fit pin is free to turn in the piston, but fits so tightly in the rod that it cannot slip through the rod and piston and hit the cylinder wall. Full floating pins are able to turn in both the piston and the rod. Circlips retain the pin at each end of the pin hole in the piston.

When the piston goes up or down in the cylinder, the angle of the rod forces the piston against one side of the cylinder. The side of the piston that receives the thrust on the power stroke is called the major thrust surface. To prevent slap, the piston pin may be offset up to $\frac{1}{16}$ inch (1.6 mm) toward the major thrust side.

Since the piston must be a slip fit in the cylinder, it requires a precision seal, in the form of piston rings, but they must seat to be effective. There are two types of rings, compression rings and oil control rings. They seal the combustion chamber, conduct heat from the piston to the cylinder wall, and scrape oil from the cylinder walls. The compression ring and oil control rail ends are separated by a small gap, necessary to account for the heat expansion of the ring.

Rings are made of cast iron or steel. They may have a face coating of chrome, molybdenum, ceramic, or another composite coating.

The connecting rod is the link between the piston and the crankshaft. Rods may be either cast or forged in an I-beam shape. Most are made of steel. Some rods are forged from iron powder.

The rod has a small end, which attaches to the piston, and a big end that fits the crankshaft rod journal. The rod may have balancing pads on the cap and at the small end of the rod.

The length of the connecting rod affects the way the piston moves and this has an effect on engine performance. Long rods dwell longer at TDC, which may improve mid-range and high-rpm power. Long rods create a better rod angle, which reduces piston and cylinder wear. Short rods create a sharper rod angle, which increases wear, but may improve leverage and low-rpm torque.

Piston speed is a prime factor in engine life. It is the average velocity of the piston at a given rpm and is usually expressed in feet per minute (or meters per minute).

Review Questions

Choose the single most correct answer.
Compare your answers with the correct answers on Page 328.

1. Which part of the piston forms the bottom of the combustion chamber?
   a. Skirt
   b. Ring groove
   c. Crown
   d. Ring land

2. Technician A says that pistons are round at room temperature. Technician B says that pistons are elliptical at room temperature.
   Who is right?
   a. A only
   b. B only
   c. Both A and B
   d. Neither A nor B

3. Modern pistons are made primarily of:
   a. Cast iron
   b. Aluminum alloy
   c. Steel
   d. Silicon

4. The best method of controlling piston expansion is:
   a. With a split skirt piston
   b. With a T-slot piston
   c. To cut a slot between the head of the piston and the skirt
   d. Cam grind the piston

5. A prime factor in the life of a piston is:
   a. Ring weight
   b. Height
   c. Piston speed
   d. Tensile strength

6. Scuffing on pistons can be caused by:
   a. Hot spots from a clogged cooling system
   b. Distorted cylinders
   c. Overheated cylinders
   d. All of the above

7. To reduce piston slap, the piston pin should be:
   a. Offset toward the minor thrust side
   b. Offset toward the major thrust side
   c. Adjusted
   d. Replaced

8. Technician A says that pistons are secured with a press-fit pin. Technician B says that pistons are secured with circlips at the pin ends.
   Who is right?
   a. A only
   b. B only
   c. Both A and B
   d. Neither A nor B

9. There are usually _____ rings on the piston in a gasoline engine.
   a. One
   b. Two
   c. Three
   d. Four

10. The bottom ring on a piston is used to:
    a. Control compression
    b. Control oil
    c. Control blowby
    d. Prevent piston slap

11. If too much oil is allowed to get into the combustion chamber the resulting carbon can:
    a. Obstruct the intake and exhaust flow
    b. Cause detonation
    c. Cause preignition
    d. All of the above

12. Technician A says that the piston rings control oil in the cylinder. Technician B says that the rings conduct heat away from the piston.
    Who is right?
    a. A only
    b. B only
    c. Both A and B
    d. Neither A nor B

13. Compression rings:
    a. Control oil
    b. Seal the combustion chamber
    c. Conduct heat from the piston
    d. All of the above

14. Technician A says that pressure from combustion seals the compression rings against the cylinder wall. Technician B says that static tension alone seals the compression rings against the cylinder wall.
    Who is right?
    a. A only
    b. B only
    c. Both A and B
    d. Neither A nor B

15. The ring gap:
    a. Allows for heat expansion of the ring
    b. Reduces blowby
    c. Improves oil control
    d. Reduces heat buildup in the ring

16. Rings are made of:
    a. Cast iron or steel
    b. Copper
    c. Aluminum
    d. Nickel

17. Connecting rods can be made of:
    a. Steel
    b. Aluminum
    c. Titanium
    d. All of the above

18. Connecting rods are usually made in a(n) _____ configuration.
    a. H-beam
    b. I-beam
    c. T-section
    d. U-section

19. To improve rod strength, some rods are:
    a. Shot-peened
    b. Plated
    c. Cam ground
    d. Drilled and relieved

20. Technician A says a short rod causes the piston to dwell longer at TDC than a long rod. Technician B says a long rod causes the piston to dwell longer at TDC.

    Who is right?
    a. A only
    b. B only
    c. Both A and B
    d. Neither A nor B

13

Engine Bearings

In the engines that you will work on, engine designers put bearings in almost all locations where there are moving parts. Correct selection, measuring, and replacing of bearings is one of the most important jobs of an automotive machinist.

Bearings are vitally important because they perform three major functions in an automotive engine: they reduce friction, support moving parts under load, and serve as a replaceable wear surface. This chapter describes bearing types and explains how bearing manufacturers select bearing materials and construct bearings to perform these functions.

BEARINGS AND FRICTION

Friction is the enemy of the automotive engine. Friction exists wherever parts rub against each other. It reduces the power produced by an engine wherever moving parts exist, since effort expended to overcome friction is subtracted from available power. Friction will also cause wear. If two metal parts continue to rub against each other, the metal will gradually wear away and the parts will loosen. The loose parts will pound against each other and eventually break or seize.

Lubrication is used to reduce friction. If there is an oil film between moving parts, wear is reduced or eliminated. An engine idling at 1,500 rpm could theoretically run forever without bearing wear. The engine is doing no work, so there is very little load on the parts. Sometimes lubrication is not enough. Even with plenty of oil between two moving parts, a heavy load on the engine can cause parts to touch. Two moving parts need a buffer or a bearing between them to reduce wear.

Consider an engine pulling a trailer up a steep grade. This engine is probably running at wide open throttle. Every intake stroke of the piston pulls in the maximum amount of mixture, and each power stroke gives the maximum push on the crankshaft. Even with proper lubrication, the rotating parts of the crankshaft and connecting rods may break through the oil film and touch, causing friction, heat, and wear.

A heavy load is not the only condition that causes parts to touch and wear to occur. When an engine is stopped, oil drains away from the internal parts and back to the oil pan. Some oil film may remain depending on how long the engine has been idle. However, during a cold start, before oil pressure has a chance to build, many parts are resting on their bearing surfaces. The greatest wear most engines ever experience is during a cold start.

Designers place bearings between any two rotating or moving parts of the engine to prevent those parts from rubbing directly against each other. When wear does eventually take place, it is usually an inexpensive bearing that is replaced instead of a more expensive crankshaft or connecting rod.

BEARING TYPES

There are two basic types of bearings: **plain** and **rolling element bearings**. A plain bearing is a solid, nonmoving bearing placed between two parts of the engine that work against each other. There are several types of plain bearings. Precision insert bearings are manufactured to exact measurements. They are made in half circles and inserted into crankcases and connecting rods, figure 13-1.

Sleeve bearings are also plain bearings. These bearings are made in a full circle and may be inserted in an engine block for a camshaft to turn in. Like insert bearings, they have a steel back for support and a special bearing material on the inside to reduce the effects of friction, figure 13-2.

Another type of plain bearing is the **bushing**, figure 13-3. A bushing is a full round sleeve or bearing. It is usually small and pressed into a hole and machined or reamed to fit a shaft. It is normally associated with slow rotation. The best example of a bushing in an engine is the pin bushing used with full-floating piston pins. It is made of bronze or is bronze lined and pressed into the small end of the rod.

A ball or roller bearing is a rolling element bearing. They have many individual balls or rollers that are held together by a cage. Ball bearings often have lubricant sealed inside them and need no regular service, figure 13-4. Roller bearings are usually not sealed, figure 13-5. Roller bearings and unsealed ball bearings require outside lubrication, but thrive on a much lower level of oil than plain bearings. Oil mist or splash is completely adequate.

The rolling friction of a ball or roller bearing is theoretically less than the sliding friction of a plain bearing. Less friction results in less wear. However, roller and ball bearings are much more costly to produce than plain bearings. Roller and ball bearings cannot take as much load and tend to be more fragile than plain bearings.

MAIN AND CONNECTING ROD BEARINGS

Designers have tried a number of different types of automotive main and connecting rod bearings. In the late 1940s and early 1950s, plain insert bearings, also called bearing shells, rapidly began replacing other types of main bearings, figure 13-6. Better lubrication technology as well as better bearing construction and materials made them practical.

Plain bearings simplify engine manufacture and repair. Since crankshaft and connecting rod bearings

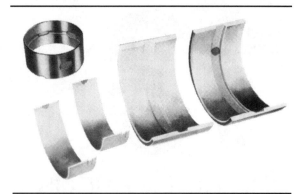

Figure 13-1. These plain bearings are used with the connecting rods and crankshaft. Each bearing has a matching insert to completely line a precision hole.

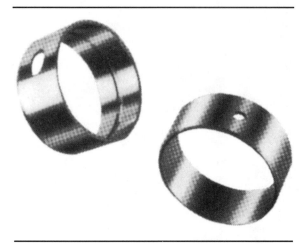

Figure 13-2. Sleeve bearings have a steel back for support and are lined with an alloy. They are circular.

Figure 13-3. Bushings are solid metal bearings without a steel backing.

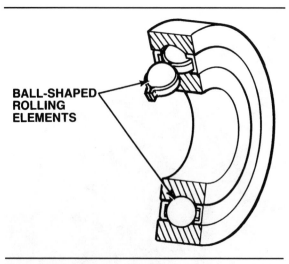

Figure 13-4. Ball bearings get their name from the shape of the rolling elements inside them.

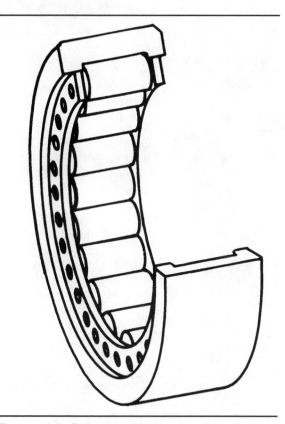

Figure 13-5. Roller bearings have rolling elements shaped like cylinders.

are made in split halves, you can easily remove and replace connecting rods or the crankshaft. They also allow a simple, strong, one-piece crankshaft. The crank is relatively easy to cast or forge and does not require an elaborate buildup by a technician. The two-piece plain bearings also require no babbitt pouring or tedious sizing and reaming.

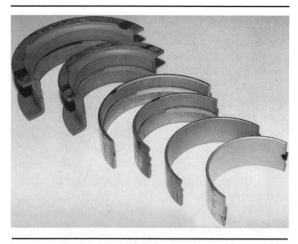

Figure 13-6. Manufacturers make main and rod bearings in an assortment of shapes and sizes.

Plain bearings are much more durable than poured babbitt and work well at high or low speeds. Although rolling element bearings have lower friction in theory, test results show that a properly lubricated plain bearing has less high-speed friction. Rolling element bearings have less initial friction, so they begin to roll more easily from a dead stop, but this is unimportant for an engine that runs no slower than idle speed. A plain bearing is also lighter and more compact for its load-carrying capacity than a rolling element bearing. For a high-volume manufacturer, the fact that plain bearings are also the cheapest type to produce is a major consideration.

Plain bearing construction
Main and connecting rod plain bearings are two-piece. Each end, where the two halves meet, is called the **parting face**, figure 13-7. Most bearing halves, or shells, do not have uniform thickness. The wall thickness of most bearings is largest at the center, or **bearing crown**, and tapers to a thinner measurement at each parting face, figure 13-8. The tapered wall

Plain Bearing: A bearing that slides or rubs on the part it supports.

Rolling Element Bearing: A bearing that rolls on the part it supports.

Bushing: A small, full-circle plain bearing.

Parting Face: The meeting points of the two halves of split plain bearings, such as main or connecting rod bearings.

Bearing Crown: The middle of a main or connecting rod bearing. This is its thickest point.

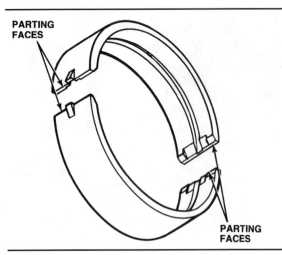

Figure 13-7. The two halves of a plain bearing meet at the parting faces.

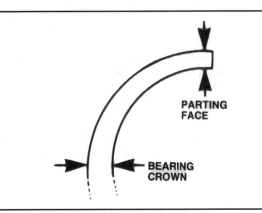

Figure 13-8. Bearing manufacturers refer to this bearing wall shape as an eccentric wall.

keeps bearing clearances close at the top and bottom of the bearing, which are the more highly loaded areas, and allows more oil flow at the sides of the bearing.

Main and rod bearings sometimes have beveled edges to allow room for the fillet on the crank journal, figure 13-9. Cranks in heavy-duty engines have larger fillets, so bearings for these engines need larger bevels.

Bearing retention

Main and rod bearings are held in place by a bearing cap, figure 13-10. To help you index the bearing in the block or rod, each bearing half has a small tang at one end which fits into a slot in the **bearing bore**, figure 13-11. Because the tangs are offset from each other, the edge of the tang bears against the lip of the other half of the bore. When you install the cap and tighten the nuts or bolts, the bearing is held firmly against the supporting metal.

Some engines have a locating dowel in the bearing bore, figure 13-12. The dowel fits into a hole in the bearing shell and prevents movement in any direction.

Figure 13-9. Beveled edges keep the bearing from riding on the crank fillet.

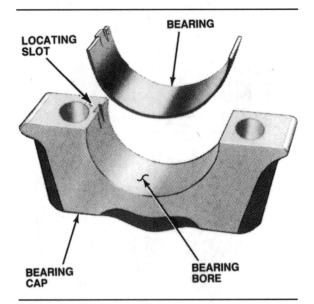

Figure 13-10. The bearing cap holds the bearing in place.

Oil grooves

As we explained in Chapter 11, rod and main bearings have an inlet hole for pressurized oil. Main bearings may also have an annular oil groove machined down the center of the bearing to help the oil flow around the entire bearing, figure 13-13. Usually only the upper bearing half has the groove because cutting the groove reduces bearing surface area and load-carrying ability.

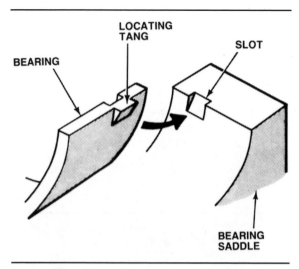

Figure 13-11. The tang and slot helps you index the bearing in its bore.

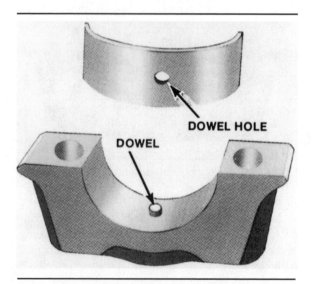

Figure 13-12. In some cases, a dowel in the bore helps index the bearing.

Some rear main bearings have an additional annular groove that functions as a drain, figure 13-14. Oil tends to collects here and the drain groove allows it to return to the oil pan, rather than put pressure on the rear main seal.

Thrust bearings
The thrust bearing is one of the crankshaft main bearings that is flanged, figure 13-15. The thrust bearing controls **crankshaft endplay** by preventing the crankshaft from sliding back and forth. The flanges rub on the crank at the sides of one of the main bearing journals, as we showed in Chapter 11.

To help assist the annular groove, some thrust bearings have a spreader groove to distribute the oil across the bearing surface. Some thrust bearings also have

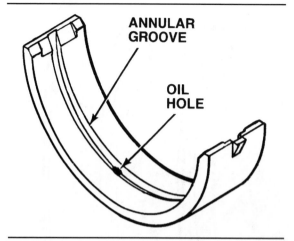

Figure 13-13. In many bearings, a groove down the middle improves oil flow around the main journal.

lateral oil grooves to help channel oil to the rear flange of the bearing. The thrust flange may contain small pockets to retain oil on the face.

Rather than put a flange on an existing main bearing, some engines use a thrust bearing insert, figure 13-16.

Plain bearing dimensions

An uninstalled bearing insert has a larger diameter than its saddle. The amount that it is greater is called the **bearing spread**, figure 13-17. Making the bearing spread dimension from 0.005 inch to 0.030 inch (0.125 to 0.750 mm) ensures that it will be tight against its bore. The bearing snaps into place when inserted correctly and will not fall out during handling.

The distance that the bearing's parting face is higher than the saddle is called **bearing crush**, figure 13-18. When you tighten the rod or main bearing cap, the bearing is actually "crushed" into the bore slightly. The crush makes sure the bearing is tight in its housing, so that it will not spin when the engine is running and so that it can shed heat to the saddle effectively.

Bearing Bore: The full circle machined surface that supports the back of a bearing. It may consist of a saddle and cap bolted together, as in rod or main bearings, or a bored hole, used for most camshafts.

Crankshaft Endplay: The end-to-end movement of the crankshaft in the bearings.

Bearing Spread: The distance the diameter of an uninstalled bearing insert is wider than the diameter of its saddle or web.

Bearing Crush: The distance that an installed bearing insert is higher than its saddle or web.

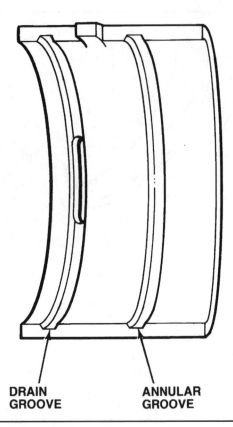

DRAIN GROOVE   **ANNULAR GROOVE**

Figure 13-14. Some rearmost main bearings have an additional drain groove to relieve oil pressure on the rear main seal.

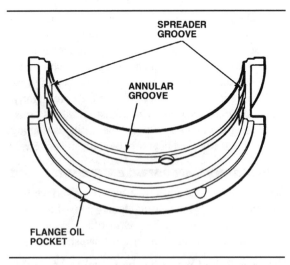

SPREADER GROOVE

ANNULAR GROOVE

FLANGE OIL POCKET

Figure 13-15. A typical thrust bearing has built-in flanges that help control crankshaft end play.

Bearings without enough crush show signs of movement in the saddle, and sometimes overheating, when the engine is torn down. Movement between the bearing and its bore will cause wear and looseness. Bearings used in heavy-duty applications have additional crush.

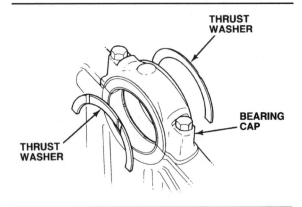

THRUST WASHER

BEARING CAP

THRUST WASHER

Figure 13-16. Thrust bearing washers are most common on European engines, while some Japanese manufacturers use them as well.

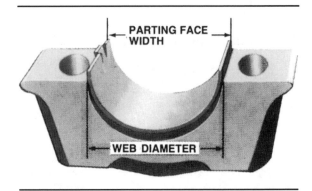

PARTING FACE WIDTH

WEB DIAMETER

Figure 13-17. Bearing spread allows the bearing to snap into place.

Because the tight fit provided by the crush is critical to maintaining good heat transfer from the bearing to the engine block, it is bad practice to oil or Loctite® main or rod bearings in place. This extra material between the bearing and saddle interferes with heat transfer.

To allow for the crush fit, plain bearing halves taper to a thinner dimension at their parting faces, figure 13-19. This **crush relief** keeps the bearing from mushrooming out and reducing clearance when you torque the bearing cap in place.

Bearing clearance

Bearing clearance is the gap or space between a bearing and its journal, figure 13-20. Its size depends on the type of bearing material and the shaft size. For example, a two-inch shaft using babbitt bearings may require 0.001 inch (0.025 mm) of oil clearance, while the same shaft using copper alloy bearings may require 0.002 inch (0.050 mm) of oil clearance. A multi-layered copper alloy bearing may require only 0.001 inch (0.025 mm) of oil clearance.

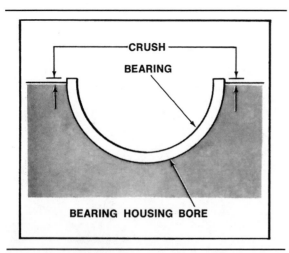

Figure 13-18. Bearing crush holds the bearing tightly in the bore and keeps it from spinning with the crank.

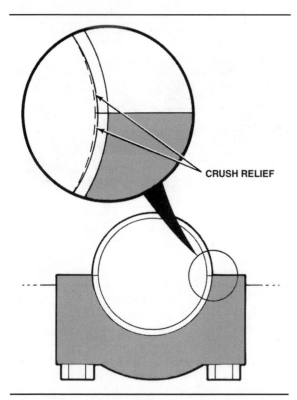

Figure 13-19. Bearings are thinner at their parting faces to provide crush relief.

Bearing clearance is usually specified as a range. For example, rod bearing clearance may be specified as 0.0015 inch to 0.003 inch (0.035 mm to 0.075 mm), with a maximum clearance in service of 0.0035 inch (0.90 mm).

Engine designers determine bearing clearance according to the diameter of the shaft and the bearing material. The greater the diameter of the shaft, the more bearing clearance it requires and there are two

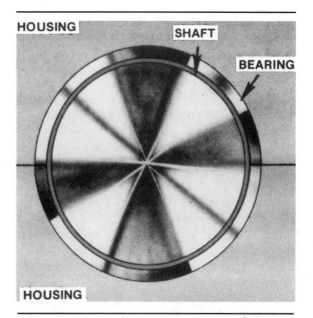

Figure 13-20. Bearing clearance is also called the oil clearance.

reasons for this. Specifications usually call for more main bearing clearance than rod bearing clearance. The reason is that main bearings are larger than rod bearings, and therefore need additional clearance.

Undersize bearings

On a modern engine, when the bearings wear, they are thrown away and replaced with new ones. If the crankshaft is not worn, you can replace the bearings with a standard size.

Often, when an engine is ready for a rebuild, the crank is worn and you must remove it from the engine and grind it to a standard undersize on a crank grinding machine. If the surface of the journal is still rough, out-of-round, or tapered, you grind it to the next available undersize. These sizes are usually 0.010 inch (0.25 mm), 0.020 inch (0.50 mm), and 0.030 inch (0.75 mm) under the standard size.

After the crank is reground, it is marked for the new bearing size it requires. For the crank to fit and operate efficiently, you must install the correct undersize bearings.

Bearings are commonly manufactured in the same 0.010 inch (0.25 mm), 0.020 inch (0.50 mm), and

Crush Relief: A thinning of the bearing at the parting face. This relief allows for crush in a bearing shell.

Bearing Clearance: The gap or space between a bearing and its shaft that fills with oil when the engine is running. Also called the oil clearance.

0.030 inch (0.75 mm) undersize amounts. An **undersize bearing** is really thicker than a standard bearing. The outside diameter of the bearing is the same; the inside diameter is undersize. The name undersize is not a description of the bearing, but of the shaft it is made to fit.

In some cases, the crankshaft will have wear, but not enough to need grinding. Micro-polishing can make it serviceable. For these cases, there are bearings available for some engines that are 0.001, 0.002, and 0.003 inch (0.025, 0.05, and 0.075 mm) undersize.

Oversize bearings

Sometimes a problem occurs with the main bearing bore when the factory machines it. To correct the bore, the factory may machine the saddle oversize. This allows them to use a block which would otherwise have to be scrapped.

The **oversize bearing** will be the standard size on the inside, but the outside will be larger. This makes no difference to engine operation. But when the bearings are changed, the engine rebuilder must be careful to spot the oversize or undersize bearing. In most cases, the bearing is marked for undersize or oversize, figure 13-21. But do not take the markings at face value; measure for yourself to determine the bearing size.

Oversize bearings are not normally made for connecting rods. Rod caps can be ground and the rod remachined to accept a standard size bearing.

CAMSHAFT BEARINGS

The camshaft is supported by several bearings. Pushrod engines use sleeve bearings, figure 13-22. Overhead-cam engines may have sleeve bearings or split bearing shells that resemble rod bearings, figure 13-23.

The load on cam bearings comes from the pressure generated by the valve springs every time a valve opens and closes. This pressure is fairly constant because of the consistent motion of the valves. Compared to main bearings, the cam bearing material is lightly loaded.

Camshaft bearings normally last at least as long as the main bearings. In some cases, a set of cam bearings will outlast two sets of main bearings. Because camshaft bearings seldom cause trouble, they are sometimes overlooked when an engine is torn down for repairs.

Camshaft bearings have oil inlet holes to allow oil to reach the camshaft journals. Camshaft sleeve bearings may also have additional outlet holes to direct oil to the valve train, figure 13-24.

Remember, cam bearings can cause trouble if they are worn enough to allow too much oil flow. This is because the same oil passageway feeds both the cam bearings and the main bearings. If the cam bearings

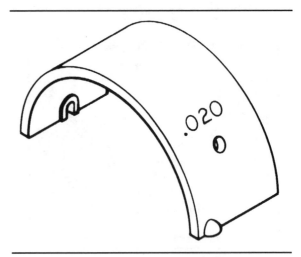

Figure 13-21. The bearing may be marked for an oversize or undersize dimension.

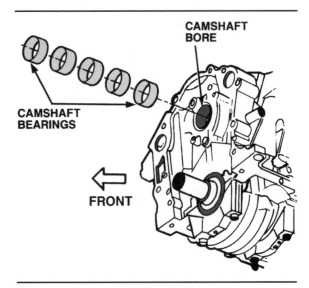

Figure 13-22. Domestic pushrod engines support the camshaft with sleeve bearings.

bleed off excessive quantities of oil, the crankshaft may be starved for oil because of a drop in oil pressure. You should always replace cam bearings when rebuilding an engine.

Engines Without Camshaft Bearings

In some overhead-camshaft aluminum cylinder heads, the camshaft runs directly on the metal of the head without any insert bearings, figure 13-25. Such heads are lightweight and inexpensive to produce and have proven very reliable. Theoretically, the camshaft runs on the oil film and an insert bearing is not absolutely necessary. Designers specify an aluminum alloy with a high silicon content for the head. This material reduces the chance of scuffing if a momentary

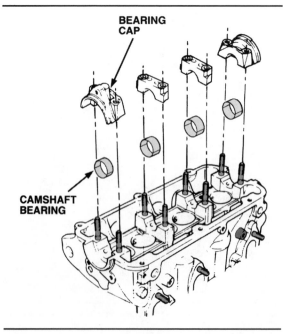

Figure 13-23. Some European overhead-camshaft engines support the camshaft with split bearing inserts.

lubrication breakdown allows the camshaft to contact the head.

Designs without cam bearings are common in most Japanese cars and motorcycles. Some European engines have cam bearing inserts, but most of the newer designs allow the cams to run in the aluminum heads without bearings.

PLAIN BEARING OPERATION

Plain bearings are theoretically unlimited in speed and load-carrying potential. However, plain bearings require adequate lubrication and oil pressure from a separate lubrication system. The supply of lubrication becomes more critical as engine speed and load rise. A rotating part with plain bearings actually rides on a film of oil while it is in motion. To understand how plain bearings work, we must explain three terms:

- Fluid friction
- Oil clearance
- Oil film theory.

Undersize Bearing: A bearing made thicker than standard. It has a smaller inside diameter to fit an undersize crankshaft.

Oversize Bearing: A bearing made thicker than standard. It has a larger outside diameter to fit an oversize bearing bore.

■ Babbitt Main and Rod Bearings

Babbitt, a mixture of tin, lead, and antimony, was the earliest practical bearing material. Isaac Babbitt invented it in 1839. It was called tin-base babbitt because 89 percent of the material was tin. When babbitt was invented, there were no automobiles, but it was useful for stationary power plants, steam engines, and many other places where shafts turned in bearings.

The lead-based babbitt was developed next and used by railroads for wheel bearings. From this use, lead-based babbitt became very popular for low-speed automotive engines. It was used for many years in connecting rods and main bearings.

Mechanics of the day poured the molten babbitt bearing material into the block and connecting rods. When the bearing material cooled, they scraped it by hand (and later line-bored it) to produce the correct bearing clearances for the crank and rods. To adjust the fit of the bearing, shims were used between the cap and the rod. Fitting bearings was an art requiring skill and experience. If the bearings were too tight, they overheated; if they were too loose, they pounded out. High rpm could cause the babbitt to smear and heavy loads could pound it out.

In the early years of automotive use, filing the caps to decrease clearance in worn big ends was considered owner maintenance. Many trips in Ford's Model T were interrupted along the way while the owner stopped on the side of the road to perform this task.

This Ford Model A connecting rod is ready to install. The babbitt has been poured and shims fitted between the cap and rod. Using a special fixture, it was bored to size so it fits the crank journal with no more than 0.001- to 0.0015-inch clearance.

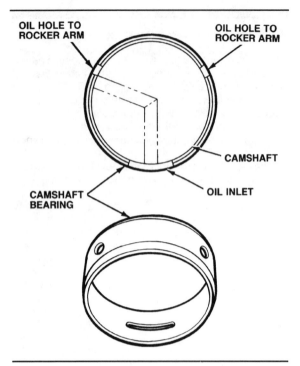

Figure 13-24. Cam bearings in pushrod engines may also have oil outlet holes so lubrication can reach the rocker arms and other parts of the valve train.

Fluid Friction

Oil reduces friction between moving parts because of a property called **fluid friction**. Like other physical objects, oil is composed of molecules and oil molecules have two important fluid friction characteristics:

- Oil molecules cling to metal surfaces more persistently than they stick to one another
- Oil molecules slide against one another freely.

Thus, when an oil film flows between two metal surfaces, the top layer of molecules clings to the top metal surface and the bottom layer of oil molecules clings to the lower metal surface. When one of the metal surfaces moves, the internal layer of oil molecules, figure 13-26, slides with much less friction than if the two metal surfaces were sliding against each other.

Oil Clearance

To employ the advantages of oil's fluid friction characteristics, manufacturers allow a space for oil between moving engine parts. The oil clearance is simply a different name for the bearing clearance, which we described earlier. The oil clearance space receives a constant flow of oil from the engine's lubrication system.

The amount of oil pressure, the bearing clearance, and bearing life are related. The oil clearance regulates the oil pressure inside the bearing as well as the

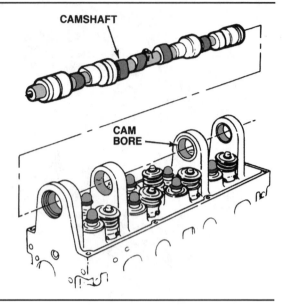

Figure 13-25. This Ford 2.3-liter OHC aluminum cylinder head has no cam bearings. The oil film in the cam bores supports the camshaft.

rate of flow of oil through the bearing. A bearing too close to the shaft will not have sufficient oil clearance. Oil pressure will be high, but oil flow through the bearing will be too slow. The result is higher bearing temperatures and increased bearing wear. If the bearing clearance is excessive, oil will pass through the bearing more quickly, and oil pressure will be lower. Low oil pressure can increase engine wear and noise. Oil thrown off the spinning crankshaft also increases when the clearance is excessive. This overloads the rings with oil, causing high oil consumption.

The larger the journal, the larger the oil clearance it needs. As we learned in Chapter 11, a larger journal has a higher surface speed for any given rpm than a small journal. The higher speed creates more heat, so a larger clearance is needed to allow a greater volume of oil, which is necessary to properly cool the part.

Oil Film Theory

When the engine is not running and oil pressure is zero, a shaft and its bearing are touching, figure 13-27. When the engine starts, rotation causes the shaft to climb up the bearing. This is why cold starts cause the most engine wear. But this period is very brief because the oil pump begins to supply oil to the bearing clearance rapidly. The combination of shaft rotation and oil pressure causes a wedge of oil to form between the shaft and bearing, figure 13-28. Since oil, like other fluids, does not compress, the shaft climbs the oil wedge, lifting the shaft from the bearing surface. Oil flows under the shaft and completely

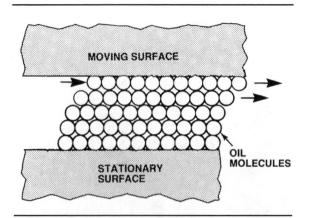

Figure 13-26. Oil molecules cling to metal surfaces, but slide against one another with little friction.

surrounds it. This event takes much more time to describe than to occur. It happens almost instantly whenever the engine is started.

When an engine is running under ideal conditions, the oil film keeps the shaft centered in the bearing. Friction and wear are insignificant. To a point, the oil film is self-perpetuating. The more pressure the shaft exerts on the oil, the more the oil resists by building a wedge.

Larger journals have less curvature and do not create an oil wedge as readily as smaller bearing journals. This is another reason why larger journals need more oil clearance: they need a greater supply or volume of oil to support the part and keep it centered in the bearing.

Boundary lubrication

Unfortunately, ideal conditions are rare in a running vehicle. The load on rotating engine parts changes often, as do the parts' speed, due to the power impulses from combustion and varying driving conditions. Although the oil does not compress, it steadily weeps out around the oil clearance between the shaft and bearing. Load sometimes squeezes the oil film very thin. In most cases, the oil film is thick enough to keep the surfaces from seizing, but not thick enough to keep them completely separated. This state is called **boundary lubrication** and some wear occurs. When the engine is under a heavy load, even boundary lubrication can fail. The piston exerts maximum force on the rod and may cause the bearing and shaft to touch momentarily, resulting in great wear or engine damage.

PLAIN BEARING MATERIALS

Using dissimilar materials where parts touch or rub reduces friction and wear. The bearings could be made of iron, but producing an entire bronze crankshaft would be expensive and not be strong enough.

For this reason, the moving parts are made of the material that provides the necessary strength, such the high-grade steel used in a crank, and the bearings are made of various alloys to provide a dissimilar surface. Over the years, engineers have identified other characteristics or qualities that a bearing must have.

■ Roller Main and Rod Bearings

Before World War II, roller bearings were very popular for high-speed engines. Some racing engines used roller bearings up to the 1960s. With these bearings, many cylindrical rollers are housed in a cage. The rod or main journal turns inside this assembly. These rolling element bearings provided low friction and had modest lubrication needs, so they were reliable at high speeds with the oils available at the time.

However, since rolling element bearings do not split, they dictate the use of a multi-piece crankshaft. Each journal is either pressed or threaded into each crankshaft cheek. Rebuilding these cranks is difficult because assembly must be very precise. Manufacturing these cranks and bearings is also expensive.

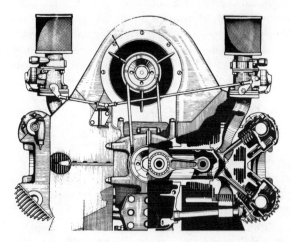

This 1954 Porsche-type 550 racing engine uses a roller-bearing crankshaft. Look closely at the big end of the rod; you can pick out the individual rollers.

Fluid Friction: The friction between the layers of molecules in a fluid.

Boundary Lubrication: Lubrication protection afforded by the remaining oil film when load has forced a part, such as a crank journal, through the oil wedge.

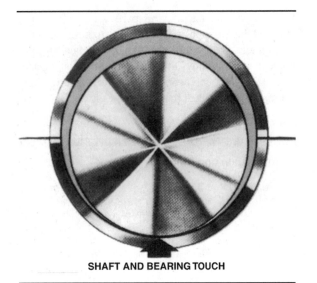

Figure 13-27. When the engine is stopped, the shaft and bearings touch.

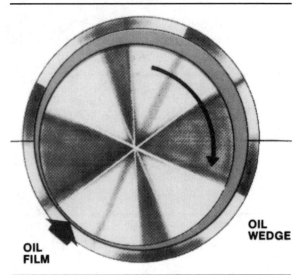

Figure 13-28. When the engine starts to turn, the shaft rolls up on a wedge of oil.

Qualities

Research and testing of bearing materials has gone on for many years in an attempt to improve the various qualities needed in bearings. A bearing must have **compatibility**. This means it must be able to slide on the journal without excessive friction or wear. Dissimilar materials have the best compatibility. For example, steel rubbing on steel does not work well because steel is not compatible with steel. In fact, the two parts would get so hot that they would weld themselves together. Copper and steel or lead and steel have good compatibility.

Another necessary quality of bearings is **fatigue strength** or the ability to carry a load. If a bearing cannot work well under a heavy load, the lining material will crack and break up or it may be wiped aside by the pressure of the shaft.

Another desirable bearing quality is **conformability**. Many crankshafts are not perfectly round or straight. They may have minute bumps or waves on their surfaces. A bearing lining material with good conformability is soft enough to change its shape slightly to conform to the shape of the crankshaft, figure 13-29. However, a lining material with too much conformability may be so soft that it cannot carry the load. Therefore, conformability can pose a problem for the bearing designer.

A necessary bearing trait is **embeddability**. Every engine has hard carbon, dirt, and small metal particles circulating with the oil. When these particles get between the bearing and the shaft, they are pushed into the bearing lining. If the particles protrude above the surface of the lining, they will scratch the shaft and

may eventually ruin it. Embeddability allows the particles to sink into the bearing lining material until they are level with the surface and can do no harm, figure 13-30. Embeddability is a very important trait because, according to bearing experts, foreign particles cause at least of half of all bearing-related failures.

There are many other less important qualities that a bearing must have. It must be able to operate at high temperatures. It must not be so hard that it will scratch or score a shaft. Low-quality bearings may use materials that are very hard on shafts.

Bearings must also be able to operate against shafts that have not been hardened or heat treated. Bearing lining must not erode or break down from acids that are formed in the engine. It must be able to conduct heat from the shaft to prevent overheating. And finally, a good bearing material must be able to be bonded to a backing shell so it can be used as a replaceable insert.

Main and Rod Bearing Materials

An automotive bearing shell has a backing made of mild steel, which provides a strong foundation. Bearing manufacturers bond other metals to one side of this backing, which becomes the inside diameter of the bearing, figure 13-31. Generally, the more layers the bearing has, the better it is for heavy-duty service — and the more expensive it is.

The first layer bearing designers add to the backing is the lining, the major functional part of the bearing. Designers usually choose babbitt, copper-lead alloy, or aluminum alloy for the lining.

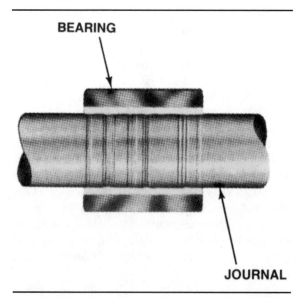

Figure 13-29. The ability of the bearing metal to change shape slightly to match irregularities in the journal is called conformability.

Figure 13-31. Modern bearings are made in layers.

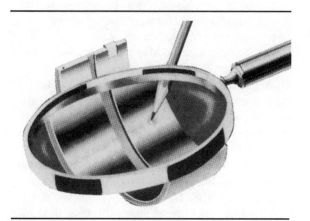

Figure 13-30. Embeddability allows a particle to sink into the bearing material.

Babbitt bearings

Babbitt bearings are two-layer: a tin- or lead-based babbitt on the steel backing. It makes a good, all-around bearing material when loads and temperatures are moderate. It operates well under such handicaps as poor lubrication and misalignment. It also has good conformability and embeddability. Most cam bearings, because they are a lighter-duty bearing, are two-layer babbitt bearings.

Multi-layer bearings

A bearing with more than two layers is considered a multi-layer bearing. For multi-layer, automotive crank, and rod bearings, designers first bond a bearing lining of copper-lead alloy or aluminum alloy to the steel backing. Engineers developed these alloys to stand up to higher speeds, temperatures, and loads

■ Full-Floating Rod Bearings

Some engines used a full-floating rod bearing. There were no tangs on the bearing insert and no notches in the rod. Also, the inserts had a negative bearing spread, so they did not snap into place. As the crankshaft rotated, the bearing insert was free to rotate with the shaft or remain stationary with the big end of the rod. The floating design used a single bearing insert for each pair of connecting rods. Wear rates were difficult to predict because the rotation of the insert was variable. The insert might rotate with one rod, with the other rod, or with the shaft.

Compatibility: The ability of a bearing lining to allow friction without excessive wear.

Fatigue Strength: The ability of a bearing to withstand the loads placed on it.

Conformability: The ability of a bearing lining to conform to irregularities in a bearing journal.

Embeddability: The ability of a bearing to absorb particle contamination into the lining of the bearing.

Overplate: A material added to the surface of the bearing lining to improve the conformability and embeddability. Only the harder, heavy-duty bearing alloys need overplates.

than babbitt bearings could tolerate. For moderate-duty bearings, this lining is followed by a flash tin plate over the entire surface of the bearing. The flash plate prevents the bearing from corroding in the engine or while it sits on the stock shelf.

Heavy-duty bearings also begin with a steel backing and a lining of copper-lead alloy or aluminum alloy. These bearings use harder alloys that provide longer life and greater resistance to fatigue, but sacrifice some measure of conformability and embeddability. To regain some of these last two qualities, bearing makers add a lead-tin-copper **overplate** onto the surface of the lining.

On the heavy-duty copper-lead alloy bearings, manufacturers usually add a nickel barrier plate between the lining and the overplate to prevent the overplate from reacting chemically with the lining material. Finally, the heavy-duty bearings also receive a flash tin plate to resist corrosion.

Aluminum alloy bearings

The aluminum alloy bearing is a combination of tin, copper, nickel, and aluminum and is cast in one piece. The finished bearing is covered with a thin plate of tin to help during the break-in period. Aluminum has a tendency to expand at high temperatures, so these bearings have a heavy wall to resist expansion. You cannot use this bearing as a replacement for a thin wall bearing. Vehicle manufacturers sometimes install aluminum bearings with hardened crankshafts for extended bearing life.

SUMMARY

Bearings perform three major functions in an automotive engine: they reduce friction, support moving parts under load, and serve as a replaceable wear surface. There are two types of bearings: plain and rolling element bearings. A plain bearing is a solid, nonmoving bearing. There are several types of plain bearings: precision insert bearings, sleeve bearings, and bushings.

A ball or roller bearing is a rolling element bearing. They have many individual balls or rollers in a cage. The rolling friction of a ball or roller bearing is theoretically less than the sliding friction of a plain bearing.

Main and connecting rod plain bearings are two-piece. Each end, where the two halves meet, is called the parting face. Most plain bearings have an eccentric wall. Main and rod bearings are held in place by a bearing cap. Each bearing half has a small tang at one end which fits into a slot in the bearing bore. The distance that the parting face of a bearing shell is greater than the diameter of its saddle is called bearing

spread. The distance that the bearing's parting face is higher than the saddle is called bearing crush.

Rod and main bearings have an inlet hole for pressurized oil. The upper main bearings may also have an annular oil groove down the center of the bearing to help the oil flow around the entire bearing.

The thrust bearing is a crankshaft main bearing that is flanged and prevents the crankshaft from sliding back and forth. Rather than put a flange on an existing main bearing, some designers use a thrust bearing insert.

Bearing clearance, also called oil clearance, is the gap or space between a bearing and its journal. Designers determine bearing clearance according to the diameter of the shaft and the bearing material. The greater the diameter of the shaft, the more bearing clearance it requires.

An undersize bearing is thicker than a standard bearing. It is designed to fit a crank that has been ground undersize. An oversize bearing is the standard size on the inside, but the outside is larger. Factories sometime fit these bearings to a block they would otherwise have to scrap because casting or machining flaws create an oversize bearing bore.

Pushrod engines use sleeve cam bearings. Some overhead-cam engines have sleeve bearings or split bearing shells that resemble rod bearings. The load on cam bearings is not as severe as the load on main bearings and a set of cam bearings may outlast the main bearings. In some overhead camshaft aluminum cylinder heads, the camshaft runs directly on the metal of the head without any insert bearings. Theoretically, the camshaft runs on the oil film and a bearing insert is not necessary.

Plain bearings work on three principles: fluid friction, oil clearance, and oil wedge. A shaft climbs the oil film to create the wedge, which lifts the shaft from the bearing surface. Oil flows under the shaft and completely surrounds it, keeping the part centered.

Plain bearing materials need four primary qualities: compatibility, fatigue strength, conformability, and embeddability. The foundation of an automotive bearing shell is a backing made of mild steel. Bearing manufacturers bond one to four other metals to this backing.

Overplate: A material added to the surface of the bearing lining to improve the conformability and embeddability. Only the harder, heavy-duty bearing alloys need overplates.

Review Questions

Choose the single most correct answer.
Compare your answers with the correct answers on Page 328.

1. There are two types of bearings, rolling element bearings and:
   a. Ball bearings
   b. Roller bearings
   c. Plain bearings
   d. None of the above

2. A ball bearing is a _____ bearing.
   a. Rolling element
   b. Plain
   c. Frictionless
   d. None of the above

3. Which of the terms below does NOT apply to engine bearings:
   a. Embeddability
   b. Roundability
   c. Conformability
   d. Compatibility

4. The bearings that support the crankshaft in the block are called:
   a. Main bearings
   b. Web bearings
   c. Support bearings
   d. Connector bearings

5. Technician A says that bearings support moving parts under load.
   Technician B says that bearings serve as a replaceable wear surface.
   Who is right?
   a. A only
   b. B only
   c. Both A and B
   d. Neither A nor B

6. The gap between the bearing and the shaft or journal is called the:
   a. Bearing clearance
   b. Oil clearance
   c. Air gap
   d. Both a and b

7. Worn camshaft bearings can:
   a. Lower oil pressure
   b. Raise the engine temperature
   c. Lower the engine temperature
   d. Raise oil pressure

8. When a bearing insert is installed in the block, the amount that the bearing's parting face is higher than its saddle is called:
   a. The float
   b. Bearing height
   c. Bearing crush
   d. Bearing level

9. A material used in bearings is called:
   a. Alloy
   b. Babbitt
   c. Layers
   d. Babbcock

10. Most bearing wear occurs when the engine is:
    a. Started cold
    b. Started hot
    c. Shut off
    d. Idling

11. Technician A says that main bearings have elliptical walls.
    Technician B says that main bearings have concentric walls.
    Who is right?
    a. A only
    b. B only
    c. Both A and B
    d. Neither A nor B

12. The greater the diameter of a shaft:
    a. The greater the oil pressure
    b. The lower the oil pressure
    c. The thicker the bearing required
    d. The more bearing clearance required

13. Main bearings are usually:
    a. The same size as the rod bearings
    b. Smaller than the rod bearings
    c. Harder than the rod bearings
    d. Larger than the rod bearings

14. Plain bearings are always made of a material _____ than the shaft on which they ride.
    a. Lighter
    b. Stronger
    c. Harder
    d. Different

15. Bearings are made in three common undersizes for most engines:
    a. 0.010, 0.020, 0.030
    b. 0.015, 0.025, 0.035
    c. 0.001, 0.002, 0.003
    d. 0.100, 0.200, 0.300

16. Technician A says camshafts require special bearings that can withstand incredibly high loads.
    Technician B says some camshafts run directly on aluminum heads without bearing inserts.
    Who is right?
    a. A only
    b. B only
    c. Both A and B
    d. Neither A nor B

17. Technician A says the theory of fluid friction explains why oil reduces friction between moving parts.
    Technician B says oil molecules cling to metal surfaces more persistently than they stick to each other.
    Who is right?
    a. A only
    b. B only
    c. Both A and B
    d. Neither A nor B

18. Technician A says that the tang on a main bearing and slot in the bore keep the bearing from spinning in a running engine.
    Technician B says bearing crush keeps a main bearing from spinning in a running engine.
    Who is right?
    a. A only
    b. B only
    c. Both A and B
    d. Neither A nor B

19. Technician A says that the foundation of most modern bearings is a babbitt backing.
    Technician B says that the foundation of most modern bearings is a mild steel backing.
    Who is right?
    a. A only
    b. B only
    c. Both A and B
    d. Neither A nor B

14

Gaskets, Seals, Sealants, Fasteners, and Engine Mounts

By reading all the previous chapters of this book, you learned the parts in the engine and how they work. The various parts that seal the engine together don't seem as significant, but they are also vitally important. The engine could not work without them. As an automotive technician, you need to understand the different kinds of seals and fasteners, and to know which ones to use for different purposes.

Gaskets, seals, and sealants are the actual compounds used in sealing engines. Adhesives hold gaskets and seals in place, cleaners prepare sealing surfaces so they form a good seal, and fasteners hold sealing surfaces together. Lubricants are sometimes needed to aid fastener installation. You must use and install all these parts and products correctly for proper engine sealing.

Engine mounts hold the engine in place and protect the rest of the car from engine vibration. They must be in good repair to do their job.

GASKETS

A gasket is used between two pieces in a stationary joint, figure 14-1, to prevent leaks. Such joints are found between the cylinder head and the block, the oil pan and the block, the intake manifold and the heads, and the exhaust manifold and heads. Smaller gaskets are used at the front cover, fuel pump, water pump, thermostat elbow, and many other places.

If two pieces fit together so perfectly that air or liquids cannot pass between them, a gasket is not necessary. A gasketless joint is difficult to make and may leak in service, so there are few gasketless joints in an automobile engine.

The purpose of a gasket, then, is to fill in the spaces between two parts. A gasket must be soft enough to fill minute grooves and holes when the joint is bolted together, but strong enough not to tear or break.

Sufficient clamping force helps prevent leaks, but the clamping force must not be so great that it distorts the gasket. Correct torque is especially important for modern, light-weight gaskets. Follow manufacturers' torque specifications.

Gasket Types

Gasket designs differ according to their intended use. High-heat, high-pressure applications, such as the cylinder head, require strong gaskets, typically with a steel core. Where fluid leakage is the main concern, as at the oil pan, a rubber gasket is most effective.

Head gaskets

The gasket between the cylinder head and engine block must seal completely to contain combustion

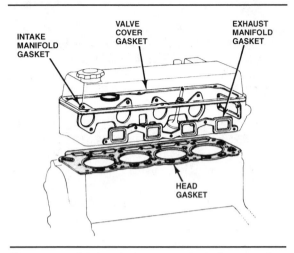

Figure 14-1. Gaskets help prevent leaks between facing surfaces. They are used in stationary joints.

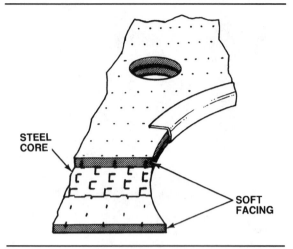

Figure 14-2. A typical head gasket has a steel core, a heat-proof, compressible facing, and metal encircling the combustion chambers.

force, and it must not burn when the air-fuel mixture ignites in the combustion chamber. The head gasket also must seal coolant passages between the head and block. In many modern engines, the block is cast iron and the head is aluminum, and the two metals expand and contract at different rates. This difference causes the head and block surfaces to slide past each other as a bi-metal engine warms up. However, the head gasket must stay in place and continue to seal as conditions change. Internal friction and wear become a problem in the head gaskets as the two sealing surfaces are stretched in different directions by the different metals.

Depending on the gasket design, you may need to retorque the head bolts because running the engine to operating temperature causes the gasket to set and relax. Using retorque gaskets usually means retorquing a third time after 300-500 miles (480-800 kilometers) of use. Read the gasket instructions to see if retorquing is required. Many modern gasket designs eliminate retorquing, but your choice is limited by what is specified for a particular engine.

Sandwich gaskets

Asbestos used to be considered a good material to use in high-temperature applications because it is compressible and non-flammable. But in the 1970s, industries became aware of health hazards related to exposure to asbestos and began eliminating it from their products. Graphite or composite materials including rubber-fiber content have become more common choices.

Today's typical aftermarket **sandwich-type head gasket** has a steel core for strength and rigidity. Both sides of the steel are covered with a soft, heat-resistant material, such as graphite or a rubber composite, figure 14-2. The softness allows compression for good

sealing and tolerates the shearing motion in bi-metal engines. Steel or another metal encircles the combustion chamber holes, to withstand heat and pressure.

Some gasket manufacturers coat the sealing surfaces with silicone, Teflon®, or another non-stick material to allow sliding motion between the gasket and head or block, figure 14-3, reducing internal friction and damage. Some gasket surfaces have a rubber bead drawn along crucial sealing points, figure 14-4. The bead increases and distributes the clamping force along the gasket when the head is torqued down, figure 14-5.

Multilayer steel gaskets

Some head gaskets are made of layers of steel, with no soft, compressible materials. **Multilayer steel head gaskets** are strong and durable, but the facing surfaces of the block and head must be extremely smooth for the gasket to seal well. In carefully controlled engine manufacturing plants, the surfaces can be made smooth enough, but most machine shops cannot provide that finish quality. Therefore, these gaskets are generally original equipment. The Ford OHC 4.6-liter V8 has a three-layer steel head gasket, for example.

Intake and exhaust manifold gaskets

The intake manifold and exhaust manifold have surfaces that mate to the cylinder head. These joints must be sealed tightly for proper engine performance. If the intake manifold-to-cylinder head joining allows air to be sucked in, the air-fuel mixture is leaner. At the least, the engine idles roughly and does not provide good acceleration. At the worst, detonation occurs. Leaks at the exhaust manifold increase exhaust turbulence and backpressure, detracting from engine performance.

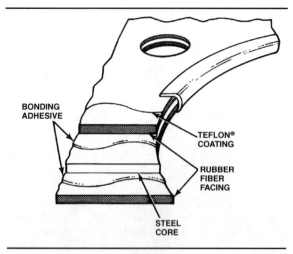

Figure 14-3. A non-stick finish between the gasket and facing surfaces allows some movement between head and block in a bi-metal engine, as the two metals expand and contract at different rates.

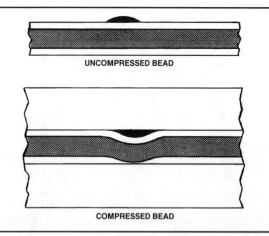

Figure 14-5. When the rubber bead is compressed between the head and block, it provides more sealing force against the gasket.

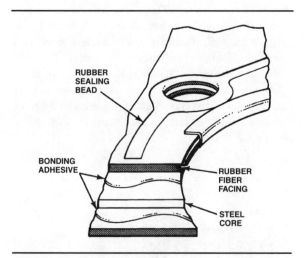

Figure 14-4. A rubber bead molded along the gasket surface adds extra sealing force.

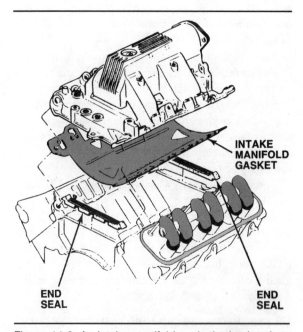

Figure 14-6. An intake manifold seals the intake air or air-fuel mixture. In a V engine, it usually also requires end strip seals.

The intake manifold gasket seals the air-fuel mixture or intake air and, depending on the engine design, may also seal coolant and oil passages. The intake manifold in a V engine generally sits in the "V" and requires strip, or end, seals between the end of the manifold and the engine block, figure 14-6. These seals are usually made of cork and rubber particles or of molded rubber. Replace them at the same time you replace the intake manifold gasket.

An exhaust manifold on an engine that has never been disassembled may seal directly to the cylinder head with no gasket. Before the manifold is used, its surface is smooth. However, if it has gone through many heating and cooling cycles and then been removed from the engine, its surface is likely to be distorted. When rebuilding an engine, machinists

usually choose to install an exhaust manifold gasket, figure 14-7, even if one was not part of the original equipment.

Sandwich-Type Head Gasket: A head gasket made in layers, typically a center layer of steel and surface layers of heat-resistant, compressible material.

Multilayer Steel Head Gasket: A head gasket made of layers of steel.

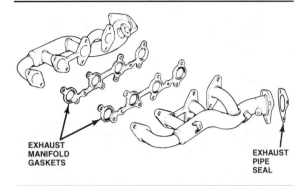

Figure 14-7. An exhaust manifold gasket seals the exhaust manifold to the cylinder head. The exhaust manifold is a high-temperature application.

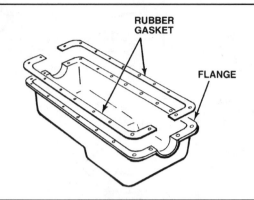

Figure 14-8. An oil pan gasket seals against fluid leaks. A flange on the pan positions the gasket.

Exhaust manifold gaskets vary in design. It is common to have a soft surface facing the engine and a steel surface against the manifold. The steel allows the manifold to move against the gasket surface when it expands as it heats and contracts as it cools. This is similar to the movement of the manifold against the engine surface when the engine is new and there is no gasket. Another design is to use steel and ceramic. Ceramic forms a good seal against a hot engine. Many heavy-duty engines have an individual shim gasket at each exhaust port. The shim gasket usually has a single embossed bead to help it seal, but in some cases there may be two beads for extra sealing.

Some in-line engines have the intake and exhaust manifolds on the same side. Such an application may require a single gasket for both manifolds. The gasket will usually have a steel core with soft facing on the engine side of the exhaust ports and both sides of the intake ports. Installing these gaskets so they form a good seal calls for special care and attention to the gasket manufacturer's instructions.

Oil pan gaskets and valve and cam cover gaskets
The gaskets that seal the oil pan to the bottom of the engine block and the valve or cam cover to the top of the cylinder head have similar requirements in that they must seal against hot, thin engine oil. The valve or cam cover is on top of the engine and usually has few bolts at low torque to hold it in place. The oil pan, on the other hand, is at the bottom of the engine and must hold the weight of the crankcase oil, so it usually has more and larger bolts holding it in place, closer together and at higher torque than the valve or cam cover has. It is important to torque bolts correctly to avoid distorting the flange or gasket.

The gaskets in these applications are usually either cork and rubber or just synthetic rubber. Cork/rubber gaskets are relatively inexpensive and provide good compressibility. The cork makes the gasket more rigid,

which makes it easier to install, but the rubber allows some flexibility. Rubber gaskets, figure 14-8, are more expensive and more difficult to install but they seal better than cork/rubber gaskets, because they deform instead of compressing. That means that when the pan pushes down on the gasket, it does not crush but it pushes back against the pan, trying to return to its original shape. Rubber's **resilience** — or tendency to return to its original shape — makes it a more effective seal.

Some valve or cam covers, or oil pans, have a molded rubber gasket permanently bonded to them. In this case, you must replace the entire cover or pan to replace the gasket.

Other gaskets
Other engine gasket applications include the:

• Timing cover
• Air cleaner
• Carburetor
• Choke tube
• EGR valve
• Water pump
• Fuel pump.

Engineers design gaskets to suit the requirements of each installation. For example, a carburetor mounting gasket is thick, often with ferruled bolt holes, to shield the carburetor from the heat of the engine. An EGR valve gasket may include metering orifices that help regulate exhaust flow.

SEALS

A seal prevents leaks past a moving part. When a rotating shaft protrudes from a piece of machinery, a seal is usually necessary, figure 14-9. A crankshaft extends from an engine at both the front and the rear; seals are used at both ends to keep oil in and dirt out. Seals are also used on other parts of the engine and accessories such as the water pump and power steering pump.

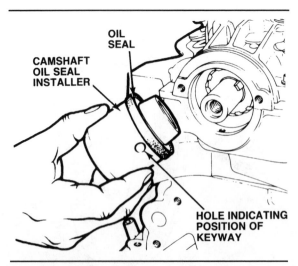

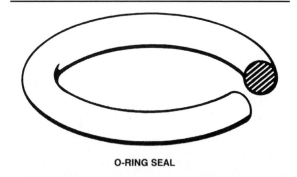

Figure 14-10. The cross section of an O-ring seal is usually shaped like an "O".

Figure 14-9. Seals are needed where shafts extend from the engine to keep fluid from leaking at these points.

Synthetic Rubber Seals

The synthetic material used for most rubber seals is neoprene. This material is not as brittle as pure rubber, and it withstands more heat and chemical attack than pure rubber.

O-ring seal

The neoprene O-ring seal is round with a circular cross section, figure 14-10. An O-ring creates a seal when it is compressed between two pieces of metal. The O-ring is fitted into a groove which is not quite as deep as the ring, squeezing the ring in its groove, thus forming a tight seal. Some valve stems have O-ring seals, which we covered in Chapter 8.

The O-ring is not used where rotational force would be placed on it, because the seal would not seat properly. The O-ring also tends to roll if it is subjected to much **axial movement**, that is, movement along the length of the shaft. For example, if an O-ring were used between the land of a piston and the inside wall of a cylinder, the piston could not be allowed to move very far inside the cylinder. If the piston moved too far, it would damage the O-ring seal.

Square-cut seal

The square-cut, or lathe-cut, neoprene seal is circular with a square cross section, figure 14-11. It is used between metal pieces subjected to slight axial movement, but not where rotational force might act on it. Rotation would tend to pull the seal away from its seat, like the O-ring seal.

Hydraulic pressure applied to a piston, such as in a disc brake caliper, moves it through the cylinder fast enough to drag and bend the inside of a square-cut seal pressed against the cylinder wall, figure 14-12. The portion of the seal pressed against the cylinder

wall does not move as far or as fast as the piston. When hydraulic pressure against the piston is released, the bent edge of the square-cut seal moves back to its original position. This helps draw the piston back in the cylinder.

A square-cut seal responds to axial movement better than an O-ring. It will not roll in its groove and is not as subject to damage.

Lip seal

The lip seal is circular like the O-ring and the square-cut seals. A lip seal is always installed with the lip facing the source of hydraulic pressure, figure 14-13.

■ When an O-Ring Is Not an O-Ring

O-rings are commonly used for sealing between moving or non-moving shafts and housings. They are usually made of rubber and designed so they are squeezed when installed, thus preventing leakage.

In racing engines, cylinder heads are sometimes O-ringed around the combustion chamber. But this is an entirely different kind of O-ring. A small groove is precisely machined into the cylinder head around the combustion chamber. Then a length of wire is laid into the groove so that approximately half of its diameter sticks up out of the groove.

When the head is installed on the engine with a head gasket, the wire exerts tremendous pressure against the gasket and cylinder block. This extra sealing pressure prevents the gasket from leaking or blowing out. Such tactics are necessary because of the high pressures developed in the combustion chambers of supercharged and turbocharged racing engines.

Resilience: The tendency of a material to return to its original shape after being bent, compressed, or deformed.

Axial Movement: Movement parallel to the axis of a shaft. The measurement of axial movement on a shaft is called endplay.

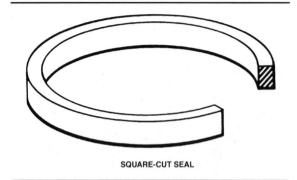

SQUARE-CUT SEAL

Figure 14-11. The cross section of a square-cut seal is square.

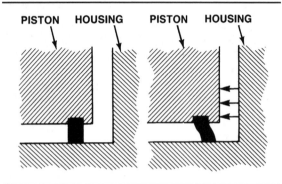

Figure 14-12. A square-cut seal deforms under pressure to keep a seal between two surfaces that move slightly in relation to each other.

Hydraulic pressure against the inside of the lip causes the lip to flare. As pressure on the lip increases, it presses harder against the shaft or cylinder it seals. When pressure against the lip is relieved, it slides easily against its shaft in either direction. In addition, many lip seals have a small circular coil spring, called a garter spring, placed behind the lip. The garter spring maintains pressure on the lip when the shaft is stopped and no oil pressure is present.

Lip seals are used where either or both rotational and axial forces are present. Because the lip that takes the rotational force is small, the seal will not distort with rotation as easily as an O-ring or square-cut seal. The front and sometimes the rear crankshaft seals are the lip type.

Other Types of Seals

Some engines have bolt-on crankshaft lip seals. The seals bolt to the rear face of the block and the main cap. A gasket between the seal and the block prevents leakage from behind the seal.

Other engines have a wick-type seal, made of rope-like material that fits inside the rear main cap and web, figure 14-14.

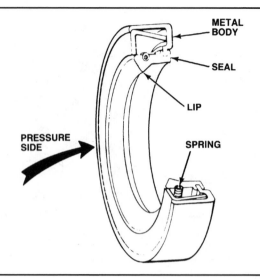

Figure 14-13. A lip seal is designed so that pressure pushes the seal against the surface it is sealing, making it seal even tighter.

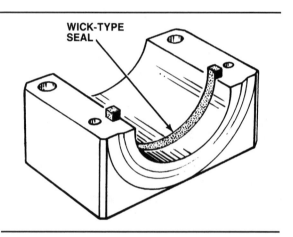

Figure 14-14. A rope-like wick seal fits into some engines' rear main bearing caps.

ADHESIVES, SEALANTS, LUBRICANTS, AND CLEANERS

Adhesives, sealants, lubricants, and cleaners used in engine assembly come packaged in tubes, figure 14-15, spray cans, figure 14-16, or small cans with a brush attached to the cap, figure 14-17. Each product serves a unique purpose.

An **adhesive** holds a gasket in place during assembly, but it does not affect sealing. A **sealant**, on the other hand, can aid a gasket's sealing ability or even act as a seal where there is no gasket. A lubricant helps loosen frozen or corroded threads during disassembly and eases assembly of threaded parts.

An **anti-seize lubricant** helps prevent bolts from becoming frozen by corrosion or seizing in an aluminum component. Heat speeds the forming of

Figure 14-15. Room-temperature vulcanizing (RTV) sealant is a common silicone sealant used just before assembly.

Figure 14-16. A light film of adhesive sprayed on the gaskets helps hold them in place during assembly.

corrosion, so parts exposed to great heat, such as exhaust manifold fastener threads, are good places to use an anti-seize lubricant.

Finally, a cleaner helps dissolve debris such as old gaskets or sealant so you can completely clean engine surfaces before assembly.

Precautions

Use moderation in applying any chemical substance to engine parts. Do not assume that if a little is good a lot is better. Read gasket installation instructions and other service literature to determine whether using these substances is necessary. Many modern gaskets, for example, work better without any sealant.

Figure 14-17. Brush adhesive onto the gasket and the facing surface and fit them together before assembly.

Never use an adhesive or sealant on any part of a seal that touches a moving shaft. However, seals that are pressed into place or clamped into position can have sealant on the back side of the seal to prevent leakage.

Some types of fittings, such as a flare nut or a ferrule and nut, are self-sealing. If they leak, change the fittings, instead of trying to stop the leak with sealant.

Do not use adhesives, sealants, or lubricants on electrical sending units, not even the threads on the mounting studs. An electrical unit must be installed dry for a good ground.

When using spray products (such as gasket cleaner) on engine surfaces, avoid overspraying into the crankcase or oil passages, as doing so will cause contamination.

Adhesives

Adhesives are usually classified as hardening and non-hardening. The hardening type sets up until it

---

Adhesive: A substance that holds a part such as a gasket in place during component assembly.

Sealant: A substance that forms a seal between two facing surfaces.

Anti-Seize Lubricant: A lubricant designed to keep threads from being frozen in place by corrosion.

becomes brittle and has to be chipped away when the parts are disassembled. The non-hardening type remains pliable. It is used if there might be some movement of the parts due to heat expansion. This movement could break a gasket coated with the hardening type.

To make engine assembly easier, apply adhesive to one side of a gasket, figures 14-16 and 14-17, and the mating side of a part. Allow the adhesive to dry until tacky and bolt the parts together with the gasket between them. Adhesive on rubber gaskets helps keep them from shifting when you tighten the flanges. It can also be used on cork/rubber gaskets.

Sealants

A sealant fills gaps and holes between mated parts. A common use for sealant is to fill irregularities between a gasket and the surface it seals. Read gasket instructions to see whether using a sealant is recommended. There are two basic types of gasket sealant:

- Aerobic
- Anaerobic.

An **aerobic sealant** cures in the presence of air (more specifically, oxygen), while an **anaerobic sealant** cures in the absence of air.

Aerobic sealants

Most aerobic sealants are silicone compounds suitable for metal or plastic parts. Silicone is thick and can seal large gaps. The temperature range of a silicone sealant usually reaches as high as 350°F (175°C), but some will also withstand the 600°F (300°C) temperatures of the exhaust system to assist a gasket in sealing.

One disadvantage of a silicone sealant is that you must fit the parts together within ten minutes of applying the sealant because it begins to cure as soon as air contacts it. Another name for the most common silicone sealant compound is **room-temperature vulcanizing (RTV) compound**, figure 14-15.

Anaerobic sealants

An anaerobic sealant cures only after mating parts are bolted together, excluding air. This is convenient for overhaul work because you can put sealant on a number of parts and leave them on your workbench until you are ready to assemble them.

Anaerobic sealant is often used on cast timing covers, intake manifolds, and rear main bearing caps. It is thinner than silicone and is not practical for flexible covers. Anaerobic sealant is recommended for use on machined surfaces only.

Some anaerobic sealants are designed to lock threaded bolts or studs into place. Machinists generally use thread-lock sealants in applications where vibration would tend to loosen a bolt.

Lubricants

When disassembling automotive components, you sometimes find fasteners frozen in place by rust or corrosion. Penetrating oil, sometimes called penetrant, helps dissolve rust and corrosion so you can remove the fastener. Corrosion and other contaminants should be cleaned out of bolt holes before a part is reassembled.

During assembly, many machinists apply anti-seize lubricant to fastener threads in places where corrosion tends to be a problem, such as in exhaust components, figure 14-18. It is common to apply anti-seize lubricant to spark plug threads in aluminum cylinder heads.

Because lubricants reduce friction, lubricating the threads of a fastener changes the torque reading when you tighten it. Check whether torque specifications are for a "wet" or a "dry" bolt. "Wet" means with lubricant applied to the threads. Either use lubricant or don't use it, according to the specifications.

Cleaners

Cleanliness is important for good engine sealing. Oil or dirt on engine surfaces prevents sealing between surfaces and keeps adhesives from working. In a bolt hole, oil or dirt interferes with torque readings, prevents sealant from forming a tight seal, and can make the fastener bottom out, preventing tight fastening. Before installing a gasket, sealant, or a fastener, make sure sealing surfaces, bolt holes, and all threads are clean and dry.

Gasket remover softens pieces of old gasket stuck to engine surfaces, figure 14-19, so the pieces scrape off easily without damage to the metal surface. Gasket removers also remove paint, so be careful where you spray them. Degreasers help remove oil film from engine surfaces and bolt threads so the dry metal can accept adhesives, seals, and sealant. Run a tap through bolt holes to clean them.

FASTENERS

Screws, bolts, and studs are the three types of threaded fasteners used in automotive applications. Unlike studs, which have no head, both screws and bolts have some type of head that a screwdriver or wrench can turn for installation and removal. Both have a threaded shank.

So what is the difference between a bolt and a screw? Some engineers say that if the head is six-pointed or hexagonal, and the shank is ¼ inch (6 mm) or larger in diameter, the fastener is a bolt. Otherwise, it's a screw: meaning that it has a shank of less than ¼ inch (6 mm) and it most probably has a slotted or Phillips head that is tightened by a screwdriver.

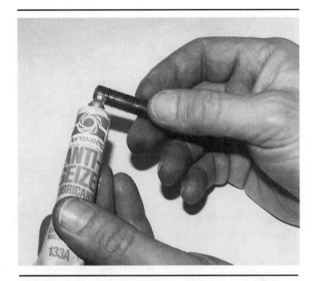

Figure 14-18. Anti-seize lubricant on exhaust system bolts can help prevent corrosion that would freeze the bolts into their holes.

Figure 14-19. Gasket remover softens pieces of old gasket stuck to the engine so they scrape off easily. Don't overspray because it is also a paint remover.

But there are exceptions. Other engineers say that a bolt is held in place by a nut and tightened by turning the nut rather than the fastener. A screw, on the other hand, is installed by turning the fastener head. By this definition, the same fastener could be a screw or a bolt depending on where you use it and how you install it. There is no consensus, but most machinists (and the writers of this book) adhere to the first definition.

Specifications

From a machinist's point of view, it's more important to use the right fastener than to call it by the right name. The important question, then, is how to check

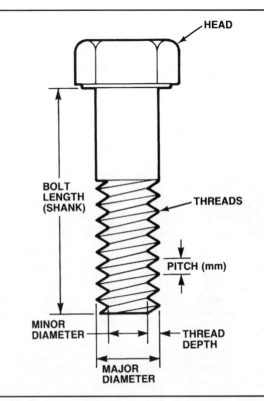

Figure 14-20. The various dimensions of a bolt or screw reflect its use.

the specifications of any given threaded fastener so that you can know whether it is right for a particular application. The things you need to know about a bolt or screw, figure 14-20, are its:

- Bolt or screw length
- Thread diameter
- Thread pitch
- Thread depth
- Head size and shape.
- Grade (strength).

The three basic parts of a screw or bolt are the shank, thread, and head. The shank is the rod part and the threads wrap around part or all of the shank. The head, usually with a flat washer, seats against the surface that the fastener is holding in place.

Aerobic Sealant: A sealant that cures in the presence of air.

Anaerobic Sealant: A sealant that cures in the absence of air.

Room-Temperature Vulcanizing (RTV) Compound: A silicone-based, aerobic sealant commonly used in automotive applications.

Bolt or screw length

Except for flathead fasteners, the length is measured from the bottom of the head to the end of the shank. The shank has to be the correct length to seat the bottom of the head firmly. If the shank is too long, the bolt will bottom out in its hole and not fasten firmly. If it is too short, it will provide less gripping strength. The head of a flathead screw is included in the length measurement because it goes under the surface when the screw is threaded into its hole.

Thread diameter

Thread major diameter is measured across the shank and includes the height of the threads. This measurement is used to designate the bolt size. In other words, a specification for a ½-inch bolt refers to a bolt whose shank measures ½ inch across the threads. Bolt sizes do not refer to the size of the head or of the wrench needed. Thread minor diameter is measured from the bottom of the threads on one side of the bolt to the bottom of the threads on the other side.

Thread pitch

The **thread pitch** refers to how closely wound the threads are, or how frequently they occur along the length of the shank. The American National measuring system specifies pitch in terms of how many threads per inch, while the metric system specifies how far apart two threads are. Both specifications indicate the same thing. Standard bolt and machine screw threads are cut at a 60-degree angle, regardless of pitch, so a larger, or coarser, pitch means deeper threads.

Thread depth

Thread depth is the height of the thread from its base to its top. The deeper the thread, the stronger it is. Threads can be cut or rolled. Rolled threads are stronger. They are formed by passing the rod through grooved dies, figure 14-21, that apply high pressure to the previously heat-treated rod and cold-form the threads. The grain of the metal follows a rolled thread and adds strength, and the metal is burnished by the rollers. Buy bolts from a reputable supplier. Some counterfeits are not heat-treated before rolling, and this produces an inferior bolt.

In contrast to rolling the threads, cutting threads into a rod chops off the grain of the metal, weakening it, and the cutting process leaves roughness that creates stress raisers. Virtually all mass-produced bolts and screws have rolled threads; you should not attempt to fabricate bolts by cutting threads on a lathe or by using a die. Internal threads, however, must be cut because there is no practical way to roll them.

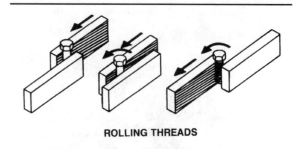

ROLLING THREADS

Figure 14-21. Cold-rolling a previously heat-treated bolt blank forms strong threads.

Head size and shape

The size and shape of the head determines what tool you must use to install the fastener. A hexagonal (six-sided) head requires a wrench, slotted heads require a flat screwdriver, and Allen heads require an Allen wrench. Figure 14-22 illustrates some bolt heads and gives their common names.

Special head bolts

Most bolts that hold down cylinder heads have either a hex-head or an Allen head. However, some Japanese cylinder heads use head bolts with external twelve-point heads, figure 14-23. Some European engines use an internal twelve-point head, also referred to as a serrated or triple-square head. Ford 2.3-liter 4-cylinder engines have head bolts with TORX® heads.

Bolt Torque

Once a bolt is in place, you must tighten it to the correct torque value. Manufacturers specify a torque value for virtually every fastener in a car.

All bolts have a range of elasticity. Up to a certain point, they stretch when tightened, and spring back to original length when loosened. Most bolt torque specifications are calculated so the bolt is in this slightly stretched elastic range when it is tightened. The stretch provides clamping force to hold the parts together under tension, and helps keeps the bolt from coming loose.

Tightening pattern

For applications such as cylinder head bolts or oil pan bolts, manufacturers also specify a tightening pattern to ensure that the part is torqued evenly. A tightening pattern avoids warping a part due to uneven clamping force. We cover the tightening pattern in more detail in Chapter 16 of the *Shop Manual*.

When following a torque pattern, do not immediately tighten each bolt to specifications. Go through the pattern three times, tightening a little bit more each time, until you are close enough to tighten to the final torque. Apply final torque on the bolts in the

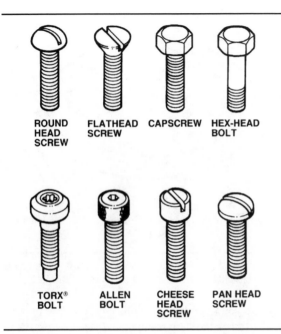

ROUND HEAD SCREW   FLATHEAD SCREW   CAPSCREW   HEX-HEAD BOLT

TORX® BOLT   ALLEN BOLT   CHEESE HEAD SCREW   PAN HEAD SCREW

Figure 14-22. Bolts and screws have different heads that determine what tools you must use to install them.

Figure 14-23. Some imported vehicles use external 12-point fasteners for head bolts.

same tightening sequence you used to bring the fasteners to that point.

Torquing bolts
There are two methods of applying final torque, and you should carefully read torque tables to see which method the manufacturer recommends. The two methods consist of tightening the bolt to a:

- Torque value
- Torque/angle.

Tightening to a torque value is fairly straightforward. Just read the torque wrench and tighten as near as possible to the specified value. A torque wrench effectively measures the friction between the bolt and the hole into which it is threaded.

At the factory, the engine parts are in perfect condition, and all of the head bolts are torqued at the same time by machine. As a result, variations in clamping force are minimal. In the field, it is a different matter. Some bolts or bolt holes may be dirtier than others, and threads can become distorted, causing certain bolts to have a higher resistance to rotation than others. Even though you torque every bolt to the same value, the final clamping forces may be quite different. For these reasons, some manufacturers believe that a torque wrench reading is not an exact enough measurement, so engineers developed the torque/angle method.

Using the torque/angle method, the exact procedure varies with the application, but the procedure for Chrysler 2.2-liter engines with 11-mm head bolts is

typical. Following the recommended sequence, tighten the bolts in three steps to the torque specified in the shop manual. This is a fairly low torque that gives the bolts an equal starting point. Then, tighten each bolt an additional ¼ turn.

In many cases, the final step is measured in degrees of rotation rather than fractions of a turn (90 degrees versus ¼ turn). Some import manufacturers require that the head bolts be turned an uneven number of degrees, such as 107. You can use the corners of the bolt head to help you calculate the number of degrees you are turning, figure 14-24. However, a torque angle gauge, available from tool companies such as Kent-Moore and Snap-On, provides an exact angle reading, figure 14-25.

Torque-to-yield-bolts
Torque-to-yield-bolts are another answer to the problem of getting bolts properly torqued in the field. These bolts are also torqued using the torque/angle method; however, their design prevents a machinist from over-torquing them and damaging the parts being clamped together.

Torque-to-yield bolts are tightened beyond their elastic range to the yield point of the bolt. This is the point where the bolt stretches so far it does not return to its original length when loosened. This is desirable

Thread Pitch: The distance between threads on a fastener; the number of threads along a certain length.

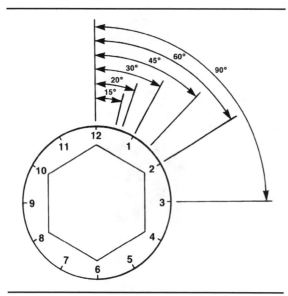

Figure 14-24. When using the torque/angle tightening method, it helps to imagine a clock face and remember that each "hour" means 30 degrees. On a hex-head, each point equals two hours.

because there is little increase in clamping force once a bolt reaches its yield point. If several identical bolts are tightened to their yield points, they all provide nearly the same amount of clamping force. This is the main reason manufacturers are switching to torque-to-yield cylinder head bolts.

Bolt-to-bolt variations in clamping force are minimal up to about 25 ft-lbs (35 Nm) of torque. Above this threshold, however, the potential variation in clamping force becomes greater and greater, even though equal torque is applied to all bolt heads. For example, a bolt torqued to 60 ft-lbs (80 Nm) may provide an actual clamping force anywhere between 4,500 and 12,500 pounds (20,000 and 55,500 N), a variation of 8,000 pounds (35,500 N). Compare this to a torque-to-yield bolt designed to apply the same clamping force. When tightened properly, the variation is only 3,500 pounds (15,600 N), less than half what can occur if bolt is torqued.

Uniform clamping force is a primary factor in assuring long head gasket life. The aluminum heads on many engines have drastically different expansion rates than the cast iron blocks they attach to. This causes significant movement between the two parts, as well as clamping force variation as the engine heats and cools. Many engines today use fewer head bolts because there is less space for them in compact, lightweight castings. As a result, every head bolt on a modern engine must be used to maximum potential.

It is important that you follow the factory recommendations for replacing torque-to-yield bolts. Some vehicle manufacturers allow a single reuse, others

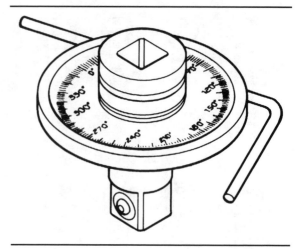

Figure 14-25. A torque angle gauge provides a reading that tells you how far you are turning the bolt.

require you to replace the bolts any time you remove them. Check the shop manual, and never reuse a head bolt unless the manufacturer specifically says to do so. If you have any doubt, install new bolts to be safe.

If you are unsure whether the engine you are working on has torque-to-yield head bolts, there are some clues to look for. Normal cylinder head bolts often have an indentation in the head; torque-to-yield bolts do not. In addition, some torque-to-yield bolts have a cut-down shank, although many do not, figure 14-26. Once again, if you are unsure about the bolts used on a specific vehicle, check the shop manual.

Securing Fasteners

Nuts are made to go on screws or bolts, securing them in place. While bolts and screws have external threads, nuts have internal threads, figure 14-27.

Some nuts, called jam nuts, are used to keep bolts and screws from loosening. Jam nuts screw on top of a regular nut and jam against the regular nut to prevent loosening. A jam nut is so called because of its intended use, rather than a special design. Some jam nuts are thinner than a standard nut.

There are also self-locking nuts of various types. Some have threads that are bent inward to grip the threads of the bolt. Some are oval-shaped at one end to fit tightly on a bolt. Fiber lock nuts have a fiber insert near the top of the nut or inside it; this type of nut is also made with a plastic or nylon insert. When the bolt turns through the nut, it cuts threads in the fiber or plastic and this puts a drag on the threads which prevents the bolt from loosening.

One of the oldest types of retaining nuts is the castle nut. It looks like a small castle, with slots for a cotter pin. A castellated nut is used on a bolt that has a hole for the cotter pin.

Flat washers are placed underneath a nut to spread the load over a wide area and prevent gouging of the material. However, flat washers do not prevent a nut from loosening.

Lockwashers are designed to prevent a nut from loosening. Spring-type lockwashers resemble a loop out of a coil spring. As the nut or bolt is tightened, the washer is compressed. The tension of the compressed washer holds the fastener firmly against the threads to prevent it from loosening. Lockwashers should not be used on soft metal such as aluminum. The sharp ends of the steel washers would gouge the aluminum badly, especially if they are removed and replaced often.

Another type of locking washer is the star washer. The teeth on a star washer can be external or internal, and they bite into the metal because they are twisted to expose their edges. Star washers are used often on sheet metal or body parts. They are seldom used on engines.

The spring steel lock washer also uses the tension of the compressed washer to prevent the fastener from loosening. The waves in this washer make it look like a distorted flat washer.

Thread and Bolt Sizes

Fastener manufacturers commonly designate nuts, bolts, and machine screws by their diameter and pitch. Automotive machinists will encounter two measuring systems for fasteners:

- American
- Metric.

We will look at what the designations in each of these systems mean.

Recognition and measuring, American sizes
In the U.S. system, bolt diameters of one-quarter inch and larger are given in fractions of an inch. The

---

■ British Whitworth

Before 1841, there was no standard for making bolts and nuts. Each fastener factory manufactured bolts and nuts its own way, so a nut would fit a bolt only if both came from the same factory. In England, Sir Joseph Whitworth thought that bolts and nuts should be interchangeable. That same year, he presented an engineering paper describing "A Uniform System of Screw-Threads."

It took about 20 years for England to adopt the new system, and by the time the automobile appeared, almost every screw thread in England was being made by the Whitworth Standard. It became known as the British Standard Whitworth or just Whitworth.

Later, another standard known as British Standard Fine was developed with finer threads. Both British Standards covered bolt diameters down to ⅛ or 3/16 of an inch. Below that, there was another standard known as the British Association. It used numbers ranging from zero to 7, but with the larger numbers for the smaller sizes. This is the opposite of bolts numbered in the American system. To avoid confusion, bolts in the British Association standard were identified by 0BA, 1BA, 2BA, etc.

Like U.S. and Metric fastener sizing systems, the British standardized their fasteners on bolt diameter, not head size. But interestingly, the wrenches were stamped this size, not the size of the head. To make this system work, a bolt had to be made with a specific size of hex head for every size of the threaded portion.

Whitworth wrenches were made this way, and identified with the size of the bolt and a "W", such as 5/16W, or ½W. The opening in the wrench was much larger than the size stamped on it. It was made to fit the head of a bolt, the threaded diameter of which was stamped on the wrench.

When the British Standard Fine bolts were made, they came out with smaller size hex heads than those on the Whitworth bolts. This was initially confusing

because a wrench marked ½W would not fit the head of a ½SF bolt. However, technicians learned that the BSF bolt heads were exactly one size smaller than the Whitworth bolt heads. So to turn a ½BSF, you would use a 7/16W wrench.

To show that the wrenches would fit both sizes, they were marked with both sizes for distribution in England. Most British cars imported into the United States used Whitworth bolts. So the American tool makers marked their wrenches with the Whitworth size only. This situation continued until 1951, when all bolt hex heads were made with the smaller BSF size. But wrenches continued to be marked with both "W" and "F" sizes.

In the middle 1960s all English vehicles started using American-sized bolts. This was the Unified system, which was formed as a result of a conference involving England, Canada, and the U.S. With American bolts on all the cars, they had to do something about wrench marking. The decision was made to use the standard U.S. marking, which gives the size of the hex head on the bolt. To show that it meant the diameter of the head, they put "AF" after the size, which means "across flats". Today, all British-made wrenches for use on American nuts or bolts have the standard fractional inch sizes with "AF" after them. Wrenches are made slightly larger than the actual size of the head so they will fit easily.

Although the customary size for bolt heads is well established, there is some variation. When ordering a hex-head bolt, the diameter, thread, and length are always specified, but the size of the head is usually not mentioned. If the head is a different size, you will have to use a different size wrench. It is important to remember that bolt sizes are always the diameter sizes. A hex-head bolt that takes a half-inch wrench is not a half-inch bolt. If you order a half-inch bolt, you will get a bolt almost twice the diameter you want.

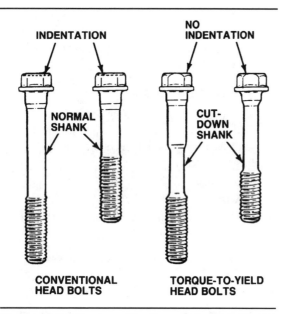

Figure 14-26. You can identify some torque-to-yield bolts by differences in the head or shank.

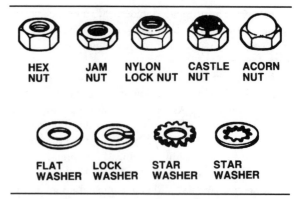

Figure 14-27. Various nuts serve different purposes. All have the common purpose of securing bolts and screws.

standard sizes are ¼ inch, ⁵⁄₁₆ inch, ⅜ inch, and up; increasing ¹⁄₁₆ inch at each step. Above ⅝ inch, the sizes increase in ⅛-inch steps.

Sizes below ¼ inch are usually designated by the American National numbering system. The numbers start at zero and go through 12. A zero size screw is 0.060 inch in diameter. A number 12 is 0.216 inch in diameter. In the progression from size zero to size 6, each size is 0.013 inch larger than the preceding one. However, there are no screws in sizes 7, 9, or 14. This means that sizes 8, 10, and 12 are each 0.026 inch larger than the next smaller size.

In the American National system, most bolts or machine screws are available in two different thread pitches, National Fine and National Coarse. For example, a number 6 screw has 32 threads per inch if it is National Coarse, and it has 40 threads per inch if it is National Fine. In writing, these two number 6 screws would be referred to as 6-32 and 6-40. Figure 14-28 is a table of all National Fine and National Coarse diameters and pitches.

In addition to National Fine and National Coarse, there is National Special. National Special thread pitch may be more or less than the National Fine for a given bolt diameter. These special thread pitches are not universally available and are seldom found on an automobile engine.

Another standard size measurement is American Standard Pipe. It applies only to pipe threads. There is only one thread pitch for each diameter of pipe. As far as pipe sizes are concerned, there are no metric measurements. American Standard is also the metric standard.

Pipe threads may be tapered or straight. The tapered threads result in a wedging action when screwed into a fitting, helping to prevent leaks. Tapered threads are standard on all sizes of pipe or fittings and are common in coolant and oil lines. Straight pipe threads are rare, and used only for special purposes.

Hex-head bolts, also called capscrews, have a customary head size relative to the bolt diameter. Quarter-inch bolts usually have a head that measures ⁷⁄₁₆ inch across the flats. ⁵⁄₁₆-inch bolts usually have a ½-inch head. The size of the head determines the size wrench to use.

Metric sizes

Measurements of metric fasteners are made in millimeters (mm). Metric bolts and screws are usually made with diameters in exact millimeter measurements. It is rare to find a metric bolt that measures in fractions of a millimeter, except in the very small sizes, figure 14-29.

Metric bolt designations list the diameter first, followed by the length and the thread pitch. The pitch is listed as the center-to-center distance in millimeters between thread crests. For example, an 8-mm bolt that is 20 mm long with a thread pitch of 1 mm may be listed as an M8 x 20-1.00. It would probably have a 12- or a 13-mm head. A capital M is often used in front of the size to indicate that it is metric. In some cases, the designation "mm" or a single "m" follows the size, as in 10-1.25 mm.

The metric pitch measurement is similar in use to threads per inch. The larger the number, the coarser the thread. Metric bolts are available with more than one thread pitch. However, the difference between the pitches is not as great as it is in the American Standard system. The terms coarse and fine are not used in the metric system. Thread pitches are referred to simply by the number of the pitch.

Size	Threads per inch		Outside Diameter Inches
	NC UNC	NF UNF	
0	. .	80	0.0600
1	64	. .	0.0730
1	. .	72	0.0730
2	56	. .	0.0860
2	. .	64	0.0860
3	48	. .	0.0990
3	. .	56	0.0990
4	40	. .	0.1120
4	. .	48	0.1120
5	40	. .	0.1250
5	. .	44	0.1250
6	32	. .	0.1380
6	. .	40	0.1380
8	32	. .	0.1640
8	. .	36	0.1640
10	24	. .	0.1900
10	. .	32	0.1900
12	24	. .	0.2160
12	. .	28	0.2160
1/4	20	. .	0.2500
1/4	. .	28	0.2500
5/16	18	. .	0.3125
5/16	. .	24	0.3125
3/8	16	. .	0.3750
3/8	. .	24	0.3750
7/16	14	. .	0.4375
7/16	. .	20	0.4375
1/2	13	. .	0.5000
1/2	. .	20	0.5000
9/16	12	. .	0.5625
9/16	. .	18	0.5625
5/8	11	. .	0.6250
5/8	. .	18	0.6250
3/4	10	. .	0.7500
3/4	. .	16	0.7500
7/8	9	. .	0.8750
7/8	. .	14	0.8750
1	8	. .	1.0000
1	. .	12	1.0000
1 1/8	7	. .	1.1250
1 1/8	. .	12	1.1250
1 1/4	7	. .	1.2500
1 1/4	. .	12	1.2500
1 3/8	6	. .	1.3750
1 3/8	. .	12	1.3750
1 1/2	6	. .	1.5000
1 1/2	. .	12	1.5000
1 3/4	5	. .	1.7500
2	4 1/2	. .	2.0000
2 1/4	4 1/2	. .	2.2500
2 1/2	4	. .	2.5000
2 3/4	4	. .	2.7500
3	4	. .	3.0000
3 1/4	4	. .	3.2500
3 1/2	4	. .	3.5000
3 3/4	4	. .	3.7500
4	4	. .	4.0000

Figure 14-28. The American National system is one method of categorizing fasteners according to size and thread pitch.

Although diameters of metric bolts are almost always in even millimeter measurements, metric pitches come in fractions of a millimeter. The usual steps are 0.25 of a millimeter, such as 1.25, 1.50, 1.75, and so on. You must make careful measurements with a pitch gauge to avoid mismatching bolts and nuts.

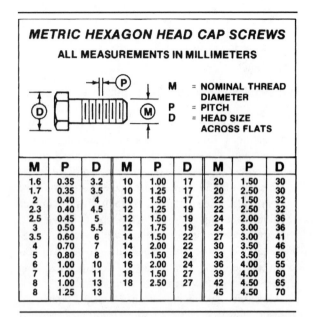

METRIC HEXAGON HEAD CAP SCREWS

ALL MEASUREMENTS IN MILLIMETERS

M = NOMINAL THREAD DIAMETER
P = PITCH
D = HEAD SIZE ACROSS FLATS

M	P	D	M	P	D	M	P	D
1.6	0.35	3.2	10	1.00	17	20	1.50	30
1.7	0.35	3.5	10	1.25	17	20	2.50	30
2	0.40	4	10	1.50	17	22	1.50	32
2.3	0.40	4.5	12	1.25	19	22	2.50	32
2.5	0.45	5	12	1.50	19	24	2.00	36
3	0.50	5.5	12	1.75	19	24	3.00	36
3.5	0.60	6	14	1.50	22	27	3.00	41
4	0.70	7	14	2.00	22	30	3.50	46
5	0.80	8	16	1.50	24	33	3.50	50
6	1.00	10	16	2.00	24	36	4.00	55
7	1.00	11	18	1.50	27	39	4.00	60
8	1.00	13	18	2.50	27	42	4.50	65
8	1.25	13				45	4.50	70

Figure 14-29. The metric system designates fasteners by diameter, length, and pitch.

■ Computer-Controlled Engine Mounts

Just when you thought vehicle manufacturers had automated everything they possibly could, automotive engineers have begun designing computer-controlled engine mounts. Believe it or not, the inconspicuous rubber engine mount may be replaced and fully automated in vehicles of the future.

"Why?" you ask. Vehicle manufacturers have determined that a consumer shopping for a vehicle considers a quiet, smooth-running engine a mark of excellence. In the past, manufacturers reduced noise and vibration by using V8s, heavy balance shafts, and filler material in body cavities to deaden sound and damp vibration. But since consumers also look for good gas mileage, engineers are looking for alternatives to large engines and fill material.

Since the early 1980s, high-output, fuel-efficient 4- and 6-cylinder engines have replaced V8s in many applications. The disadvantage of a smaller engine is that it does not run as smoothly because it does not have the overlapping power strokes that tend to smooth out engine vibration. But customers still dislike vibration — and this is why design engineers are developing computer-controlled engine mounts.

A computer-controlled engine mount system uses sensors to detect vibration between the chassis and engine. The sensors send this data to the computer, which determines the distance and frequency at which the engine mounts should move to counter this vibration. In essence, the engine mounts are electronically controlled shock absorbers. The computer sends a message to an actuating mechanism that moves the engine mounts and smooths out the vibration.

The size of a metric bolt is taken as the size of the threaded part of the bolt, the same as in American Standard. Hex-head sizes in metric bolts are measured in millimeters.

Wrench length relationship to bolt head size

Look at any set of open end, box end, or combination wrenches. With wrenches of the same design, the smaller the wrench opening, the shorter the wrench. This is to give the right feel to the wrench when the nut or bolt is tightened and to prevent overtightening.

The load the bolt must carry determines bolt size and strength. The strength of the bolt must be higher than the force imposed on it. When you tighten a bolt, you turn it with more force than it will bear in service. For example, if a bolt is going to support a load of 100 pounds, you might tighten it to 150 foot-pounds. This ensures that the bolt will work without failing.

Machinists and technicians use a torque wrench to measure how tight a nut or bolt is. Torque is described in Chapter 2. With a torque wrench you can tighten a nut or bolt to a specific amount of torque. Shop manuals give the torque specifications for every important bolt or nut on the engine. The specifications are carefully calculated so that the bolts will carry their loads and not become loose.

Most torque specifications are written for fasteners that are clean and dry. As mentioned earlier, if fasteners are lubricated, the pressure applied to the nut or bolt will be greater because there is less friction to overcome. If the torque specification chart does not mention any lubricant, then the threads should be clean and dry.

Fastener Strength

Nuts, bolts, and screws must be strong enough to hold parts together as designed, and to withstand the maximum tension put upon them. Fasteners are made in different grades of strength, and are marked so you can easily determine this. Figure 14-30, Table A, shows U.S. Customary bolt grade designations, materials and treatment, and grade identification markings. Notice that in the U.S. system grades are indicated by numbers. The larger the number, the greater the **tensile strength** of the bolt. Decimals are used to represent variations of the same strength level.

U.S. Customary nuts are marked according to figure 14-30, Table B. Grade 2 nuts and some kinds of grade 5 and grade 8 nuts are not required to be marked. There are three possible styles of grade identification. Style A is applicable to all nuts. Style B is optional depending on the supplier, the purchaser, and the size of the nut. Style C is for nuts which are cut from a hexagon-shaped steel bar.

TABLE A—U.S. CUSTOMARY BOLTS, SCREWS & STUDS

SAE Grade Designation	Material & Treatment	Grade Identification Marking
1, 2	Low or medium carbon steel	
4	Medium carbon cold drawn steel	
5	Medium carbon steel, quenched and tempered	
5.1	Low or medium carbon steel, quenched and tempered	
5.2	Low carbon martensite steel, quenched and tempered	
7	Medium carbon alloy steel, quenched and tempered	
8	Medium carbon alloy steel, quenched and tempered	
8.1	Elevated temperature drawn steel, medium carbon alloy	
8.2	Low carbon martensite steel, quenched and tempered	

TABLE B—U.S. CUSTOMARY NUTS

SAE Grade Designation	Grade Identification Marking		
	Style A	Style B	Style C
2	None	None	None
5	Dot and radial or circumferential line 120° CCW from dot	Dot at one corner of nut and radial line at the corner 120° CCW from dot	One notch at each hexagon corner
8	Dot and radial or circumferential line 60° CCW from dot	Dot at one corner of nut and radial line at the corner 60° CCW from dot	Two notches at each hexagon corner

TABLE C—METRIC NUTS & BOLTS

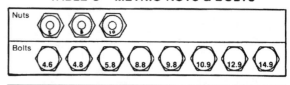

Figure 14-30. The U.S. Customary (as well as the metric) system indicates fastener strength.

Metric fasteners generally have the strength classification embossed on the head of each bolt, or the face of each nut. The number indicates the tensile strength in kilograms of force per square millimeter. This means the larger the number, the stronger the bolt or nut, figure 14-30, Table C.

When replacing a fastener, always use one of the correct grade. In some cases, manufacturers recommend replacing specific nuts and bolts instead of reusing them, even though they might appear to be in perfect condition. Always comply with these requirements, because a bolt that has been weakened by being torqued to a high value could fail if reused.

ENGINE MOUNTS

Engine mounts insulate the vehicle body from engine vibrations and hold the engine firmly in place. An engine mount is a bracket that bolts between the engine and the frame or lower body of the vehicle. Similar mounts are also used for locating the transmission. For the purpose of mounting the powertrain in the vehicle, the engine and transmission can be treated as a single unit. You can refer to the mount at a transmission as a transmission mount or an engine mount. Front-engine, rear-wheel-drive vehicles typically have three mounts — one at each side of the engine and one at the back of the transmission, figure 14-31. Front-wheel-drive vehicles also usually have three mounts — two on each side of the block and one at the front of the engine, figure 14-32. One of the mounts may also help support the transmission. Most front-wheel-drive vehicles have a torque arm that limits engine movement. It usually bolts to the cylinder head and connects to the body.

Rear-engine and mid-engine vehicles sometimes have different problems in isolating engine vibrations. There have been rear-engine vehicles with engine mounts at the top of the engine. An upper mount does not really support the weight of the engine but acts as a steadying rest. Most rear- and mid-engine vehicles have a three-point mounting system similar to front-engine cars.

A simple engine mount consists of two metal brackets bonded to a block of rubber. Even these simple engine mounts are "tuned" to absorb vibrations from the engine through choosing the best shape of the mount, the right amount of rubber, and the proper stiffness of rubber. In time, the bonding between the rubber and the metal weakens, and the engine moves around in the frame. A separated engine mount can allow the engine to hit other parts, such as the steering mechanism or radiator. Some engine mounts have a bracket or tang to prevent the engine from moving enough to cause damage if the rubber fails. The extra bracket provides metal-to-metal contact when the rubber bonding fails.

Some vehicles use more sophisticated, fluid-filled rubber engine mounts, figure 14-33. The fluid flows between upper and lower chambers through an orifice, providing hydraulic damping like a mini-shock absorber.

Electric Engine Mounts

The 1990 Honda Accord introduced electronically controlled engine mounts. They were used only as rear (closest to the firewall) mounts on Accords equipped with automatic transmissions. These mounts smooth low-frequency vibration that can occur during

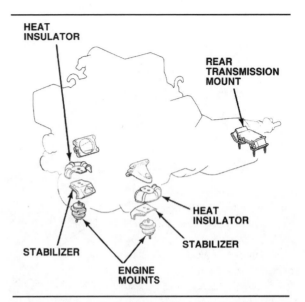

Figure 14-31. A typical front-engine, rear-wheel-drive car has two engine mounts at each side of the block and a rear mount that supports the transmission.

low-rpm, high-load running — such as idling while in gear with the air conditioning running.

The mounts contain two fluid-filled chambers separated by a barrel valve, figure 14-34. Below 850 rpm the barrel valve is open so that the full volume of both chambers can dampen vibration. The computer control senses when 850 rpm is reached by monitoring ignition pulses. At 850 rpm the computer signals a vacuum actuator to close the barrel valve. Thus, one of the fluid-filled chambers is closed, firming the engine mount.

SUMMARY

Gaskets prevent leaks between facing surfaces in stationary joints. Common types include head gaskets, intake and exhaust manifold gaskets, oil pan gaskets, and valve and cam cover gaskets. Gasket designs differ to suit the various applications. You must prep surfaces before gasket installation for good sealing. Seals prevent leaks past moving parts. Common types of seals are O-rings, square-cut seals, lip seals, and positive or umbrella-type valve seals. Chemical adhesives, sealants, lubricants, and cleaners help in the task of engine assembly. Adhesives tack gaskets in place during assembly. Sealants form a seal between two

Tensile Strength: A metal's or, in this case, a bolt's resistance to a force pulling it apart. This force is measured in pounds per square inch (psi) or millions of pascals (MPa).

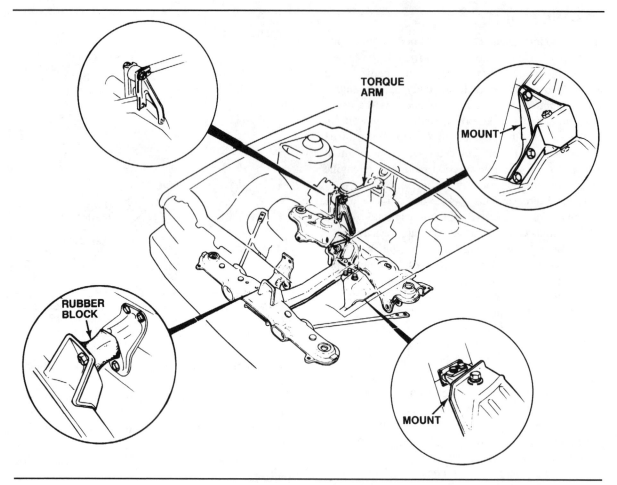

Figure 14-32. A typical front-wheel-drive vehicle may have three or four engine mounts and a torque arm.

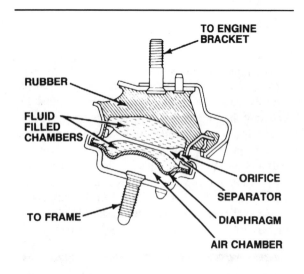

Figure 14-33. This Lexus LS400 engine mount uses fluid-filled rubber to provide hydraulic damping of engine vibration.

surfaces. Lubricants reduce friction between parts, and cleaners help you prepare parts for assembly.

Threaded fasteners are categorized by length, diameter, thread pitch and depth, grade, and head size and shape. Manufacturers specify a torque value for virtually every fastener in a vehicle. For applications such as cylinder head bolts or oil pan bolts, they also specify a tightening pattern. The two methods of torquing bolts are tightening the bolt to a torque value, or tightening to a torque value then turning the bolt a certain number of degrees.

Torque-to-yield bolts are tightened beyond their elastic range to the yield point of the bolt to provide more uniform amounts of clamping force on all bolts. This is the main reason manufacturers are switching to torque-to-yield cylinder head bolts. Different kinds of nuts may be used to secure screws and bolts: standard nuts, jam nuts, castle nuts, and various types of self-locking nuts.

The two measuring systems for fasteners are American and metric. In the U.S. system, bolt diameters of one-quarter of an inch and larger are given in

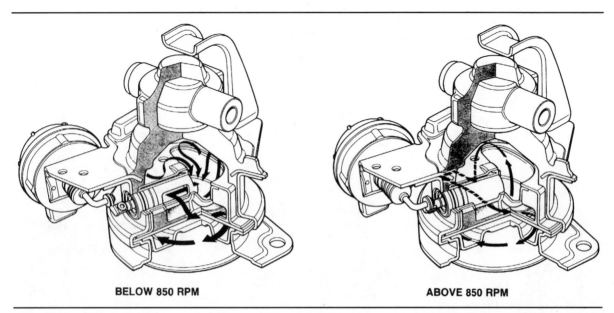

BELOW 850 RPM   **ABOVE 850 RPM**

Figure 14-34. These Honda Accord engine mounts change fluid volume electronically to filter out both high- and low-frequency vibrations.

fractions of an inch. The standard sizes are ¼ inch, ⁵⁄₁₆ inch, and up; increasing ¹⁄₁₆ inch at each step. Above ⅝ inch, the sizes increase in ⅛-inch steps. Sizes below ¼ inch are usually designated by the American National numbering system. The numbers start at zero and go through 12. Metric bolts and screws are usually made with diameters in exact millimeter measurements. Metric bolt designations list the diameter first, followed by the length and thread pitch.

The U.S. Customary system indicates fastener strength, measured in tensile strength. Metric bolts also have their heads embossed with a tensile strength rating. On both, the larger the number, the stronger the bolt.

Engine mounts secure the engine and transmission to the frame and shield the frame from engine vibration. The mounts may use rubber, fluid, or some kind of electronic control.

Review Questions
Choose the single most correct answer.
Compare your answers with the correct answers on Page 328.

1. Technician A says that you should let the engine cool completely before beginning to disassemble it.
   Technician B says that if the head gasket is in very good condition, you can reuse it.
   Who is right?
   a. A only
   b. B only
   c. Both A and B
   d. Neither A nor B

2. A sandwich-type head gasket usually has a core made of:
   a. Copper
   b. Iron
   c. Steel
   d. Aluminum

3. Technician A says it is not necessary to use an exhaust manifold gasket when reassembling an engine.
   Technician B says that if you do use an exhaust manifold gasket, the steel side should face the engine.
   Who is right?
   a. A only
   b. B only
   c. Both A and B
   d. Neither A nor B

4. The cross section of a lathe-cut seal is:
   a. Round
   b. Square
   c. Oval
   d. Triangular

5. An O-ring seal should be used where there is:
   a. Rotational movement
   b. Axial movement
   c. Little movement of any kind
   d. Absolutely no movement

6. A lip seal often contains a:
   a. Garter spring
   b. Lock washer
   c. Spring washer
   d. Teflon® surface

7. Torque-to-yield bolts:
   a. Are used to install most exhaust manifolds
   b. Can be reused indefinitely
   c. Are sometimes used as head bolts
   d. Are not found on domestic vehicles

8. Technician A says that an adhesive helps form a seal when applied between two facing surfaces.
   Technician B says that an adhesive helps hold a gasket in place during engine assembly.
   Who is right?
   a. A only
   b. B only
   c. Both A and B
   d. Neither A nor B

9. An anaerobic sealant:
   a. Must be applied just before assembly
   b. Is also known as RTV
   c. Cures in the absence of air
   d. Is good to use on flexible surfaces

10. Anti-seize lubricants are commonly used on:
    a. Exhaust system fastener threads
    b. Intake system fastener threads
    c. Rear main seal grooves
    d. Cylinder head bolt threads

11. Which of the following does *not* indicate a thread pitch?
    a. National Fine
    b. National Coarse
    c. National Standard
    d. National Special

12. The size of the wrench to use depends on the:
    a. Bolt shank
    b. Specified torque
    c. Bolt head
    d. Thread pitch

13. Pitch for a metric bolt is the:
    a. Center-to-center distance between threads in millimeters
    b. Thread angle divided by the diameter in millimeters
    c. Number of threads per centimeter
    d. Number of threads per millimeter

14. Technician A says you should always lubricate the threads of a bolt before installing it.
    Technician B says you should clean out the bolt hole with a tap before installing the bolt.
    Who is right?
    a. A only
    b. B only
    c. Both A and B
    d. Neither A nor B

15. Technician A says that engine mounts made of metal brackets with rubber insulators are tuned to reduce engine vibration.
    Technician B says that some engine mounts use hydraulic damping.
    Who is right?
    a. A only
    b. B only
    c. Both A and B
    d. Neither A nor B

ASE Technician Certification Sample Test

This sample test is similar in format to the series of tests given by the National Institute for Automotive Service Excellence (ASE). Each of these exams covers a specific area of automotive repair and service. The tests are given every fall and spring throughout the United States.

For a technician to earn certification in a particular field, he or she must successfully complete one of these tests, and have at least two years of "hands on" experience (or a combination of work experience and automobile mechanics training). A person who successfully completes all of the tests is a certificated ASE Master Automotive Technician.

The questions in this sample test follow the format of the ASE exams. Learning to take this kind of test will help you if you plan to apply for certification later in your career. You can find the answers to the questions in this sample exam on page 328 of this Classroom Manual.

For test registration forms or additional information on the automobile technician certification program, write to:

National Institute for
AUTOMOTIVE SERVICE EXCELLENCE
13505 Dulles Technology Drive
Herndon, VA 22071-3415

1. The micrometer reading shown corresponds to:

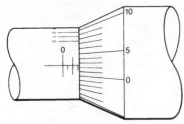

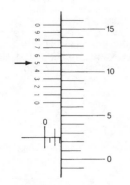

   a. 0.1055 inch
   b. 0.0775 inch
   c. 0.155 mm
   d. 1.525 mm

2. Engines are sometimes fitted with undersize or oversize bearings, lifters, or other components at the factory. These engines can be identified by:
   a. Casting marks
   b. The VECI sticker
   c. The VIN code
   d. None of the above

3. A compression test was performed on a four-cylinder engine with the following results: Cylinder #1 = 138 psi, cylinder #2 = 127 psi, cylinder #3 = 115 psi, cylinder # 4 = 135 psi.
   Technician A says low compression in cylinders 2 and 3 indicates that the cylinder head gasket is leaking between these two cylinders.
   Technician B says the compression test is not conclusive enough to determine the problem. Uneven readings may be caused by burnt, sticking, or improperly adjusted valves or carbon deposits on the valve faces.
   Who is right?
   a. A only
   b. B only
   c. Both A and B
   d. Neither A nor B

4. A vacuum gauge is connected to an engine. The gauge reads 18 in-Hg at idle. When engine speed is increased, the gauge momentarily drops to near zero, then slowly rises and stabilizes at 9 in-Hg.
   What could cause these readings?
   a. Weak piston rings
   b. Incorrect fuel mixture
   c. Restricted exhaust
   d. Retarded ignition timing

5. A cylinder leakage test, or leak-down test, is a diagnostic procedure that can be used to detect:
   a. Broken piston rings
   b. Leaking cylinder head gaskets
   c. Burnt valves
   d. All of the above

6. Which of the following problems can cause low engine oil pressure?
   a. Worn main bearings
   b. Oil pressure relief valve stuck closed
   c. Worn oil control rings
   d. All of the above

7. An engine checks out good on all of these diagnostic tests, manifold vacuum, compression, cylinder leakage, and cylinder power balance but suffers from high oil consumption even though the valve guide seals were recently replaced.
   Technician A says that the problem may be caused by a clogged PCV valve or worn valve guides.
   Technician B says clogged oil drainback passages or worn pushrod tips can increase oil consumption.
   Who is right?
   a. A only
   b. B only
   c. Both A and B
   d. Neither A nor B

8. The dial indicator in this illustration is set up to read:

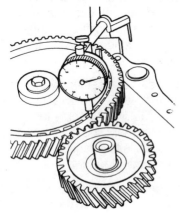

  a. Camshaft endplay
  b. Cam gear backlash
  c. Cam gear runout
  d. Both A and B

9. Technician A says that when installing a new timing belt, you should turn the camshaft slightly if the belt teeth do not line up with the cogs on the sprockets. Technician B says that you should rotate the crankshaft slightly to align the belt teeth to the sprockets when installing a new timing belt.
  Who is right?
  a. A only
  b. B only
  c. Both A and B
  d. Neither A nor B

10. Technician A says that since the camshaft rotates at half crank speed and does not receive the force of combustion, there is no need to measure the cam journals or lobes before reinstalling a used camshaft into a rebuilt engine.
  Technician B says that you must measure both the journals and lobes of a used camshaft. If the cam checks out, install a new set of lifters.
  Who is right?
  a. A only
  b. B only
  c. Both A and B
  d. Neither A nor B

11. A common way to repair a cast-iron cylinder head that has a crack in a nonstressed area is:
  a. Welding
  b. Stop-drilling
  c. Pinning
  d. All of the above

12. Which of the following block machining operations is not generally performed when rebuilding a pushrod engine?
  a. Boring or honing the main bearing bores
  b. Boring or honing the cylinder bores
  c. Boring or honing the cam bearing bores
  d. Deck resurfacing

13. Technician A says that worn cylinders in a hypereutectic engine block should not be bored to oversize. The only proper repair is to sleeve the cylinders and install standard pistons. You can't bore it out because there's no way to get oversize pistons for these engines.
  Technician B says that you can open the bores of a hypereutectic casting by honing to accept oversize pistons, which are readily available.
  Who is right?
  a. A only
  b. B only
  c. Both A and B
  d. Neither A nor B

14. After boring or honing the main bearing bores, you must:
  a. Install oversize main bearings
  b. Install undersize main bearings
  c. Grind the crankshaft journals to match the bores
  d. None of the above

15. Technician A says that a torque plate should be installed and tightened to specified torque in order to tension the engine block when you bore and hone the cylinders to oversize. Technician B says a torque plate and head gasket should be installed to tension the engine block when you finish hone the cylinders.
  Who is right?
  a. A only
  b. B only
  c. Both A and B
  d. Neither A nor B

16. Which of the following is the standard sequence of operations for reconditioning a cylinder block?
  a. Recondition the cylinder bores, recondition the main bearing bore, recondition the deck surface
  b. Recondition the main bearing bore, recondition the deck surface, recondition the cylinder bores
  c. Recondition the deck surface, recondition the main bearing bore, recondition the cylinder bores
  d. None of the above

17. A cylinder bore measures from 3.072 inches at the bottom and 3.077 below the ridge at the top, the standard factory bore is 3.065. What would be the proper repair?
  a. Knurling the pistons, deglazing the cylinder walls, and installing new cast-iron rings
  b. Boring and honing the cylinder to 3.097 inches and installing 0.030 over pistons with new rings
  c. Boring and honing the cylinder to 3.075 inches and installing knurled pistons with new rings
  d. Boring and honing the cylinder to 3.095 inches and installing 0.030 over pistons with new rings

18. Technician A says that dressing the valve faces increases the valve spring installed height and you must correct it by grinding the valve tip. Technician B says that dressing the valve faces decreases the valve spring installed height and you can correct it by installing shims under the spring.
  Who is right?
  a. A only
  b. B only
  c. Both A and B
  d. Neither A nor B

19. After facing, valves must have a margin of at least:
  a. 1/64 inch
  b. 1/32 inch
  c. 1/8 inch
  d. None of the above

20. To avoid valve-to-piston interference on a high-compression engine, you must measure:
  a. Installed height
  b. Cylinder head height
  c. Deck height
  d. Cam lobe height

21. Which of the following is the proper sequence for repairing a warped aluminum cylinder head that requires new valve guides?
    a. Remove the valve guides, straighten the head, install new guides, grind the seats, resurface the head, then line bore the cam bearing bores
    b. Install new valve guides, grind the seats, straighten the head, line bore the cam bearing bores, then resurface the head
    c. Remove the valve guides, straighten the head, install new guides, grind the seats, line bore the cam bearing bores, then resurface the head
    d. Straighten the head, replace the valve guides, grind the seats, resurface the head, then line bore the cam bearing bores

22. Resurfacing an OHC cylinder head will:
    a. Increase valve lift
    b. Alter cam timing
    c. Change valve lash
    d. All of the above

23. Which of the following statements is NOT true?
    a. Worn stamped-steel rocker arm should be replaced
    b. The contact pads on cast-iron rocker arms can often be resurfaced
    c. New bushings are often installed to repair shaft-mounted cast-iron rocker arms
    d. Excessive wear on rocker shafts or pivot fulcrums usually results from lack of lubrication

24. You measure valve spring installed height from:
    a. Spring seat to valve tip
    b. Valve guide top to valve keeper groove
    c. Valve retainer top to valve stem end
    d. Spring seat to valve retainer

25. What is the most common cause of premature bearing failure in a rebuilt engine?
    a. Improper oil clearance
    b. Particle contamination
    c. Poor crankshaft journal finish
    d. Inadequate bearing crush

26. The tool in the illustration is used to:

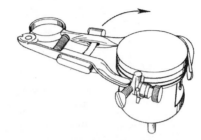

    a. Measure piston diameter
    b. Undercut a worn ring groove to install a shim
    c. Expand piston rings
    d. Clean ring grooves

27. When balancing an in-line engine, which of the following do you not need?
    a. Crankshaft
    b. Wrist pins
    c. Rod bearings
    d. Connecting rods

28. Technician A says that when calculating the bob weight for most V8 engines, you must double the reciprocating weight. Technician B says it is not necessary to double the reciprocating weight because of the balance factor most commonly used for V8 engines. Who is right?
    a. A only
    b. B only
    c. Both A and B
    d. Neither A nor B

29. Technician A says that bad integral valve seats in a cast-iron head must be removed by counter boring. Technician B says that bad insert seats in an aluminum head can usually be driven out with a long punch or lifted out with a pry bar. Who is right?
    a. A only
    b. B only
    c. Both A and B
    d. Neither A nor B

30. Technician A says to knurl valve guides that are worn more than 0.005 inch (0.13 mm) oversize. Technician B says that valve guide inserts usually have an interference fit into the cylinder head of about 0.001 to 0.002 inch (0.025 to 0.050 mm). Who is right?
    a. A only
    b. B only
    c. Both A and B
    d. Neither A nor B

31. Technician A says that when grinding valve seats, it is best to leave a wider intake seat than exhaust seat for improved heat transfer.
    Technician B says that when grinding the valve seats, it is best to leave a wider exhaust seat than intake seat for improved airflow.
    Who is right?
    a. A only
    b. B only
    c. Both A and B
    d. Neither A nor B

32. Engine block core plugs that are larger than standard size are usually stamped _____ on the cup face.
    a. OS
    b. SO
    c. OO
    d. B

33. Technician A says to install cam bearings with a few drops of motor oil so that they slide into place easier and do not damage their edges.
    Technician B says to install cam bearings dry because any lubricant on the outside diameter could increase their chances of spinning in a running engine.
    Who is right?
    a. A only
    b. B only
    c. Both A and B
    d. Neither A nor B

34. A customer says his car, which has 90,000 miles, is running poorly and is low on power. The ignition and fuel systems are in good condition and set to specifications. Compression and cylinder leakage tests were performed with the following results:

	Compression	Leakage
Cylinder #1	142 psi	17%
Cylinder #2	147 psi	15%
Cylinder #3	144 psi	14%
Cylinder #4	150 psi	12%
Cylinder #5	145 psi	14%
Cylinder #6	148 psi	13%

The engine manufacturer's specifications call for compression to be 155 to 185 psi with no more than 30 psi difference between cylinders. What could be the cause of the poor performance?
    a. Carbon deposits on the valves and combustion chambers
    b. Worn camshaft lobes or collapsed hydraulic lifters
    c. Leaking port fuel injectors or injector seals
    d. Camshaft out of time

35. An engine has been steam cleaned, removed from the vehicle, and disassembled. The engine appears that it will require extensive machine work to bring it within specification. Technician A says that your next steps should be to check the castings for cracks by magnetic particle inspection, thoroughly clean the casting, and then perform the necessary machine work.
Technician B says you should perform the machine work, clean the castings, and then check for cracks using a combination of magnetic particle inspection and pressure testing.
Who is right?
a. A only
b. B only
c. Both A and B
d. Neither A nor B

36. An 0.018 inch feeler gauge will slip easily under a straightedge on an aluminum cylinder head from an OHV engine. The block deck did not require surfacing and the factory specifications allow you to remove up to 0.030 inch from both surfaces.
Technician A says the head should be straightened before it is surfaced because it will take about a 0.025-inch cut to clean up the head and even though it is within specification, camshaft timing will be affected.
Technician B says that you could simply machine the head flat, but that changes the compression ratio. The best repair would be to straighten the head, first, to keep things as close to stock as possible.
Who is right?
a. A only
b. B only
c. Both A and B
d. Neither A nor B

37. Technician A says to check crankshaft endplay by first prying the crank as far back in the block as it will go and zeroing the dial indicator. Then, pry the crank as far forward as it will go and note the reading on the indicator.
Technician B says to check crank endplay by zeroing the dial indicator with the crankshaft centered in the block. Then, pry the crank as far forward in the block as it will go and note the reading on the indicator.
Who is right?
a. A only
b. B only
c. Both A and B
d. Neither A nor B

38. A hot tank filled with a caustic soda solution is best for cleaning which type of material?
a. Aluminum and magnesium
b. Non-ferrous metals
c. Iron and steel
d. Bronze

Glossary of Technical Terms

Adhesive: A substance that holds a part such as a gasket in place during component assembly.

Aerobic Sealant: A sealant that cures in the presence of air.

Air-Fuel Ratio: The ratio of the weight of air to gasoline in the air-fuel mixture drawn into an engine.

Align Boring: A boring process that aligns main bearing seats or camshaft bearings to ensure that they are concentric. Also called line boring.

Alloy: A mixture of two or more metals. Brass is an alloy of copper and zinc. Stainless steel is an alloy of steel and chromium.

Anaerobic Sealant: A sealant that cures in the absence of air.

Annealing: The process of heating a metal to a high temperature and allowing it to cool very slowly. Annealing makes metal softer and easier to machine.

Anodizing: The process of oxidizing a metal, such as aluminum or magnesium, by running electric current through a special solution containing the metal anode.

Anti-Seize Lubricant: A lubricant designed to keep threads from being frozen in place by corrosion.

Antioxidants: Chemicals or compounds added to motor oil to reduce oil oxidation, which leaves carbon and varnish in the engine.

API Service Classification: A system of letters signifying an oil's performance. It is assigned by the American Petroleum Institute.

Atmospheric Pressure: The pressure on the earth's surface caused by the weight of air in the atmosphere. At sea level, this pressure is 14.7 psi (101 kiloPascals) at 32°F (0°C).

Atomization: Breaking a liquid down into small particles or a fine mist.

Axial Movement: Movement parallel to the axis of a shaft. The measurement of axial movement on a shaft is called endplay.

Backpressure: The resistance, caused by turbulence and friction, that is created as a gas or liquid is forced through a restrictive passage.

Base Circle: An imaginary circle that can be drawn on the profile of a cam lobe that intersects the very bottom of the lobe. It is the lowest point on the cam lobe in relation to the valve train.

Bearing Bore: The full circle machined surface that supports the back of a bearing. It may consist of a saddle and cap bolted together, as in rod or main bearings, or a bored hole, used for most camshafts.

Bearing Clearance: The gap or space between a bearing and its shaft that fills with oil when the engine is running. Also called the oil clearance.

Bearing Crown: The middle of a main or connecting rod bearing. This is its thickest point.

Bearing Crush: The distance that an installed bearing insert is higher than its saddle or web.

Bearing Spread: The distance the diameter of an uninstalled bearing insert is wider than the diameter of its saddle or web.

Bifurcated: Separated into two parts. A bifurcated exhaust manifold has four primary runners that converge into two secondary runners; these converge into a single outlet in the exhaust system.

Billet: An unfinished block or bar of metal that is ready for machining.

Bimetal Temperature Sensor: A sensor or switch that reacts to changes in temperature. It is made of two strips of metal welded together that expand differently when heated or cooled, causing the strip to bend.

Blowby: Combustion gases that get past the piston rings into the crankcase; these include water vapor, acids, and unburned fuel.

Bob Weight: A weight added to the crankshaft of a V engine prior to spinning the crank on a balancing machine. The bob weight simulates a percentage of the reciprocating mass.

Bore: The diameter of an engine cylinder. Also, to enlarge or finish the surface of a drilled hole.

Boring: A process by which a machinist enlarges a hole or cylinder.

Bottom Dead Center (BDC): The exact bottom of a piston stroke.

Boundary Lubrication: Lubrication protection afforded by the remaining oil film when load has forced a part, such as a crank journal, through the oil wedge.

Brake Horsepower: The power available at the flywheel of an engine for doing useful work, as measured on a dynamometer.

British Thermal Unit (Btu): The amount of heat required to raise the temperature of one pound of water one degree Fahrenheit.

Brittle: A metal is brittle if a blow can easily crack or break it.

Bushing: A small, full circle plain bearing.

calorie: The amount of heat required to raise the temperature of one gram of water by one degree Celsius. The lowercase "c" is not a misprint. When capitalized, it stands for 1,000 calories.

Cam Follower: A one-piece valve train part that follows the motion of the cam lobe and transfers it to the valve stem. A follower has a leverage ratio similar to a rocker arm. Also called a finger follower.

Cam Lobes: Egg-shaped lobes that open and close the valves. A series of cam lobes on a shaft is a camshaft.

Carbon Monoxide: An odorless, colorless, tasteless, poisonous gas. A major pollutant given off by an internal combustion engine.

Carburizing: A process that causes the cam lobe surfaces to absorb additional carbon. Then, the surfaces can be hardened to a greater degree than the parent metal.

Cast Iron: Pig iron remelted and poured into a product shape.

Cavitation: An undesirable condition in the cooling system caused by the water pump blades. As the blades turn, they form low-pressure (vacuum) bubbles behind them.

Centrifugal Force: A force exerted by a rotating object, tending to move the object toward the outer edge of the circle of rotation.

Circlips: Round spring-steel retainers that fit into grooves at the end of the piston pin bore to hold the pin in place.

Clearance Volume: The combustion chamber volume with the piston at TDC.

Closing Ramp: The part of the cam lobe that closes the valve. The closing ramp lobe reduces its rate of letdown so the valve closes very gently against the seat.

Coil Bind: The condition in which all the coils of a spring are touching one another so that no further compression is possible.

Compatibility: The ability of a bearing lining to allow friction without excessive wear.

Composite: A man-made material that consists of two or more physically or chemically different components tightly bound together. The composite material has properties that neither component possesses alone.

Compression Ratio: The ratio of the total cylinder volume when the piston is at BDC compared to the volume of the combustion chamber when the piston is at TDC.

Conformability: The ability of a bearing lining to conform to irregularities in a bearing journal.

Counterweights: The weights opposite the rod journals on a crankshaft. They balance the weight of the reciprocating mass: the journal, bearing, connecting rod, and piston assembly.

Crankpins: Another name for rod bearing journals.

Crankshaft Counterweights: Weights cast into, or bolted onto, a crankshaft to balance the weight of the piston, piston pin, connecting rod, bearings, and crankpin.

Crankshaft Endplay: The end to end movement of the crankshaft in the bearings.

Crankshaft Throw: A part of the crankshaft measured from the crank centerline to the crankpin centerline. It is equal to one-half of the crankshaft stroke.

Cross-drilled Crankshaft: A crankshaft with an additional oil drillway in the main bearing journal that improves lubrication to the rod journals.

Crossflow Cylinder Head: A head with the intake port and exhaust port exiting from opposite sides of the head. The design provides an almost straight path for the mixture to flow across the top of the piston at overlap, for scavenging.

Crush Relief: A thinning of the bearing at the parting face. This relief allows for crush in a bearing shell.

Cylinder Sleeve: The liner for a cylinder which provides a good surface for the piston rings. Some sleeves can be replaced when worn to provide a new cylinder surface.

Detergent: A chemical compound added to motor oil that removes dirt or soot particles from surfaces, especially piston rings and grooves.

Detonation: An unwanted explosion of a small portion of the air-fuel mixture caused by a sharp rise in combustion chamber pressure. The noise it makes is commonly called "knocking" or "pinging."

Dispersant: A chemical added to motor oil that keeps sludge and other undesirable particles picked up by the oil from gathering and forming deposits in the engine.

Displacement: The measurement of engine volume. It is calculated by multiplying the piston displacement of one cylinder by the number of cylinders. The total engine displacement is the volume displaced by all the pistons.

Dry Sleeve: A sleeve that does not come in direct contact with the engine coolant.

Ductility: The ability to be drawn or stretched without breaking. Ductile metals are not brittle.

Duration: The valve timing period that a valve is open. Measured in degrees of crankshaft rotation.

Dynamometer: A machine used to measure the power of an engine or motor.

Eccentric: A circle mounted off-center on a shaft. It is used to convert rotary motion to reciprocating motion.

Electronic Fuel Injection (EFI): A computer-controlled fuel-injection system which gives precise mixture control under all operating conditions at all speed ranges.

Embeddability: The ability of a bearing to absorb particle contamination into the lining of the bearing.

Energy: The ability to do work by applying force.

External Combustion Engine: An engine, such as a steam engine, in which fuel is burned outside the engine.

Externally Balanced Engines: Engines that have crank counterweights but require additional weights on the flywheel and/or vibration damper to complete the crank balance.

Fatigue Strength: The ability of a bearing to withstand the loads placed on it.

Fillet: The curved portion of the valve where the stem meets the head. Also, a radius ground on the crankshaft where a journal joins a cheek. The fillet strengthens this joint.

Firing Order: The sequence by cylinder number in which combustion occurs in the cylinders of an engine.

Flanks: The ramps on either side of the cam lobe, between the nose and base circle.

Flat Tappet: A valve lifter or tappet where the base of the lifter that rides against the cam lobe appears flat. Either solid or hydraulic lifters can be flat tappets.

Float Bowl: Gasoline is delivered by the fuel pump from the fuel tank to the carburetor float bowl, where it is ready for use.

Fluid Friction: The friction between the layers of molecules in a fluid.

Force: A push or pull acting on an object; it may cause motion or produce a change in position. Force is measured in pounds in the U.S. system and in Newtons in the metric system.

Four-Stroke Engine: The Otto-cycle engine. An engine in which a piston must complete four strokes to make up one operating cycle. The strokes are: intake, compression, power, and exhaust.

Gross Brake Horsepower: The flywheel horsepower of a basic engine, without accessories, measured on a dynamometer.

Gross Valve Lift: The distance that the valve lifts off its seat. It is a theoretical number because it does not take into account lift lost from deflections or clearances in the valve train.

Hardness: The resistance to being dented or penetrated. Hard metals are usually brittle.

Heel: The part of the cam lobe directly opposite the nose. All valve adjustment is done with the cam in this position.

Helical Gears: Gears with teeth cut at an angle to the shaft instead of parallel to it.

Honing: A process by which a machinist enlarges or finishes the interior surface of a cylinder.

Horsepower: A measure of the rate at which work is done, equal to 33,000 foot-pounds of work per minute, or 0.746 kilowatts.

Hydraulic Lifter: A valve lifter or tappet that uses oil pressure to automatically adjust for wear or play in the valve train. Hydraulic lifters always maintain zero clearance in the valve train so that they need no periodic adjustment and operate quietly.

Hydraulic Lifter Leak Down: Hydraulic leakage between the lifter plunger and body. It allows for heat expansion in the valve train parts and helps reduce lifter pump-up during valve float.

Hydrocarbon: Any chemical compound made up of hydrogen and carbon. A major pollutant given off by an internal combustion engine. Gasoline is a hydrocarbon compound.

Ignition Interval: The number of degrees of crankshaft rotation between ignition sparks. Also called firing interval.

Inertia: The tendency of an object at rest to remain at rest. Also, the tendency of an object in motion to remain in motion.

Injection Pump: A pump used on diesel engines to deliver fuel under high pressure at precisely timed intervals to the fuel injectors.

Interference Angle: The angle or difference between the valve face angle and the valve seat angle.

Internal Combustion Engine: An engine, such as a gasoline or diesel engine, in which fuel is burned inside the engine.

Internally Balanced Engines: Engines that rely on the crankshaft counterweights for all crankshaft balance.

Jet: A metal orifice in a carburetor which meters fuel flow. Larger jets allow more fuel to flow than smaller jets. Jets are usually replaceable and available in different sizes.

Kilowatt (kW): A measure of the rate at which work is done, equal to 1.3405 horsepower.

Kinetic Energy: The energy of mass in motion. All moving objects possess kinetic energy.

Lash Ramp: The initial portion of the cam lobe that meets the lifter as the cam rotates off its base circle. The lash ramp compensates for small defections and begins to take up slack in the valve train without actually lifting the valve.

Lifter Pump-up: The condition in which a hydraulic lifter adjusts or compensates for the additional play in the valve train created by valve float.

Lobe Lift: The lift provided by the cam lobe. Best measured with a dial indicator.

Lobe Separation Angle: On a camshaft, the angle between the centerline of the intake lobe and the centerline of the exhaust lobe of a single cylinder.

Main Bearing Journals: The parts of the crankshaft on which it rotates in the cylinder block.

Major Thrust Surface: The side of the piston that pushes against the cylinder wall during the power stroke.

Mass: The measure of the amount of matter in an object.

Matter: What all material elements are made of; anything that occupies space and is perceptible to the senses in some way.

Microinch: One millionth of an inch, abbreviated μin. The standard measurement for surface finish in the American customary system.

Micro-meter: One millionth of a meter, abbreviated μm. The standard measurement for surface finish in the metric system.

Multigrade: An oil that meets viscosity requirements at more than one test temperature, and so has more than one SAE viscosity number.

Multilayer Steel Head Gaskets: A head gasket made of layers of steel.

Mushroom Lifter: A valve lifter or tappet with a foot or base larger in diameter than its body.

Net Brake Horsepower: The flywheel horsepower of a fully equipped engine, with all accessories in operation, measured on a dynamometer.

Net Valve Lift: The actual amount the valve lifts off its seat. It is slightly less than gross lift because of the bending of pushrods, rocker arms, and flex in the camshaft itself.

Newton-meter (Nm): The metric unit of torque. One Newton-meter equals 1.356 foot-pounds. One foot-pound equals 0.736 Newton-meter.

Nitriding: A form of casehardening where nitrogen is placed into the surface of a metal.

Normalize: The process of heating metal to its normalizing temperature, holding it there for about one hour for each inch of thickness, and allowing it to cool in the open air. Normalizing removes internal stresses in the metal.

Nose: The highest part of the cam lobe that holds the valve in the fully open position.

Nose Circle: An imaginary circle that can be drawn on the profile of a cam lobe that intersects the very top of the lobe. It is the highest point on the cam lobe in relation to the valve train.

Octane Rating: The measure of a gasoline's resistance to detonation.

Opening Ramp: The part of the cam lobe that does the actual opening of the valve.

Overlap: A relatively short valve timing period during which both the intake and exhaust valves are open. Measured in degrees of crankshaft rotation.

Overplate: A material added to the surface of the bearing lining to improve the conformability and embeddability. Only the harder, heavy-duty bearing alloys need overplates.

Oversize Bearing: A bearing made thicker than standard. It has a larger outside diameter to fit an oversize bearing bore.

Oxidation Reaction: A chemical reaction in which electrons are added to a compound, such as when oxygen molecules combine with other molecules to form a new compound.

Oxides of Nitrogen: Chemical compounds of nitrogen given off by an internal combustion engine. They combine with hydrocarbons to produce smog.

Particulates: Liquid or solid particles such as lead and carbon that are given off by an internal combustion engine as pollution.

Parting Face: The meeting points of the two halves of split plain bearings, such as main or connecting rod bearings.

Pig Iron: Small solidified blocks of iron poured from the blast furnace into channels dug in sand. Used to make steel and cast iron.

Pin Boss: The area surrounding the hole in the piston for the piston pin.

Piston Crown: The top of the piston. It forms the bottom of the combustion chamber.

Piston Pin: The hollow metal rod that attaches the piston to the connecting rod. Also called a wrist pin.

Piston Skirt: The part of the piston below the pin bosses. The skirt stabilizes the piston in the bore to prevent the piston from rocking.

Piston Slap: A noise caused by a loose-fitting piston skirt hitting against the cylinder wall.

Piston Speed: The average speed of the piston, in feet or meters per minute, at a specified engine rpm.

Plain Bearing: A bearing that slides or rubs on the part it supports.

Plateaued Finish: A surface finish in which the highest parts of a surface have been honed to flattened peaks. A plateaued finish is the best finish possible for new cylinder bores.

Plenum: A chamber that stabilizes the air-fuel mixture and allows it to rise to a pressure slightly above atmospheric pressure.

Poppet Valve: A valve that plugs and unplugs its opening by linear movement.

Port Fuel Injection: A fuel-injection system in which individual injectors are installed in the intake manifold at a point close to the intake valve. Air passing through the manifold mixes with the injector spray just as the intake valve opens.

Positive Crankcase Ventilation (PCV): Late-model crankcase ventilation systems that return blowby gases to the combustion chambers.

Potential Energy: Energy stored but not being used. A wound-up spring and gallon can of gasoline both have potential energy.

Pour-Point Depressants: Chemical compounds added to motor oil to help the oil flow at colder temperatures.

Power: The rate or speed of doing work.

Preignition: An unwanted, early ignition of the air-fuel mixture.

Pressure Differential: The difference in pressure between two points.

Primary Vibration: Strong, low-frequency vibration produced in a single plane. It is caused by the up-and-down movement of the piston and connecting rod.

Quench Area: An area in the combustion chamber that has only a few thousandths of an inch clearance from the piston at TDC. The close proximity of the cool cylinder head and piston crown prevent the end gases between them from becoming hot enough to autoignite.

Ramps: The part of the cam lobe that lifts and closes the valve.

Reciprocating Engine: An engine in which the pistons move up and down or back and forth as a result of the combustion of an air-fuel mixture at one end of the piston cylinder. Also called piston engine.

Reduction Reaction: A chemical reaction in which electrons are removed from a compound, such as when oxygen is removed from a compound.

Reed Valve: A one-way check valve. A reed, or flap, opens to admit a fluid or gas under pressure from one direction, while closing to deny movement from the opposite direction.

Resilience: The tendency of a material to return to its original shape after being bent, compressed, or deformed.

Resonate: An effect produced when the natural vibration frequency of a part is greatly amplified by reinforcing vibrations at the same or nearly the same frequency by another part.

Ring Flutter: Rapid up and down movement of a piston ring that breaks the seal between the ring and the cylinder wall.

Ring Grooves: Grooves in the upper part of the piston that hold the piston rings.

Ring Lands: The part of the piston between the ring grooves. The lands strengthen and support the ring grooves.

Ring Tension: The natural spring tension of a piston ring. It helps the ring seal against the cylinder wall during the intake and exhaust strokes and the beginning of the compression stroke, when combustion gases are not forcing the rings against the wall.

Road Draft Tube: The earliest type of crankcase ventilation; it vented blowby gases to the atmosphere.

Rocker Arm Geometry: The angle of the rocker arm in relation to the valve stem. Rocker arm geometry is correct when the center of the rocker arm tip contacts the centerline of the valve stem when the valve train is at 30 to 50 percent of its maximum lift.

Rocker Arm Ratio: The measure of the rocker arm's lever or mechanical advantage. On a rocker arm, the distance from the pivot point to the pushrod seat (or cam lobe) is shorter than the distance from the pivot point to the valve stem pad. The rocker arm ratio is usually around 1.5:1.

Rocking Couple: Vibration or unbalance set up by an alternating rising and falling of two masses that are out of phase.

Rod Bearing Journals: The parts of the crankshaft that the connecting rods attach to. They are also called crankpins.

Roller Chain: A chain made of links, pins, and rollers. Each link has two plates, held together by pins, and separated by a roller around the pin. The length of the rollers determines chain width.

Roller Tappet: A valve lifter or tappet with a roller that rides on the camshaft lobe. Either solid or hydraulic lifters can be roller tappets.

Rolling Element Bearing: A bearing that rolls on the part it supports.

Room-Temperature Vulcanizing (RTV) Compound: A silicone-based, aerobic sealant commonly used in automotive applications.

Rotary Valve: A valve that rotates to cover and uncover the intake port of a two-stroke engine at the proper time. Rotary valves are usually flat discs driven by the crankshaft. Rotary valves are more complex than reed valves, but are effective in broadening a two-stroke engine's power band.

Runners: The passages or branches of an intake manifold that connect the manifold's plenum chamber to the engine's inlet ports.

SAE Viscosity Grade: A system of numbers signifying an oil's viscosity at a specific temperature. It is assigned by the Society of Automotive Engineers.

Sandwich-Type Head Gasket: A head gasket made in layers, typically a center layer of steel and surface layers of heat-resistant, compressible material.

Scuffing: The transfer of metal between two rubbing parts. It is usually caused by a lack of lubrication.

Sealant: A substance that forms a seal between two facing surfaces.

Secondary Vibration: Relatively weak, high-frequency vibration. It is caused by the difference in upward and downward force of the piston due to the rod-to-stroke ratio.

Shot-Peening: A surface treatment process in which the surface of a metal part is blasted with round, steel shot in a controlled manner. Shot-peening reduces the chances of surface cracks developing.

Siamesed Cylinder Bores: A bore design where the cylinder walls of adjacent cylinders are attached to each other.

Silent Chain: A chain made of flat links that pivot around pins. Each link is a series of plates held together by the pins. The number of plates determines chain width.

Single Grade: An oil that has been tested at only one temperature, and so has only one SAE viscosity number.

Sludge: A thick, black deposit caused by the mixing of blowby gases and oil.

Solid Lifter: A valve lifter or tappet that is one-piece, lightweight, and usually made of cast iron. Solid lifters cannot compensate for wear so the valve clearance must be checked and adjusted at regular intervals.

Spring Height: The dimension from the spring seat to the bottom of the retainer. Automotive machinists measure it when springs are installed.

Squish Area: A narrow space between the piston and the cylinder head. As the piston approaches top dead center, the mixture squishes or shoots out of the squish area across the combustion chamber. Mixture turbulence results and improves combustion.

Stellite: The brand name for a very hard alloy of chromium, cobalt, and tungsten applied to valve faces for longer wear.

Stoichiometric Ratio: The chemically correct air-fuel mixture for combustion in which all oxygen and all fuel will be completely burned.

Stress Riser: A groove, scratch, or other imperfection in the metal that could develop into a crack.

Stroke: One complete top-to-bottom or bottom-to-top movement of an engine piston.

Sulfur Oxides: Chemical compounds given off by processing and burning gasoline and other fossil fuels. As they decompose, they combine with water to form sulfuric acid.

Surface Roughness Indicator: An electrical instrument that measures surface roughness by running a diamond stylus over the surface of a metal.

Surface-to-Volume Ratio: The ratio of the surface area of a three-dimensional space to its volume.

Synthetic Motor Oils: Lubricants formed by artificially combining molecules of petroleum and other materials.

Tempering: The process of reheating steel to a temperature between 300°F to 1,100°F (150°C to 600°C), then allowing it to cool slowly. Tempering makes the steel more workable and improves toughness.

Tensile Strength: A metal's resistance to a force pulling it apart. The force is measured in pounds per square inch (psi) or millions of pascals (MPa).

Tensioners: A wheel or pad, pressed against a belt or chain, to maintain its tension.

Tetraethyl Lead: A gasoline additive used to help prevent detonation.

Thread Pitch: The distance between threads on a fastener. Also, the number of threads along a certain length.

Throttle-Body Fuel Injection (TBI): A fuel-injection system in which one or two injectors are installed in a carburetor-like throttle body mounted on a conventional intake manifold. Fuel is sprayed at a constant pressure above the throttle plate to mix with the incoming air charge.

Top Dead Center (TDC): The exact top of a piston stroke. Also a specification used when tuning an engine.

Torque: The tendency of a force to produce rotation around an axis. It is equal to the distance to the axis times the amount of force applied expressed in foot-pounds or inch-pounds (U.S. customary system) or in newton meters (metric system).

Torsional Vibration: Vibration forces that place a twisting stress on a part.

Toughness: The ability of a metal to withstand a sudden blow. The test for toughness uses a falling weight onto the test metal.

Tuftriding: A tradename for a form of salt-bath nitriding. General Motors favors this process because it works well in mass production.

Two-Stroke Engine: An engine in which a piston makes two strokes to complete one operating cycle.

Undersize Bearing: A bearing made thicker than standard. It has a smaller inside diameter to fit an undersize crankshaft.

Vacuum: Air pressure lower than atmospheric pressure.

Valve Face: A precision ground surface of the valve that seals against the valve seat.

Valve Float: The condition in which the valve continues to open or stays open after the cam lobe has moved from under the lifter. This happens when the inertia of the valve train at high speeds overcomes the valve spring tension.

Valve Margin: A small margin around the edge of the valve head between the end of the head and the valve face. The margin adds strength and heat resistance to the valve.

Valve Reliefs: Small pockets cut into the piston crown to provide clearance for the valves.

Valve Train: The assembly of parts that transmits force from the camshaft lobe to the valve, including the camshafts and valves themselves.

Vaporization: Changing a liquid, such as gasoline, into a vapor (gas) by evaporation or boiling.

Vaporize: To change from a solid or liquid into a vapor.

Variable Rate Spring: A spring that changes its rate of pressure increase as it is compressed. This is achieved by unequal spacing of the spring coils.

Varnish: An undesirable deposit, usually on the engine pistons, formed by oxidation of fuel and of motor oil.

Venturi: A restriction in an airflow, such as in a carburetor, that speeds the airflow and helps create a pressure differential.

Viscosity: The tendency of a liquid such as oil to resist flowing.

Viscosity Index (VI) Improvers: Chemical compounds added to motor oil to help the oil resist thinning at high temperatures.

Volatility: The ease with which a liquid changes from a liquid to a gas or vapor. Gasoline is more volatile than water because it evaporates quicker.

Volumetric Efficiency: The comparison of the actual volume of air-fuel mixture drawn into an engine to the theoretical maximum volume that could be drawn in. Written as a percentage.

Water Jackets: Passages in the head and block that allow coolant to circulate throughout the engine.

Weight: The measure of the earth's gravitational pull on an object.

Wet Sleeve: A sleeve that comes in contact with the engine coolant.

Work: The application of energy through force to produce motion.

Answers to Review and ASE Questions

Chapter 1: Engine Operation and Construction

1-A 2-B 3-D 4-C 5-A 6-B 7-D 8-B
9-B 10-C 11-A 12-D 13-C 14-B
15-D 16-C 17-D 18-A 19-D 20-D
21-A 22-C

Chapter 2: Engine Physics and Chemistry

1-D 2-A 3-C 4-B 5-C 6-D 7-D 8-B
9-A 10-D 11-C 12-D 13-A 14-D
15-B 16-C 17-D 18-D 19-C 20-B

Chapter 3: Engine Materials and Manufacture

1-D 2-C 3-B 4-A 5-B 6-C 7-D 8-A
9-C 10-D 11-B 12-D 13-C 14-A
15-C 16-C 17-D 18-D 19-A 20-B

Chapter 4: Cooling Systems

1-B 2-A 3-B 4-B 5-C 6-D 7-A 8-C
9-D 10-C 11-D 12-D 13-B 14-C
15-C 16-A 17-C 18-C 19-A 20-B

Chapter 5: Intake and Exhaust Systems

1-A 2-D 3-A 4-A 5-D 6-C 7-A
8-A 9-C 10-B 11-C 12-B 13-C
14-D 15-C 16-D 17-B

Chapter 6: Engine Lubrication

1-C 2-D 3-B 4-D 5-D 6-B 7-C 8-A
9-D 10-B 11-D 12-A 13-D 14-C
15-A 16-C 17-C 18-B

Chapter 7: Cylinder Blocks and Heads

1-B 2-B 3-C 4-A 5-B 6-A 7-B 8-C
9-D 10-B 11-D 12-A 13-D 14-C
15-B 16-D 17-C 18-A 19-B 20-D

Chapter 8: Valves, Springs, Guides, and Seats

1-D 2-D 3-B 4-A 5-D 6-B 7-D 8-D
9-C 10-C 11-B 12-C 13-B 14-B
15-C 16-C 17-A 18-D 19-C 20-A

Chapter 9: Camshafts, Camshaft Drives, and Valve Timing

1-A 2-D 3-D 4-B 5-A 6-C 7-B 8-C
9-D 10-C 11-D 12-B 13-D 14-C
15-D 16-B 17-A 18-D 19-A 20-C

Chapter 10: Lifters, Followers, Pushrods, and Rocker Arms

1-B 2-D 3-D 4-C 5-C 6-A 7-A 8-A
9-C 10-B 11-D 12-B 13-A 14-C
15-B 16-B 17-D 18-C

Chapter 11: Crankshafts, Flywheels, Vibration Dampers, and Balance Shafts

1-B 2-B 3-B 4-D 5-A 6-D 7-C 8-C
9-D 10-C 11-B 12-B 13-C 14-D
15-A 16-C 17-C 18-D 19-A 20-A

Chapter 12: Piston, Rings, and Connecting Rods

1-C 2-B 3-B 4-D 5-C 6-D 7-B 8-C
9-C 10-B 11-D 12-C 13-D 14-A
15-A 16-A 17-D 18-B 19-A 20-B

Chapter 13: Engine Bearings

1-C 2-A 3-B 4-A 5-C 6-D 7-A 8-C
9-B 10-A 11-A 12-D 13-D 14-D
15-A 16-B 17-C 18-B 19-B

Chapter 14: Gaskets, Seals, Sealants, Fasteners, and Engine Mounts

1-A 2-C 3-D 4-B 5-C 6-A 7-C 8-C
9-C 10-A 11-C 12-C 13-A 14-B
15-C

Answers to ASE Technician Certification Sample Test

1-B 2-D 3-B 4-C 5-D 6-A 7-C
8-C 9-A 10-D 11-C 12-C 13-B
14-D 15-B 16-B 17-D 18-D
19-B 20-C 21-A 22-B 23-C
24-D 25-B 26-D 27-C 28-B
29-C 30-B 31-D 32-A 33-B
34-D 35-D 36-B 37-A 38-C

Index

NOTES

NOTES

NOTES

NOTES

NOTES

NOTES

NOTES

NOTES

NOTES

NOTES

NOTES

NOTES

NOTES

NOTES

NOTES

NOTES